Field-Effect Sensors: From pH Sensing to Biosensing

Field-Effect Sensors: From pH Sensing to Biosensing

Editors

Michael J. Schöning
Sven Ingebrandt

MDPI • Basel • Beijing • Wuhan • Barcelona • Belgrade • Manchester • Tokyo • Cluj • Tianjin

Editors
Michael J. Schöning
Aachen University of Applied Sciences
Germany

Sven Ingebrandt
RWTH Aachen University
Germany

Editorial Office
MDPI
St. Alban-Anlage 66
4052 Basel, Switzerland

This is a reprint of articles from the Special Issue published online in the open access journal *Sensors* (ISSN 1424-8220) (available at: https://www.mdpi.com/journal/sensors/special_issues/field_effect_sensors).

For citation purposes, cite each article independently as indicated on the article page online and as indicated below:

LastName, A.A.; LastName, B.B.; LastName, C.C. Article Title. *Journal Name* **Year**, *Volume Number*, Page Range.

ISBN 978-3-0365-5513-3 (Hbk)
ISBN 978-3-0365-5514-0 (PDF)

Cover image courtesy of Sven Ingebrandt

Contents

About the Editors

Michael J. Schöning

Michael J. Schöning (Dr.-Ing.) received his diploma degree in electrical engineering (1989) and his Dr.-Ing. degree in the field of semiconductor-based microsensors for the detection of ions in liquids (1993), both from the University of Karlsruhe (now, KIT). In 1989, he joined the Institute of Radiochemistry at Research Centre Karlsruhe. Since 1993, he has been with the Institute of Thin Films and Interfaces at Research Centre Jülich (now, Institute of Biological Information Processing). In 1999, he was appointed full professor at Aachen University of Applied Sciences, Campus Jülich. Since 2006, he has served as Founding Director of the Institute of Nano- and Biotechnologies (INB) at the Aachen University of Applied Sciences. His main research subjects concern silicon-based chemical and biological sensors, thin-film technologies, solid-state physics, microsystem, and nano(bio-)technology.

Sven Ingebrandt

Sven Ingebrandt (Dr. rer. nat.) received his diploma degree in physics (1998) and his Dr. rer. nat. degree in the field of physical chemistry (2001) both from the Johannes Gutenberg University Mainz. During his doctoral studies, he worked at the Max-Planck-Institute for Polymer Research in Mainz, where he developed silicon field-effect transistors for the coupling to neuronal cells. From 2001 until 2002, he worked as a postdoctoral researcher at the Frontier Research System RIKEN, Wako, Tokyo, Japan. In 2002, he joined the Institute of Thin Films and Interfaces at Research Centre Jülich (now, Institute of Biological Information Processing) as a group leader. In 2008, he was appointed full professor of biomedical instrumentation at the University of Applied Sciences Kaiserslautern, Campus Zweibrücken. Since 2018, he has served as Chair of Micro- and Nanosystems and is Director of the Institute of Materials in Electrical Engineering 1 at RWTH Aachen University. His main research interests are transistor-based chemical and biological sensors, thin-film technologies, the integration of nanomaterials into microsystems, and cell-based biosensors.

Preface to "Field-Effect Sensors: From pH Sensing to Biosensing"

The groundbreaking work of Piet Bergveld in 1970 ("Development of an ion-sensitive solid-date device for neurophysiological measurements", *IEEE Trans. on Biomed. Eng.*) has stimulated and attracted the interest of a multitude of (young) scientists within the last five decades who work with ion-sensitive field-effect (ISFET) devices for chemical sensing and biosensing, distinctly enhancing the device structures, materials, (bio)receptor layers including living cells or microorganisms, electronic amplifier circuits, system integration, and sensor performance. In this half-century, the basic theoretical concepts were elaborated on, the devices were miniaturized and the used materials and the areas of applications were gradually broadened. During our scientific careers, we have been greatly inspired by this ISFET concept and are still actively working within the field.

To retrace Piet's idea, the following three main types of (bio-)chemical field-effect sensors are discussed in the literature: (i.) ISFETs (ion-sensitive field-effect transistors), nowadays referred to as nanowire devices with nanometer dimensions; (ii.) LAPS (light-addressable potentiometric sensors), where visualization of the chemical specimen can be achieved by means of a two-dimensional mapping with an additional movable light pointer or an array of light sources; and (iii.) capacitive EIS (electrolyte–insulator–semiconductor) sensors, representing the simplest set-up for a field-effect (bio-)chemical sensor, which are used to study and model interface phenomena at the interface "analyte/receptor layer".

This Special Issue "Field-Effect Sensors: From pH Sensing to Biosensing" is devoted to the different types and the scope of their applications, compiling examples of state-of-the-art technologies. The 12 articles included focus on the following topics:

- Device concepts for field-effect sensors for (bio-)chemical sensing (*Yoshinobu & Miyamoto, Tintelott et al., Nii et al.*);
- Modelling and theory of field-effect sensors (*Khodadadian et al., Poghossian et al., Medina-Bailon et al.*);
- Nanomaterial-modified field-effect (bio-)chemical sensors (*Tintelott et al., Shukla et al.*);
- Field-effect sensors for biomedical analysis, food control, and environmental monitoring (*Miyamoto et al., Sakata*);
- Using field-effect sensors to record neuronal and cell-based signals (*Wu et al., Goda*).

The collected articles include one perspective, three reviews, four articles, and two communications. Both editors would like to thank all authors of this volume for their contributions. The book offers graduate students, academic researchers, and industry professionals insight into different up-to-date examples of field-effect sensors for (bio-)chemical sensing. We hope that this collection of articles encourages readers to adapt the described concepts or to develop new ideas and applications of ISFETs.

Michael J. Schöning and Sven Ingebrandt

Editors

Perspective

Technical Perspectives on Applications of Biologically Coupled Gate Field-Effect Transistors

Toshiya Sakata

Department of Materials Engineering, School of Engineering, The University of Tokyo, 7-3-1 Hongo, Bunkyo-ku, Tokyo 113-8656, Japan; sakata@biofet.t.u-tokyo.ac.jp; Tel.: +81-3-5841-1842

Abstract: Biosensing technologies are required for point-of-care testing (POCT). We determine some physical parameters such as molecular charge and mass, redox potential, and reflective index for measuring biological phenomena. Among such technologies, biologically coupled gate field-effect transistor (Bio-FET) sensors are a promising candidate as a type of potentiometric biosensor for the POCT because they enable the direct detection of ionic and biomolecular charges in a miniaturized device. However, we need to reconsider some technical issues of Bio-FET sensors to expand their possible use for biosensing in the future. In this perspective, the technical issues of Bio-FET sensors are pointed out, focusing on the shielding effect, pH signals, and unique parameters of FETs for biosensing. Moreover, other attractive features of Bio-FET sensors are described in this perspective, such as the integration and the semiconductive materials used for the Bio-FET sensors.

Keywords: biosensing; potentiometric biosensor; biologically coupled gate field-effect transistor (Bio-FET); ionic and biomolecular charge; Debye length; measurement solution; pH response; subthreshold slope; semiconductive material; integrated device

Citation: Sakata, T. Technical Perspectives on Applications of Biologically Coupled Gate Field-Effect Transistors. *Sensors* **2022**, *22*, 4991. https://doi.org/10.3390/s22134991

Academic Editors: Sven Ingebrandt, Michael J. Schöning and Antonio Di Bartolomeo

Received: 30 May 2022
Accepted: 30 June 2022
Published: 1 July 2022

Publisher's Note: MDPI stays neutral with regard to jurisdictional claims in published maps and institutional affiliations.

1. Introduction

Ionic or biomolecular charges induce a change in potential at the electrolyte solution/electrode interface. As a type of potentiometric biosensor, biologically coupled gate field-effect transistors (Bio-FETs), which are originally based on solution-gated FETs, are attracting attention worldwide [1–6]. This is probably because various types of biomolecules with charges can be directly detected as electrical signals with the Bio-FETs in a label-free and real-time manner, and various semiconductive materials can also be applied to biosensing [7–14]. Furthermore, the integrated Bio-FET chip based on a complementary metal oxide semiconductor (CMOS) technology enables the simultaneous detection of multiple samples [15].

However, some critical issues constrain such advantages of the Bio-FETs, such as the shielding effect due to counter ions (Debye length limit) and the fabrication process. The Debye length limit is controlled by changing the ionic strength in a measurement solution, that is, diluted measurement solutions are useful for improving the detection sensitivity of the Bio-FETs to charged biomolecules because of the reduction in the shielding effect by counter ions [16–30]. Although the dilution of measurement solutions contributes to the improvement, it is not useful for real samples with high ionic strengths such as blood in a real-time measurement [30], depending on the application. On the other hand, solution-gated FETs are promising for the detection of changes in pH owing to the equilibrium reaction between hydrogen ions with the smallest size and hydroxy groups at an oxide gate insulator, in accordance with the Nernstian response. That is, the detection of changes in pH induced by biological phenomena may be straightforward and effective for biosensing with solution-gated FETs [31–41], although various receptor molecules should be modified on the gate electrode to specifically and selectively detect target biomolecules and to broaden the applications of Bio-FETs as a platform technology for biosensing, considering the Debye length limit.

Moreover, the Bio-FETs are not simple potentiometric biosensors. In other words, their features can be effectively utilized for biosensing. For instance, the subthreshold slope (SS) near the thermal limit contributes to a large shift in drain current (I_D) at a constant gate voltage (V_G) in the SS region, indicating a high sensitivity with a low limit of detection (LOD) [14,42,43]. Alternatively, the capacitive components of functional polymer membranes on the gate electrode are electrically changed by the interaction with noncharged biomolecules [41,44,45].

Considering the above, the technical issues of the Bio-FETs are pointed out in this perspective, focusing on the shielding effect, pH signals, and the unique parameters of FETs for biosensing.

2. How Is the Measurement Solution Used?

Around two decades ago, a nonoptical and label-free DNA analytical method was proposed on the basis of Bio-FET technology [17–28]. Not only were DNA molecules an easy target for Bio-FETs owing to their molecular charges based on phosphate groups, but the development of label-free DNA chips was also actively pursued as one of the post-genome technologies. Single-stranded DNA probes were chemically tethered on the gate electrode, and then the complementary DNA targets were hybridized with the probes, the immobilization density of which was at least on the order of, ca., $10^{11}/cm^2$ [26], inducing the change in the density of negative charges on the gate electrode (Figure 1). Moreover, extension reactions were performed for nonhybridized sequences of target DNA partly complementary to the probe on the gate electrode, resulting in the increase in the density of negative charges. Indeed, these reactions were successfully detected for DNA molecules with a few tens of bases on the basis of the principle of Bio-FETs, whereas longer DNA sequences could not be electrically detected [27]. However, relatively long DNA molecules of approximately 5–10 nm in length could be detected with the Bio-FETs as expected. This expectation was based on the detection of DNA molecular recognition events in a measurement buffer solution with a relatively low ionic strength (i.e., relatively large Debye length) after the bound/free (B/F) molecule separation for each reaction. That is, targeted molecules are specifically bound to substrates, whereas molecules nonspecifically and unexpectedly adsorbed there are washed out. Note that the same buffer solution should be used for each measurement after the B/F molecule separation because the effect of buffer concentration on signal drifts could be neglected. This means that the DNA chip for applications such as single-nucleotide polymorphism (SNP) genotyping, which is based on the hybridization or extension reaction, is tolerant to the B/F separation in every measurement. Thus, the diluted measurement solution can be used for reducing the shielding effect by counter ions. In addition, the B/F separation may be needed to wash out the gate electrode and reduce the nonspecific adsorptions of interfering species with charges. Then, the same measurement solution should be used before and after the reactions to maintain the Debye length. Their applications do not necessarily require the in situ measurement of real samples containing more counter ions. Similarly, the above consideration is also applicable to antigen–antibody reactions and so forth [29].

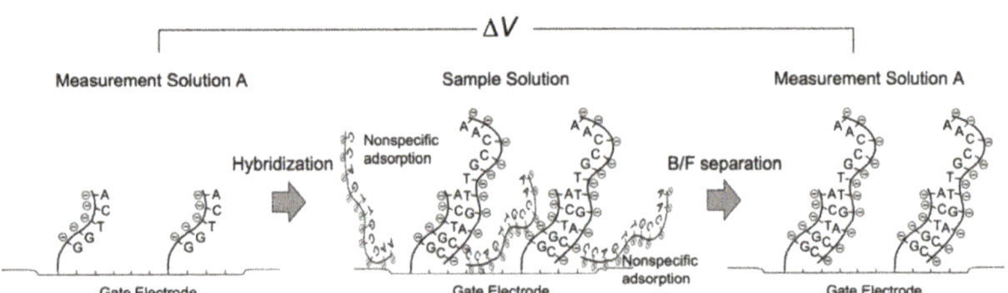

Figure 1. Schematic illustration of measurement process with Bio-FET.

3. Straightforward Mechanism in Bio-FETs

A general cell culture medium includes various ions and chemicals such as serum and glucose. As described in Section 2, in such a medium, the shielding effect caused by counter ions is a problem because Bio-FETs are very insensitive to the changes in the density of molecular charges based on biomolecular recognition events on the gate electrode in the cell culture medium. In other words, nonspecific electrical signals can be prevented from interfering with species in the cell culture medium because some proteins contained in it have been nonspecifically adsorbed on the gate electrode during preculture. Then, what specific targets are detected by the Bio-FETs under this condition? Hydrogen ions, in particular, which have the smallest size, induce changes in pH. Actually, cellular respiration activities can be easily and continuously monitored for any living cells using Bio-FETs with an oxide gate electrode in the cell culture medium [32,34–40]. Some proteins in the cell culture medium are adsorbed at the oxide gate surface during preculture, resulting in the adhesion of cells at the substrate. These macromolecules prevent targeted ionic charges from coming into contact with the gate, but hydrogen ions can easily attach to the oxide gate surface, where the equilibrium reaction between hydroxyl groups and hydrogen ions contributes to the change in the charge density at the oxide gate electrode (Figure 2). Moreover, hydrogen ions are concentrated in the closed nanogap space between the cell membrane and the oxide gate electrode [36,38]. This detection mechanism is very simple, that is, living cells are simply cultured on the oxide gate electrode of the original solution-gated FET (i.e., pH-responsive ion-sensitive FET (ISFET)) for monitoring cellular respiration, although there is a report that the action potential of nerve cells can be monitored in less than one second on the basis of the capacitive coupling model of the cell membrane and the oxide gate electrode [46]. In addition, the cell culture medium with high ionic strength contributes to the reduction in the effect of other ionic and biomolecular charges on the output signal by minimizing the Debye length. This is a straightforward mechanism in the pH-responsive ISFET. As a similar case, we had a breakthrough in label-free DNA sequencing with arrayed ISFET devices based on the CMOS process, which resulted in massively parallel DNA sequencing followed by a cost-effective and high-speed gene analysis [15]. This method was based on the detection of ionic charges, that is, not negative charges of extended base pairs mentioned in Section 2 but positive charges of hydrogen ions generated by enzymatic reactions as byproducts [31]. This means that the pH-responsive ISFET was principally utilized for label-free DNA sequencing, which makes the Debye length limit almost negligible. Thus, it is also important to reconsider the intrinsic features of Bio-FETs, which allow the stable monitoring without additional modifications of the gate electrode.

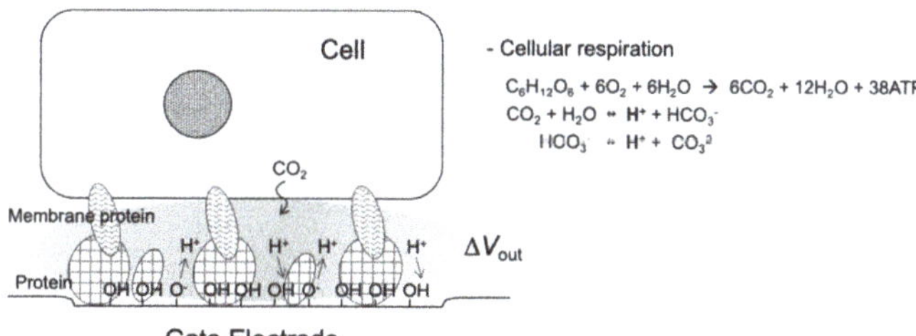

Figure 2. Schematic illustration of nanogap interface between cell and Bio-FET.

4. Features of Transistor for Biosensing

In general, biomolecular recognition events can be analyzed from transistor characteristics such as a V_G–I_D transfer characteristic (e.g., ΔV_G at a constant I_D regarded as a threshold voltage shift (ΔV_T)) (Figure 3). Mostly, ΔV_G at a constant I_D before and after various biomolecular recognition events (e.g., DNA hybridization) is estimated in the linear region of Bio-FETs. This evaluation method is appropriate for potentiometric biosensors.

Indeed, pH-responsive ISFETs ideally show the Nernstian response (59.2 mV/pH at 25 °C) on the basis of ΔV_G at a constant I_D (ΔV_T). On the other hand, ultrasensitive recognition of biomolecules is expected in the subthreshold regime of Bio-FETs (Figure 3). For instance, the solution-gated FET with a 20 nm thick indium tin oxide (ITO) channel exhibited a markedly steep SS, which was very close to the thermal limit (60 mV/dec at 300 K) and may result in a steep SS of less than 60 mV/dec in two-dimensional (2D)-FETs [14]. As a result, the electrical signals measured in the subthreshold regime were about 10 times larger than those measured in the linear regime, which could contribute to the ultrasensitive detection of biomolecules. Moreover, the sensitivity of one-dimensional (1D) nanowire-FET sensors was exponentially enhanced in the subthreshold regime [43]. Thus, the intrinsic features of Bio-FETs should be further improved for biosensing. Note that the Bio-FETs with steeper SS should also be developed not only as simple potentiometric biosensors, although their electrical stabilities have to be improved for the measurements in electrolyte solutions. With these features, 1D and 2D semiconductive materials (1D, e.g., silicon nanowire and carbon nanotube; 2D, e.g., graphene and molybdenum disulfide (MoS_2)) are attractive for the development of novel Bio-FETs owing to their high responsiveness [7–11,14,43].

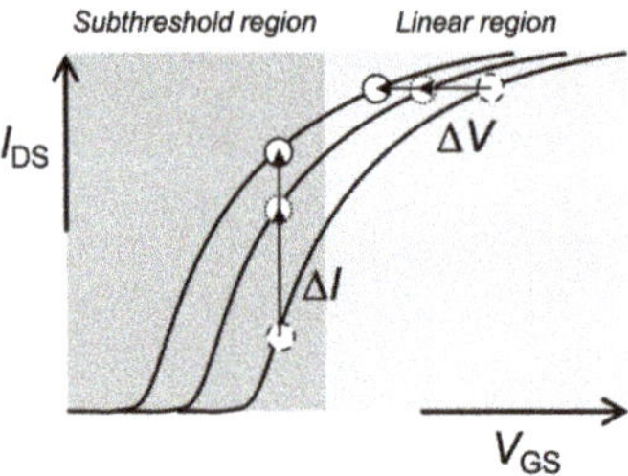

Figure 3. Schematic illustration of V_{GS}–I_{DS} transfer curve of Bio-FET.

Moreover, ΔV_T in the solution-gated FETs is based on the change in the density of ionic and molecular charges at the gate electrode. As mentioned in Section 1, the equilibrium reaction between hydrogen ions and hydroxyl groups at the oxide gate electrode contributes to the change in the charge density at the gate electrode surface, which depends on pH. pH-responsive ISFETs with the oxide gate electrode (e.g., Ta_2O_5) ideally follow the Nernstian response because the site density of hydroxy groups at the Ta_2O_5 surface is expected to be about 10^{15}/cm^2 [47], which is sufficiently high. That is, regardless of the area of the oxide gate electrode, which comes in direct contact with electrolyte solutions, such pH-responsive ISFETs must show the Nernstian response with the change in pH if the change in the charge density is identical. In accordance with this concept, the smaller the area of the gate electrode, the fewer the number of biomolecules reacting at the gate electrode surface. This indicates that a single-biomolecule measurement may be realized using Bio-FETs with a smaller area of the gate electrode on a molecular scale. Actually, the nanowire-based Bio-FETs appear to show an ultrasensitive biomolecular recognition [7]. In addition, the pH responsivity may be increased beyond the Nernst limit using dual-gate FETs with nanowires on the basis of the capacitive coupling effect between the liquid and bottom gates [48–50]. Note that the amplification of electrical signals based on the detection principle may include that of background noise derived from interfering species, leakage, photoinduced fluctuations, and the temperature effect, as well as that of specific signals expected from targeted biomolecules. That is, some treatments such as surface modifications of functional membranes at the active gate electrode are required for increasing the signal-to-noise ratio (S/N).

5. Conclusions

In this perspective, the significant features of Bio-FETs and the important points for measuring using the Bio-FETs were indicated, focusing on the measurement solution, their basic and reliable pH dependence, and the transistor parameters. In addition, the arrayed-gate Bio-FETs should be necessarily applied for multibiosensing, as mentioned in Section 1. This may be actually the most unique feature of FETs because other biosensors (e.g., surface plasmon resonance (SPR) sensors and quartz crystal microbalance (QCM) sensors) hardly enable the integration of electrodes as in CMOS sensors. Moreover, FETs are commonly used in various electric devices such as smartphones and body thermometers because FETs in themselves are miniaturized and included in such devices. Moreover, new semiconductive materials, the functionalities of which are controlled on the nanometer order, must expand the possible applications of Bio-FETs in the future. Note that functional membranes at the electrolyte solution/gate electrode interface should be continuously developed for detecting selectively specific target biomarkers [51–56], considering the prevention/filtering of nonspecific signals based on interfering species [57,58]. Moreover, such functional membranes (e.g., lipid membrane) may extend the Debye length to improve the detection limit for biosensing [59].

Funding: This research received no external funding.

Institutional Review Board Statement: Not applicable.

Informed Consent Statement: Not applicable.

Data Availability Statement: Not applicable.

Acknowledgments: I would like to thank Y. Miyahara of Tokyo Medical and Dental University and members of Sakata Laboratory for their help and useful discussion.

Conflicts of Interest: The author declares no conflict of interest.

References

1. Bergveld, P. Development of an Ion-Sensitive Solid-State Device for Neurophysiological Measurements. *IEEE Trans. Biomed. Eng.* **1970**, *BME-17*, 70–71. [CrossRef] [PubMed]
2. Matsuo, T.; Wise, K.D. An Integrated Field-Effect Electrode for Biopotential Recording. *IEEE Trans. Biomed. Eng.* **1974**, *BME-21*, 485–487. [CrossRef]
3. Esashi, M.; Matsuo, T. Integrated Micro-Multi-Ion Sensor Using Field Effect of Semiconductor. *IEEE Trans. Biomed. Eng.* **1978**, *BME-25*, 184–192. [CrossRef] [PubMed]
4. Caras, S.; Janata, J. Field Effect Transistor Sensitive to Penicillin. *Anal. Chem.* **1980**, *52*, 1935–1937. [CrossRef]
5. Schöning, M.J.; Poghossian, A. Recent Advances in Biologically Sensitive Field-Effect Transistors (BioFETs). *Analyst* **2002**, *127*, 1137–1151. [CrossRef]
6. Sakata, T. Biologically coupled gate field-effect transistors meet in vitro diagnostics. *ACS Omega* **2019**, *4*, 11852–11862. [CrossRef]
7. Stern, E.; Klemic, J.F.; Routenberg, D.A.; Wyrembak, P.N.; Turner-Evans, D.B.; Hamilton, A.D.; LaVan, D.A.; Fahmy, T.M.; Reed, M.A. Label-free immunodetection with CMOS-compatible semiconducting nanowires. *Nature* **2007**, *445*, 519–522. [CrossRef]
8. Ohno, Y.; Maehashi, K.; Yamashiro, Y.; Matsumoto, K. Electrolyte-Gated Graphene Field-Effect Transistors for Detecting pH and Protein Adsorption. *Nano Lett.* **2009**, *9*, 3318–3322. [CrossRef]
9. Rezek, B.; Krátká, M.; Kromka, A.; Kalbacova, M. Effects of Protein Inter-Layers on Cell–Diamond FET Characteristics. *Biosens. Bioelectron.* **2010**, *26*, 1307–1312. [CrossRef]
10. Sarkar, D.; Liu, W.; Xie, X.J.; Anselmo, A.C.; Mitragotri, S.; Banerjee, K. MoS_2 Field-Effect Transistor for Nextgeneration Label-Free Biosensors. *ACS Nano* **2014**, *8*, 3992–4003. [CrossRef]
11. Pachauri, V.; Ingebrandt, S. Biologically Sensitive Field-Effect Transistors: From ISFETs to NanoFETs. *Essays Biochem.* **2016**, *60*, 81–90. [CrossRef] [PubMed]
12. Sakata, T.; Nishimura, K.; Miyazawa, Y.; Saito, A.; Abe, H.; Kajisa, T. Ion Sensitive Transparent-Gate Transistor for Visible Cell Sensing. *Anal. Chem.* **2017**, *89*, 3901–3908. [CrossRef] [PubMed]
13. Chu, C.H.; Sarangadharan, I.; Regmi, A.; Chen, Y.W.; Hsu, C.P.; Chang, W.H.; Lee, G.Y.; Chyi, J.I.; Chen, C.C.; Shiesh, S.C.; et al. Beyond the Debye Length in High Ionic Strength Solution: Direct Protein Detection with Field-Effect Transistors (FETs) in Human Serum. *Sci. Rep.* **2017**, *7*, 5256. [CrossRef] [PubMed]
14. Sakata, T.; Nishitani, S.; Saito, A.; Fukasawa, Y. Solution-gated ultrathin channel indium tin oxide-based field-effect transistor fabricated by one-step procedure that enables high-performance ion sensing and biosensing. *ACS Appl. Mater. Interfaces* **2021**, *13*, 28569–38578. [CrossRef] [PubMed]

15. Rothberg, J.M.; Hinz, W.; Rearick, T.M.; Schultz, J.; Mileski, W.; Davey, M.; Leamon, J.H.; Johnson, K.; Milgrew, M.J.; Edwards, M.; et al. An Integrated Semiconductor Device Enabling Non-optical Genome Sequencing. *Nature* **2011**, *475*, 348–352. [CrossRef] [PubMed]

16. Schasfoort, R.B.M.; Bergveld, P.; Kooyman, R.P.H.; Greve, J. Possibilities and Limitations of Direct Detection of Protein Charges by Means of an Immunological Field-Effect Transistor. *Anal. Chim. Acta* **1990**, *238*, 323–329. [CrossRef]

17. Souteyrand, E.; Cloarec, J.P.; Martin, J.R.; Wilson, C.; Lawrence, I.; Mikkelsen, S.; Lawrence, M.F. Direct Detection of the Hybridization of Synthetic Homo-Oligomer DNA Sequences by Field Effect. *J. Phys. Chem. B* **1997**, *101*, 2980–2985. [CrossRef]

18. Berney, H.; West, J.; Haefele, E.; Alderman, J.; Lane, W.; Collins, J.K. A DNA Diagnostic Biosensor: Development, Characterisation and Performance. *Sens. Actuators B* **2000**, *68*, 100–108. [CrossRef]

19. Fritz, J.; Cooper, E.B.; Gaudet, S.; Sorger, P.K.; Manalis, S.R. Electronic Detection of DNA by Its Intrinsic Molecular Charge. *Proc. Natl. Acad. Sci. USA* **2002**, *99*, 14142–14146. [CrossRef]

20. Sakata, T.; Kamahori, M.; Miyahara, Y. Immobilization of Oligonucleotide Probes on Si_3N_4 Surface and Its Application to Genetic Field Effect Transistor. *Mat. Sci. Eng. C* **2004**, *24*, 827–832. [CrossRef]

21. Pouthas, F.; Gentil, C.; Cote, D.; Bockelmann, U. DNA Detection on Transistor Arrays Following Mutation-Specific Enzymatic Amplification. *Appl. Phys. Lett.* **2004**, *84*, 1594–1596. [CrossRef]

22. Uslu, F.; Ingebrandt, S.; Mayer, D.; Böcker-Meffert, S.; Odenthal, M.; Offenhäusser, A. Labelfree Fully Electronic Nucleic Acid Detection System Based on a Field-Effect Transistor Device. *Biosens. Bioelectron.* **2004**, *19*, 1723–1731. [CrossRef] [PubMed]

23. Sakata, T.; Miyahara, Y. Potentiometric Detection of Single Nucleotide Polymorphism Using Genetic Field Effect Transistor. *ChemBioChem* **2005**, *6*, 703–710. [CrossRef] [PubMed]

24. Sakata, T.; Kamahori, M.; Miyahara, Y. DNA Analysis Chip Based on Field Effect Transistors. *Jpn. J. Appl. Phys.* **2005**, *44*, 2854–2859. [CrossRef]

25. Sakata, T.; Miyahara, Y. Detection of DNA Recognition Events Using Multi-Well Field Effect Transistor. *Biosens. Bioelectron.* **2005**, *21*, 827–832. [CrossRef]

26. Sakata, T.; Miyahara, Y. DNA Sequencing Based on Intrinsic Molecular Charges. *Angew. Chem. Int. Ed.* **2006**, *45*, 2225–2228. [CrossRef]

27. Sakata, T.; Miyahara, Y. Direct Transduction of Primer Extension into Electrical Signal Using Genetic Field Effect Transistor. *Biosens. Bioelectron.* **2007**, *22*, 1311–1316. [CrossRef]

28. Ingebrandt, S.; Han, Y.; Nakamura, F.; Poghossian, A.; Schöning, M.J.; Offenhäusser, A. Label-Free Detection of Single Nucleotide Polymorphisms Utilizing the Differential Transfer Function of Field-Effect Transistors. *Biosens. Bioelectron.* **2007**, *22*, 2834–2840. [CrossRef]

29. Stern, E.; Wagner, R.; Sigworth, F.J.; Breaker, R.; Fahmy, T.M.; Reed, M.A. Importance of the Debye Screening Length on Nanowire Field Effect Transistor Sensors. *Nano Lett.* **2007**, *7*, 3405–3409. [CrossRef]

30. Kim, A.; Ah, C.S.; Park, C.W.; Yang, J.H.; Kim, T.; Ahn, C.G.; Park, S.H.; Sung, G.Y. Direct label-free electrical immunodetection in human serum using a flow-through-apparatus approach with integrated field-effect transistors. *Biosens. Bioelectron.* **2010**, *25*, 1767–1773. [CrossRef]

31. Sakurai, T.; Husimi, Y. Real-Time Monitoring of DNA Polymerase Reactions by a Micro ISFET pH Sensor. *Anal. Chem.* **1992**, *64*, 1996–1997. [CrossRef] [PubMed]

32. Wolf, B.; Brischwein, M.; Baumann, W.; Ehret, R.; Kraus, M. Monitoring of cellular signalling and metabolism with modular sensor-technique: The PhysioControl-Microsystem (PCM®). *Biosens. Bioelectron.* **1998**, *13*, 501–509. [CrossRef]

33. Nishida, H.; Kajisa, T.; Miyazawa, Y.; Tabuse, Y.; Yoda, T.; Takeyama, H.; Kambara, H.; Sakata, T. Self-Oriented Immobilization of DNA Polymerase Tagged by Titanium-Binding Peptide Motif. *Langmuir* **2015**, *31*, 732–740. [CrossRef]

34. Sakata, T.; Saito, A.; Mizuno, J.; Sugimoto, H.; Noguchi, K.; Kikuchi, E.; Inui, H. Single embryo-coupled gate field effect transistor for elective single embryo transfer. *Anal. Chem.* **2013**, *85*, 6633–6638. [CrossRef] [PubMed]

35. Yang, H.; Honda, M.; Akiko, A.; Kajisa, T.; Yanase, Y.; Sakata, T. Non-optical detection of allergic response with a cell-coupled gate field-effect transistor. *Anal. Chem.* **2017**, *89*, 12918–12923. [CrossRef]

36. Satake, H.; Saito, A.; Sakata, T. Elucidation of interfacial pH behaviour at cell/substrate nanogap for in situ monitoring of cellular respiration. *Nanoscale* **2018**, *10*, 10130–10136. [CrossRef]

37. Sakata, T.; Saito, A.; Sugimoto, H. In situ measurement of autophagy under nutrient starvation based on interfacial pH sensing. *Sci. Rep.* **2018**, *8*, 8282. [CrossRef]

38. Sakata, T.; Saito, A.; Sugimoto, H. Live Monitoring of Microenvironmental pH Based on Extracellular Acidosis around Cancer Cells with Cell-Coupled Gate Ion-Sensitive Field-Effect Transistor. *Anal. Chem.* **2018**, *90*, 12731–12736. [CrossRef]

39. Saito, A.; Sakata, T. Sperm-Cultured Gate Ion-Sensitive Field-Effect Transistor for Nonoptical and Live Monitoring of Sperm Capacitation. *Sensors* **2019**, *19*, 1784. [CrossRef]

40. Satake, H.; Sakata, T. Estimation of extracellular matrix production using cultured-chondrocyte-based gate ion-sensitive field-effect transistor. *Anal. Chem.* **2019**, *91*, 16017–16022. [CrossRef]

41. Sakata, T.; Nishitani, S.; Yasuoka, Y.; Himori, S.; Homma, K.; Masuda, T.; Akimoto, A.M.; Sawada, K.; Yoshida, R. Self-oscillating chemoelectrical interface of solution-gated ion-sensitive field-effect transistor based on Belousov-Zhabotinsky reaction. *Sci. Rep.* **2022**, *12*, 2949. [CrossRef] [PubMed]

42. Sze, S.M.; Kwok, K. Ng *Physics of Semiconductor Devices*, 3rd ed.; Wiley: New York, NY, USA, 2007.

43. Gao, X.P.A.; Zheng, G.; Lieber, C.M. Subthreshold Regime has the Optimal Sensitivity for Nanowire FET Biosensors. *Nano Lett.* **2010**, *10*, 547–552. [CrossRef] [PubMed]
44. Matsumoto, A.; Sato, N.; Sakata, T.; Kataoka, K.; Miyahara, Y. Chemical-to-Electrical-Signal Transduction Synchronized with Smart Gel Volume Phase Transition. *Adv. Mater.* **2009**, *21*, 4372–4378. [CrossRef] [PubMed]
45. Masuda, T.; Kajisa, T.; Akimoto, A.M.; Fujita, A.; Nagase, K.; Okano, T.; Sakata, T.; Yoshida, R. Dynamic Electrical Behaviour of Thermoresponsive Polymer in Well-Defined Poly(N-isopropylacrylamide)-Grafted Semiconductor Devices. *RSC Adv.* **2017**, *7*, 34517–34521. [CrossRef]
46. Fromherz, P.; Offenhäusser, A.; Vetter, T.; Weis, J. A Neuron-Silicon Junction: A Retzius Cell of the Leech on an Insulated-Gate Field-Effect Transistor. *Science* **1991**, *252*, 1290–1293. [CrossRef]
47. Akiyama, T.; Ujihira, Y.; Okabe, Y.; Sugano, T.; Niki, E. Ion-Sensitive Field-Effect Transistors with Inorganic Gate Oxide for pH Sensing. *IEEE Trans. Electron Devices* **1982**, *29*, 1936–1941. [CrossRef]
48. Knopfmacher, O.; Tarasov, A.; Fu, W.; Wipf, M.; Niesen, B.; Calame, M.; Schönenberger, C. Nernst Limit in Dual-Gated Si-Nanowire FET Sensors. *Nano Lett.* **2010**, *10*, 2268–2274. [CrossRef]
49. Go, J.; Nair, P.R.; Reddy, B., Jr.; Dorve, B.; Bashir, R.; Alam, M.A. Coupled Heterogeneous Nanowire–Nanoplate Planar Transistor Sensors for Giant (>10 V/pH) Nernst Response. *ACS Nano* **2012**, *6*, 5972–5979. [CrossRef]
50. Ahn, J.-H.; Choi, B.; Choi, S.-J. Understanding the signal amplification in dual-gate FET-based biosensors. *J. Appl. Phys.* **2020**, *128*, 184502. [CrossRef]
51. Iskierko, Z.; Sosnowska, M.; Sharma, P.S.; Benincori, T.; D'Souza, F.; Kaminska, I.; Fronc, K.; Noworyta, K. Extended-Gate Field-Effect Transistor (EG-FET) with Molecularly Imprinted Polymer (MIP) Film for Selective Inosine Determination. *Biosens. Bioelectron.* **2015**, *74*, 526–533. [CrossRef]
52. Nishitani, S.; Sakata, T. Potentiometric Adsorption Isotherm Analysis of a Molecularly Imprinted Polymer Interface for Small-Biomolecule Recognition. *ACS Omega* **2018**, *3*, 5382–5389. [CrossRef] [PubMed]
53. Kajisa, T.; Li, W.; Michinobu, T.; Sakata, T. Well-Designed Dopamine-Imprinted Polymer Interface for Selective and Quantitative Dopamine Detection among Catecholamines Using a Potentiometric Biosensor. *Biosens. Bioelectron.* **2018**, *117*, 810–817. [CrossRef] [PubMed]
54. Kajisa, T.; Sakata, T. Molecularly Imprinted Artificial Biointerface for an Enzyme-Free Glucose Transistor. *ACS Appl. Mater. Interfaces* **2018**, *10*, 34983–34990. [CrossRef] [PubMed]
55. Nakatsuka, N.; Yang, K.-A.; Abendroth, J.M.; Cheung, K.M.; Xu, X.; Yang, H.; Zhao, C.; Zhu, B.; Rim, Y.S.; Yang, Y.; et al. Aptamer-Field-Effect Transistors Overcome Debye Length Limitations for Small-Molecule Sensing. *Science* **2018**, *362*, 319–324. [CrossRef]
56. Sakata, T.; Nishitani, S.; Kajisa, T. Molecularly imprinted polymer-based bioelectrical interface with intrinsic molecular charges. *RSC Adv.* **2020**, *10*, 16999–17013. [CrossRef]
57. Nishitani, S.; Sakata, T. Polymeric Nanofilter Biointerface for Potentiometric Small-Biomolecule Recognition. *ACS Appl. Mater. Interfaces* **2019**, *11*, 5561–5569. [CrossRef]
58. Himori, S.; Nishitani, S.; Sakata, T. Aptamer-based nanofilter interface for small-biomarker detection with potentiometric biosensor. *Electrochim. Acta* **2021**, *368*, 137631. [CrossRef]
59. Lee, D.; Jung, W.H.; Lee, S.; Yu, E.-S.; Lee, T.; Kim, J.H.; Song, H.S.; Lee, K.H.; Lee, S.; Han, S.-K.; et al. Ionic contrast across a lipid membrane for Debye length extension: Towards an ultimate bioelectronic transducer. *Nat. Commun.* **2021**, *12*, 3741. [CrossRef]

 *sensors*

Review

Chemically Induced pH Perturbations for Analyzing Biological Barriers Using Ion-Sensitive Field-Effect Transistors

Tatsuro Goda

Department of Biomedical Engineering, Faculty of Science and Engineering, Toyo University, 2100 Kujirai, Kawagoe, Saitama 350-8585, Japan; goda@toyo.jp; Tel.: +81-49-239-1746

Abstract: Potentiometric pH measurements have long been used for the bioanalysis of biofluids, tissues, and cells. A glass pH electrode and ion-sensitive field-effect transistor (ISFET) can measure the time course of pH changes in a microenvironment as a result of physiological and biological activities. However, the signal interpretation of passive pH sensing is difficult because many biological activities influence the spatiotemporal distribution of pH in the microenvironment. Moreover, time course measurement suffers from stability because of gradual drifts in signaling. To address these issues, an active method of pH sensing was developed for the analysis of the cell barrier in vitro. The microenvironmental pH is temporarily perturbed by introducing a low concentration of weak acid (NH_4^+) or base (CH_3COO^-) to cells cultured on the gate insulator of ISFET using a superfusion system. Considering the pH perturbation originates from the semi-permeability of lipid bilayer plasma membranes, induced proton dynamics are used for analyzing the biomembrane barriers against ions and hydrated species following interaction with exogenous reagents. The unique feature of the method is the sensitivity to the formation of transmembrane pores as small as a proton (H^+), enabling the analysis of cell–nanomaterial interactions at the molecular level. The new modality of cell analysis using ISFET is expected to be applied to nanomedicine, drug screening, and tissue engineering.

Keywords: potentiometry; label-free; cell membranes; tight junctions; cytotoxicity

Citation: Goda, T. Chemically Induced pH Perturbations for Analyzing Biological Barriers Using Ion-Sensitive Field-Effect Transistors. *Sensors* **2021**, *21*, 7277. https://doi.org/10.3390/s21217277

Academic Editors: Michael J. Schöning and Sven Ingebrandt

Received: 15 October 2021
Accepted: 29 October 2021
Published: 1 November 2021

Publisher's Note: MDPI stays neutral with regard to jurisdictional claims in published maps and institutional affiliations.

1. Introduction

Proton (hydronium ion) is involved in many essential biological reactions, such as the equilibrium of carbonate ions, glycolysis, and enzymatic reactions. Systemic pH level has long been recognized as a typical sign of the homeostatic condition. In fact, altered pH is closely related to pathological conditions, such as tumor growth, bacterial infection, and dental caries [1–3]. As a result, various sensing techniques have been developed for biological pH measurements, including implantable sensors, positron emission tomography (PET) imaging, magnetic resonance imaging (MRI), and electron paramagnetic resonance (EPR) imaging [4–8]. These techniques have shown the ability to provide semi-quantitative information during in vivo studies, although the accuracy has been difficult to confirm [9]. In recent years, pH sensing has been applied for bioassays and bioanalytical systems [10,11].

The pH monitoring in culture media or extracellular space is useful for noninvasively determining the conditions of cells. The microenvironmental pH gradually decreases with cellular metabolites, including carbon dioxide and lactate; therefore, the acidification rate in the extracellular medium represents the degree of cellular activity [12]. For accurate quantitative sensing, a potentiometric ion-sensitive field-effect transistor (ISFET) array with a perfusion system was developed to measure the acidification rate of extracellular pH for determining the respiration and glycolysis of tumor cells adhered to the gate insulator [13–15]. ISFETs have been used for pH sensing and biosensing for decades [16–19]. The Nernst response at the solution/insulator interface is the main mechanism for potentiometric pH-sensitivity. ISFETs attract attention as a compact, label-free, real-time, high-throughput, and non-cytotoxic biosensing platform because they are manufactured by the complementary MOS (CMOS) process. Therefore, ISFET-based approaches are straightforward for miniaturized multi-parallel biosensing on a small chip [20]. Moreover, the incorporation of a selective layer on the gate insulator surface can extend the applicability to other biosensing targets [21–23]. A surface coating with an ion-selective membrane can provide the signal selectivity to physiological ions [24–26]. The gate potential of ISFET also responds to microenvironmental changes at the solution/gate interface caused by cell detachment or cell morphology changes on the gate insulator. As a result, an acute cellular response to cytotoxic reagents was estimated using ISFET [14,27].

Recently, a new pH-sensing method was developed to evaluate biological barriers, such as biomembranes and intercellular junctions. Analyzing the biological barriers with high sensitivity, specificity, and spatiotemporal resolutions is essential in the advancement of bioengineering and nanomedicine. The review paper describes a novel potentiometric pH sensing method with the aid of external chemical stimuli and their applications regarding the label-free sensing of unique cellular processes, such as biomembrane injury and epithelial barrier breakdown.

2. Analysis of Cell Barriers for Nanomedicine

Spatiotemporal control of the delivery of therapeutic agents to a specific site is going to be realized by a nanobiotechnology-based drug delivery system (DDS), i.e., nanomedicine [28]. DDS applications include cancer immunotherapy, gene therapy, and nucleic acid-based vaccination, such as messenger RNA (mRNA) vaccines against coronavirus disease 2019 (COVID-19) [29]. Anticancer drugs or therapeutic agents are administered in complexation with nanocarriers for enhancing the safety, stability, and targetability in biological conditions [30]. In other words, nanocarriers are intentionally designed to carry payload for maximizing the therapeutic efficacy while minimizing side effects. To this end, an important challenge is overcoming biological barriers without compromising body defense systems [31].

A eukaryotic cell protects itself by its self-assembled lipid bilayer plasma membrane with a thickness of 6–10 nm. A biomembrane is semipermeable; water and small neutral molecules are permeable by passive diffusion; however, charged species and electrolytes are impermeable because of the hydrophobic core of the bilayer [32]. Macromolecules are usually taken up by cells in a series of energy-dependent mechanisms, called endocytosis [33]. Representative nanocarriers are liposomes, polymeric micelles, and inorganic nanoparticles from natural and synthetic origins, whose typical size ranges from tens to a few hundred nanometers. These nanocarriers are usually endocytosed by cells and entrapped in endosomal compartments after internalization. Therefore, therapeutic agents, which are designed to function in the cell organelle, have to escape through the endosomal biomembranes to enhance the drug efficacy. Many efforts have been made for facilitating endosomal escape using the microenvironmental changes associated with lysosomal digestion known as cell autophagy. Another opportunity is that nanocarriers may bypass the endocytic pathways to reach the cytosol by directly permeating through plasma membranes. Some cationic and amphiphilic nanocarriers permeabilize biomembranes by making tiny transmembrane pores or altering the lipid bilayer polarity during internalization [34]. The permeation mechanisms by different nanocarriers are not completely understood because of the lack of sensing techniques for analyzing the interaction between biomembranes and nanocarriers with high spatiotemporal resolutions.

Biological barriers are also found in epithelial/endothelial tissues. Epithelial cells can exert a rigorous barrier function by forming a multi-protein network in the cell gaps called tight junctions (TJs) [35–37]. Proteins, such as claudin, occludin, and zonula occludens (ZO), are the main constituents of TJs. These junctions are underpinned by the cytoskeleton via transmembrane proteins in the lateral biomembranes. TJs seal peripherals in the top of apical domain so that solutes and water molecules cannot freely permeate the basolateral side through the paracellular pathways. Epithelial barriers are essential for vertebrates to protect the interior against viral and microbial challenges from the external world. TJs form a selective channel for small ions and water by altering the subtype of the constituent proteins, allowing homeostatic maintenance in epithelial tissues, and nutrition uptake in digestive tracts. Therefore, TJs could be a potential drug discovery target for curing malabsorption, dermatitis, and inflammatory diseases [38]. In DDS, epithelial tissues, such as skin and the mucous membrane, are a convenient drug administration route. Epithelial barriers need to be partially and temporarily breached to increase drug permeability, while also creating safety issues. *Clostridium perfringens* enterotoxin (CPE)-derived TJ-binder was used for promoting mucosal absorption and for cancer targeting in nanomedicine [39]. Neural tissues have a clear boundary to blood circulation, namely the blood-brain barrier (BBB). The interface is composed of endothelial cells with cell–cell junctions including TJs. Overcoming the BBB is essential for nanomedicines to cure brain pathologies and neural diseases. Receptor-mediated transcytosis or induced TJ-loosening is a major route for nanocarriers to translocate across the BBB [40]. Moreover, most malignant tumors originate from epithelial cells. During cancer progression, epithelial cells undergo phenotypic changes termed epithelial-mesenchymal transition (EMT) [41]. EMT includes the loss of epithelial cell–cell junctions including TJs, enabling the cells to invade into neighboring tissues and initiate metastasis. Revealing the regulatory mechanisms of the epithelial barrier transitions in tumor microenvironments will guide the development of new cancer therapies.

There are several ways for evaluating biological barriers in vitro. Measuring the leakage of a biomembrane-impermeable indicator from cell cytosol after challenges by nanocarriers is a common method for investigating biomembrane injuries. Indicators include small fluorescence dyes (e.g., calcein) and proteins (e.g., lactate dehydrogenase (LDH), hemoglobin) [42,43]. Leakage assays are frequently used as cytotoxicity assays because large-scale biomembrane lysis leads to acute necrosis. Although the assays are simple, are applicable to various cell types, and have high throughput, they are difficult to use to characterize the permeation mechanisms of nanocarriers. Specifically, traditional indicators cannot pass through smaller transmembrane pores because of the molecular sieve effect. This is crucial because some nanocarriers are suspected to enter cytoplasm by creating pores at molecular levels. Moreover, some indicators can permeate a biomembrane whose hydrophobic/hydrophilic balance is altered by interacting with nanocarriers [44]. These phenomena cause false-positive and false-negative signals for analyzing nanocarrier-biomembrane interactions. The patch clamp technique can electrically determine cell barrier properties by monitoring ionic currents across the biomembrane at the suction area of a single cell using a micropipette [45]. This method is sensitive because the current represents the diffusion of physiological electrolytes of low molecular weight through damaged biomembranes. However, the method can only analyze a single cell at a time and it requires skilled technicians and custom equipment. Therefore, a novel method that can determine biological barriers with high sensitivity, selectivity, and throughput is needed.

A new analytical technique that measures the leakage of proton through a damaged biomembrane was proposed [46]. The pH changes in the cell microenvironment were caused by nanocarrier-induced biomembrane damage. Details on this method are described in the latter sections of this paper.

3. Weak Acid/Base-Induced pH Perturbation in a Cell Microenvironment

The pH-responsive ISFET-based sensors have been successfully used for noninvasively determining the growth, metabolism, and physiological conditions of cells and bacteria [47–50]. The acidification rate of the extracellular microenvironment is an important indicator of live cell state [15,51]. The pH changes with the flux of acidic/basic substances via membrane transporters on the oocyte were estimated by the potentiometric responses of an ISFET-microfluidics system [52].

Although passive potentiometric sensing of extracellular pH is simple, it has some drawbacks. First, ISFET-based potentiometric sensing inevitably has a gradual signal drift over time, which needs to be compensated for by a reference signal for accurate measurement [21]. Second, the pH changes are caused by many factors including respiration, metabolism, transporters, adhesion/detachment, and morphological changes. As a result, it is laborious to identify the cause of the signal. To address these issues, an active method of pH sensing, in which the dynamic response in pH is measured following external physical or biochemical stimuli to the cells, occasionally using a fluidic system, has been reported [53,54]. Compared with passive pH sensing, active pH sensing can acquire a specific signal of interest without interference by the homeostatic activities of cells. In addition, the active method avoids signal drift because the pH perturbation upon external stimuli occurs in a short period of time, typically < 1 min. Adaptation of the active sensing into cellular pH measurements introduces a method for better understanding live cell functions. The effect of an amiloride inhibitor for sodium/hydrogen exchangers (NHE) expressed on the surface of live mammalian cells cultured on an ISFET was successfully evaluated by recording dynamic pH changes induced by intervals of ammonia loading and unloading [55].

A weak acid or base, such as a buffer solution containing ammonium chloride or sodium acetate, are effective pH oscillators in the cell microenvironment (Figure 1) [46,55]. In physiology, a weak acid is a traditional manipulator of cytosolic pH via the proton sponge effect [56]. An extracellular pH gradient was generated by a temporary non-equilibrium state of acid-base reaction as a result of semi-permeable mass transport between the cell interior and exterior. For example, upon exposing cells to an ammonium chloride (i.e., weak acid) solution, neutral ammonia in the extracellular space diffuses into the cytosol across the cell membrane in a concentration-gradient manner, while impermeable ammonium ions remain in the cell exterior. Consequently, a proton is generated in the extracellular microenvironment for rebalancing the NH_3/NH_4^+ equilibrium. This is recorded with the ISFET as a negative overshoot in pH. When the cell microenvironment reaches a steady-state condition, the extracellular pH transient disappears. A positive pH transient occurs after flushing the ammonium chloride solution from the culture media because cell-charged ammonia is diffused out by the concentration-gradient, followed by temporarily rebalancing the NH_3/NH_4^+ equilibrium by consuming protons in the extracellular space. The differential of the potentiometric pH signal is distinct (ΔpH~1 at 10 mM NH_4Cl) and reproducible (RSD < 5%) during the intervals of weak acid/base loading and unloading. Repeated exposure of ammonium chloride or sodium acetate solutions (~10 mM) cause no apparent acute cytotoxicity during the assay for up to several hours [57]. A variety of cell types can be applied to the pH perturbation assay. Measurements with floating cells, such as T lymphocytes, are possible by functionalizing the ISFET surface with a biomembrane-anchoring molecule [58]. The signal time course is slightly influenced by the cell type, cell adhesion area, cell density, and formation of cell–cell contacts. On the other hand, the pH perturbation assay has some drawbacks toward a wide range of applications. First, long-term measurements with superfusion have increased risks for cell detachment. Second, it is not clear that the pH perturbation can be obtained using thick samples of cell multilayers and ex vivo tissues. Third, numerical analysis is required for quantitatively understanding the perturbation signal.

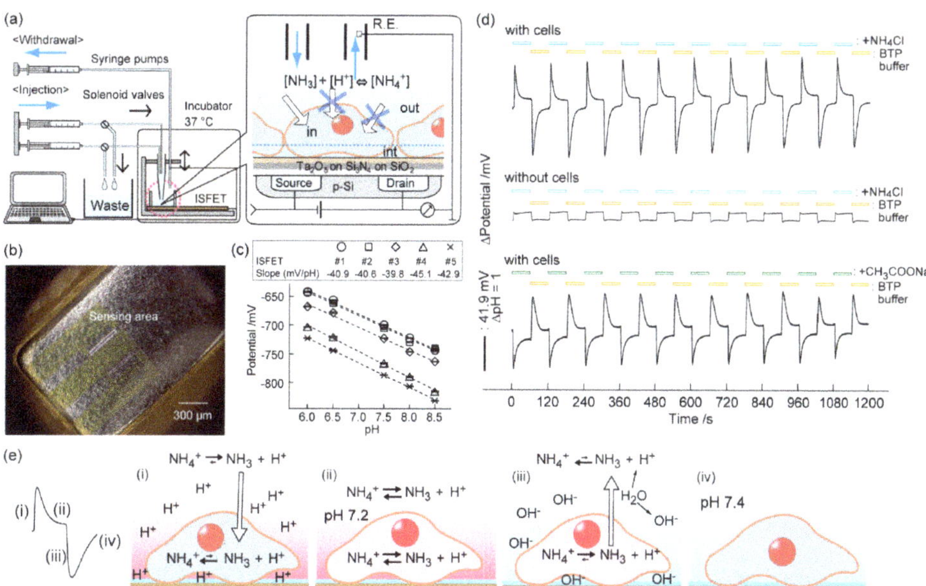

Figure 1. Cell/ISFET system. (**a**) HepG2 cells were cultured on the poly-L-lysine-coated gate insulator of the pH-sensing transistor at subconfluent levels. An automated fluidic system comprising syringe pumps and solenoid valves achieved instant exchange of the solutions surrounding the cells. The inset shows the mass transfer of NH_4^+, NH_3, and H^+ between the bulk phase, cells/ISFET interspace, and intracellular compartment, respectively, and the ammonia equilibrium reaction in each phase. The semi-permeability of the cell membranes prevented the passive diffusion of charged species into the cells. (**b**) Phase contrast image of HepG2 cells on the gate insulator. (**c**) The Nernst pH response of the ISFET with HepG2 monolayers (*n* = 5). (**d**) Time course of the ISFET potential during periodic flushes (1 min each) of isotonic buffers containing 10 mM (NH_4Cl), (CH_3COONa), or 20 mM (sucrose). A pH overshoot occurred when the buffer solution surrounding the cells was exchanged stepwise. The direction was opposite for CH_3COONa. No pH overshoots occurred in the absence of cells on the gate insulator. (**e**) Schematic illustrations explaining the mechanism of local pH changes during the periodic flushes of NH_4Cl in the extracellular space. Reproduced with modification from [46] with permission by Elsevier.

4. Detection of Pore Formation on Biomembranes

Considering the pH overshoots occur by the semi-permeability of healthy biomembranes at the point of loading and unloading of weak acid/base in the cell microenvironment, a new method for detecting leaky biological membranes was developed (Figure 2) [46]. Namely, ISFET-based active pH sensing was applied for the evaluation of cell membrane damages induced by surfactants or a nanocarrier. A model study using HepG2 cell cultures on an ISFET demonstrated an irreversible decrease in the pH overshoot at the point of ammonia exchange following exposure of the cells to poly(ethyleneimine) (PEI), which a common gene transfer reagent. Cationic PEI forms a pore in anionic biomembranes by pulling the polar headgroup toward the hydrophobic core in the bilayer [59,60]. Therefore, the reduced pH overshoot can be interpreted as elimination of the imbalanced NH_3/NH_4^+ equilibrium because of the free permeation of NH_4^+ and H^+ through the pores on PEI-treated biomembranes. The normalized ISFET signal was determined by the degree of pH perturbation before (ΔV_0) and after one-, two-, and three-time exposures (ΔV_i) of a reagent as: (ISFET signal) = $(\Delta V_0 - \Delta V_i)/\Delta V_0$. The interpretation is supported by a simulation, in which the pH overshoot disappears when the total pore area exceeds 0.1% the whole cell surface.

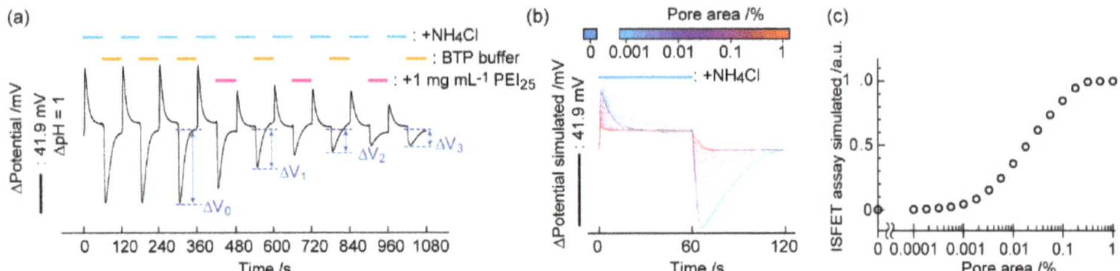

Figure 2. Biomembrane injuries decreased the pH overshoots. (**a**) Time course of the ISFET potential during the intervals of NH_4Cl loading and unloading with three 1-min exposures of 1 mg/mL [PEI] to the cells. (**b**) Simulation of the ISFET signal during the NH_4Cl treatment at various ion-accessible pore area percentages (0–1%) of the plasma membranes. (**c**) The reduction rate of the pH overshoot $(1-\Delta V/\Delta V_0)$ as a function of the pore area on the plasma membranes. Reproduced from [46] with permission by Elsevier.

The ammonia-induced active pH sensing for a biomembrane toxicity assay has sensitivity and specificity in the detection of molecularly sized transmembrane pores on the cell membranes because small proton and ammonium ions with a Stokes radii (R_H) < 0.33 nm are an indicator of the leakiness of the biomembrane. This is in sharp contrast with the conventional indicators for membrane toxicity assays, such as calcein dye of R_H~0.74 nm, hemoglobin of R_H > 3.1 nm, and lactate dehydrogenase (LDH) of R_H > 4.2 nm [61,62]. In fact, the ISFET-based assay detects subtle damage of biomembranes that was not detected by LDH leakage assays for membrane toxicity (Figure 3). The results spur us to determine the biomembrane leakages at molecular levels caused by interactions with the nanocarrier and nanomaterial. Information about cell–nanocarrier interactions with molecular definiteness will aid the development of efficient and safe nanocarriers.

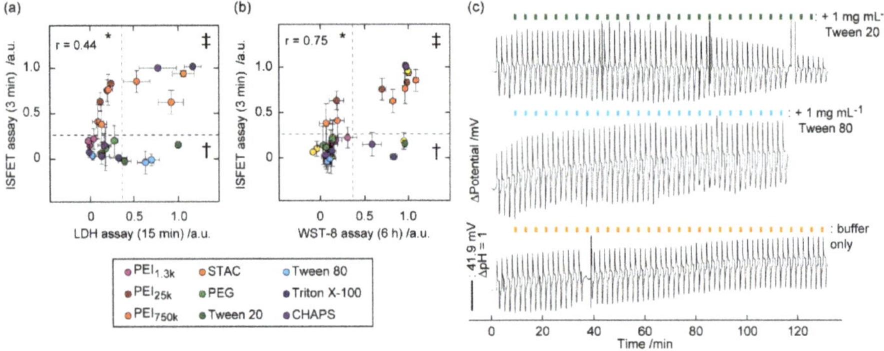

Figure 3. Cluster analysis using the ISFET vs. conventional assays and cell apoptosis detection. (**a**) Scatter plots between the ISFET (3 min) and LDH assays (15 min). *, †, and ‡ represent the $ISFET^+/LDH^-$, $ISFET^-/LDH^+$, and $ISFET^+/LDH^+$ regimes, respectively. Data points identify the two signals at set concentrations with mean ± SD (n = 3). LDH signals were normalized by those obtained at 1 mg/mL Tween 20 for 15 min. Colored symbols show chemical species. Dashed lines represent the thresholds. Correlation coefficient: r. (**b**) A scatter plot between the ISFET (3 min) and WST-8 (6 h) assays. *, †, and ‡ represent the $ISFET^+/WST\text{-}8^-$, $ISFET^-/WST\text{-}8^+$, and $ISFET^+/WST\text{-}8^+$ regimes, respectively. (**c**) Time course of the ISFET signal during intervals of NH_4Cl exchanges with/without multiple exposures of the cells to 1 mg/mL Tween 20 and Tween 80. Reproduced with modification from [57] with permission by the Royal Society of Chemistry.

The signal for ammonia-induced pH perturbation is robust against buffering agents in the cell microenvironment [46]. The pH overshoots (ΔpH~1) are not affected by the buffering effect of chemical reagents surrounding cells or proton transporter activities on the cell membranes. Therefore, the ISFET assay has wide applicability to various reagents and cell types. Moreover, the features on high sensor resolution (~40 μV for ΔpH~1.0 × 10^{-3}), downsizing, and integration for metal oxide semiconductor-based transistors with the aid of microfabrication technologies could introduce multi-parallel cytotoxicity testing with single-cell resolution. Potentiometric pH measurements using available commercial ISFETs with a superfusion system require only a small number of cells (~10 whole cells), because of the small sensing area (10 μm × 340 μm). Alternatively, existing cytotoxicity assays require thousands of cells per well of a microtiter plate.

5. Identification of Biomembrane Injury Type and Cell Death

Biomembrane toxicity assays are frequently used for characterizing the safety of engineered molecules and materials. Cell membrane injuries are typically pore formation, polarity alteration, and disruption (i.e., membrane lysis) [44,63–65]. Identification of biomembrane damage is crucial for understanding cell–nanomaterial interactions. However, no existing method could classify biomembrane injuries. Conventional biomembrane toxicity assays, including the calcein/LDH release assays and hemolysis assay, only measure the cumulative amount of indicators or biomarkers released from the cytosol across the leaky plasma membrane of dying or dead cells for an extended period.

As mentioned above, the degree of pH perturbations specifically responds to the pore-forming activity by exogenous reagents because hydrated ions (NH_4^+ and H^+) only pass through transmembrane pores [57]. On the other hand, amphiphilic protein indicators (LDH and hemoglobin) also bypass damaged biomembranes by fusion mechanisms and leakage through large pores. Therefore, the combination of the ISFET assay with conventional membrane toxicity assays was able to classify the type of biomembrane injuries. For example, the scatter plots from the ISFET and LDH assays were categorized into four regimes by setting thresholds (Figure 3a). The double-negative and double-positive regimes indicate intact biomembranes and membrane disruption or lysis, respectively. The $ISFET^+/LDH^-$ regime, which was assigned to cationic reagents, indicates membrane permeability to small NH_4^+ and H^+ ($R_H < 0.33$ nm), and membrane impermeability to large LDH ($R_H > 4.2$ nm). This is because cationic reagents form pores sizes smaller than LDH on anionic biomembranes via electrostatic interaction [59,60]. While, the $ISFET^-/LDH^+$ regime, which was assigned to non-ionic or anionic surfactants, was interpreted as membrane impermeability to hydrated NH_4^+ and H^+, and membrane permeability to amphiphilic LDH. The phenomenon is understood as LDH leakage by the fusion mechanism without forming transmembrane pores. The fusion is driven by polarity changes of the biomembranes because of the partitioning of non-ionic surfactants into the lipid bilayers. A similar explanation can be used for the combined analysis of results from the ISFET and calcein assays. Notably, the simple combination of existing assays (without ISFET) was unable to classify the biomembrane injuries.

6. Identification of Type of Cell Death

In addition to classifying cell membrane injuries, it is important to understand how cells die as a result of external stimuli, which is key to designing and fabricated safety nanocarriers. Acute cell death mainly occurs by necrosis or apoptosis at different time scales. A severe biomembrane injury leads to detrimental necrosis followed by proinflammatory responses. Apoptosis is categorized in programmed cell death and leads to anti-inflammatory responses. However, conventional cytotoxicity assays only report the results following cell exposures to exogenous reagents and stimuli at a certain endpoint. To address this issue, scatter plots combining the results of the ISFET and commercial cytotoxicity (WST-8) assay were used (Figure 3b) [57]. A detailed analysis using the correlation diagram revealed the cytotoxicity mechanisms. The $ISFET^+/WST\text{-}8^+$ cluster was assigned to necrotic cell death induced by irreversible membrane leaking [66]. The $ISFET^-/WST\text{-}8^+$ cluster indicates the cytotoxicity induced by damages to subcellular compartments without inducing the biomembrane leakage. The $ISFET^+/WST\text{-}8^-$ cluster indicates minor biomembrane damages that can be recovered without causing cytotoxicity.

The process toward cell apoptosis was detected by ISFET monitoring of the pH perturbation for an extended period (~2 h). Apoptosis is the time-dependent programmed cell death mediated by caspase enzymes in the cytosol, leading to gradual disordering of biomembranes (eat-me signaling), followed by eventual clearance by phagocytes. This is in sharp contrast with instant membrane injury during necrosis [67]. The ISFET signal started to respond after 1 h of exposure of cells to an apoptotic inducer (Tween 20), corresponding to increased activity of caspase-3 as an apoptosis marker (Figure 3c) [57]. In contrast, cell exposures to Tween 80, which is a molecular analog to Tween 20, but not an apoptotic inducer, did not cause a time-dependent ISFET signal. Therefore, the ISFET assay for an extended period is effective for distinguishing apoptosis-mediated biomembrane disruptions from direct biomembrane injuries.

Therefore, the ISFET assay complements biomembrane and cytotoxicity assays because of the sensitive and specific detection of small transmembrane pores on the cell surface in real time. The unique features enable us to identify the type of biomembrane injury and the cause of cell death when the analysis is combined with conventional techniques (Figure 4).

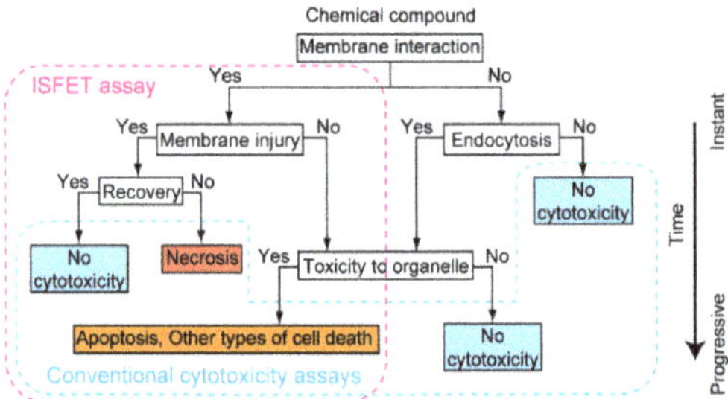

Figure 4. Flow chart showing an invasion of a toxic chemical compound into various cytosolic compartments and the assays that monitor different aspects of time-dependent cytotoxic processes. An invasion of toxicant to the plasma membranes or organelle leads to eventual cell death via different pathways. The ISFET assay detects plasma membrane leakage instantaneously or apoptosis-mediated membrane disorder in several hours. Conventional assays monitor the consequence of toxic activity in minutes to days. Reproduced from [57] with permission by the Royal Society of Chemistry.

7. Understanding the Permeation Mechanism of Nanocarrier

Nanomaterials are promising for their use as nanocontainers for DDS. In recent years, some nanomaterials were identified to enter the cell cytosol without energy-dependent cellular uptake mechanisms [68,69]. Oligopeptides, such as the transactivating transcriptional activator (TAT: GRKKRRQRRRPQ) and octa-arginine (R8: RRRRRRRR), are known as cell-penetrating peptides (CPPs) because of their high permeability to live cell cytosols [70]. A water-soluble amphiphilic random copolymer comprising 30 mol% 2-methacryloyloxyethyl phosphorylcholine (MPC) and 70 mol% *n*-butyl methacrylate (BMA), poly(MPC$_{30}$-*r*-BMA$_{70}$) (PMB30W) was found to penetrate the cell membranes in an energy-independent manner without acute cytotoxicity [71,72]. MPC is a methacrylate monomer bearing a zwitterionic phosphorylcholine group in the side chain. PMB30W is a phospholipid-mimicking polymer because of its similar chemical structure to phosphatidylcholine as a main constituent of the lipid bilayers [73]. The energy-independent translocation may improve the efficacy of DDS by bypassing endosomal entrapments during the endocytic processes [74].

Direct penetration of nanomaterials may occur by the creation of transient pores on the cell surface or via fusion with the lipid bilayers [75]. Alternatively, the direct interaction with cell membranes may cause cytotoxic or biocidal effects by breaching the barrier functions. Therefore, an in depth understanding of the underlying mechanisms of non-endocytic internalization of these nanomaterials is important for developing safe and efficient nanocarriers. However, the mechanisms remain elusive because of the lack of sensing methods for detecting the molecularly sized nanopores on the biomembranes. As a result, the ISFET-based pH perturbation assay was used for the analysis of cell-nanomaterial interaction. The ability to sense the formation of pores as small as a proton ($R_H < 0.33$ nm) can provide solid experimental evidence for the transport mechanism of nanomaterials [46]. Additionally, the combination of the ISFET assay with conventional membrane toxicity and cytotoxicity assays help identify the type of biomembrane injury and cell death as mentioned above [57].

Energy-independent internalization of PMB30W was analyzed using an ISFET-based pH perturbation assay [76]. The ISFET signal was stationary following exposures of membrane-permeable PMB30W, a transfection reagent Lipofectamine®, membrane-impermeable PMPC, and poly(ethylene glycol) (PEG) (Figure 5). The results indicate PMB30W did not form pores. Alternatively, the ISFET signal responded to TAT and R8 exposures, which indicates proton leaks through small pores. Notably, the pore formations by TAT and R8 were not detected by conventional techniques such as the LDH assay and electrochemical impedance spectroscopy. The scattered plots between the ISFET and LDH assays were classified into the four previously described regions. The ISFET$^+$/LDH$^-$ (2) regimes for TAT and R8 indicate the formation of molecularly sized pores, which are permeable for H$^+$ ($R_H < 0.33$ nm) and impermeable to LDH ($R_H > 4.2$ nm). This is the same response as the PEI exposure. In contrast, ISFET$^-$/LDH$^+$ (3) for PMB30W, Lipofectamine, and PEG was interpreted as chemically induced structural disorders or polarity alterations of biomembranes. PMB30W has an *n*-butyl group, which is less hydrophobic than the fatty acid groups in phospholipids. Therefore, the polarity of the cell membrane could be altered when PMB30W hydrophobically interacts with the lipid bilayer cores. Permeability is expressed as the product of the diffusion coefficient and solubility coefficient. The solubility coefficient of the altered cell membranes differs depending on the solute. Hydrated ions are more hydrophilic than proteins, so they did not dissolve in the altered biomembranes. Therefore, the polarity change allows for permeation of LDH as an amphiphilic enzyme while maintaining the ion-barrier functions of biomembranes. This interpretation aligns with the simulation results [77]. In conclusion, PMB30W entered cells by the amphiphilicity-induced membrane fusion mechanism, not by pore formation.

The same analytical technique was used for other nanocarriers of sulfobetaine polymers that can directly penetrate into cells. Zwitterionic sulfobetaine polymers have bio-inertness and stimuli-responsiveness against temperature, pH, and salts, leading to their application for DDS. Four sulfobetaine polymers with different main chains and cationic moieties were compared [78]. The cluster analysis results indicate that the permeation mechanisms depend on small differences in the chemical structure of the four polymers. Specifically, the intramolecular and intermolecular interactions, such as dipole–dipole, hydrophobic, π–π, NH−π, and cation−π interactions, between the polymer chains are the main drivers for non-endocytic internalization with different mechanisms.

The method combining the ISFET and LDH assays was further used to explore the effects of lipid-based and polymer-based transfection reagents on the permeability of model endosomal membranes [79]. Commercial Lipofectin™ and in vivo JetPEI® transfection reagents exhibited pH-dependent pore-forming activity under physiological and endosomal pH conditions. Lipid-based Lipofectin™ created proton-permeable small pores. In contrast, polymer-based in vivo JetPEI® caused LDH-permeable large pores. These results are consistent with previous findings that polymer-based cationic nanocarriers achieve endosomal escape through pore formation rather than the proton sponge effect [80,81]. In summary, the ISFET-based pH perturbation assay is expected to help reveal translocation mechanisms of a wide range of nanocarriers.

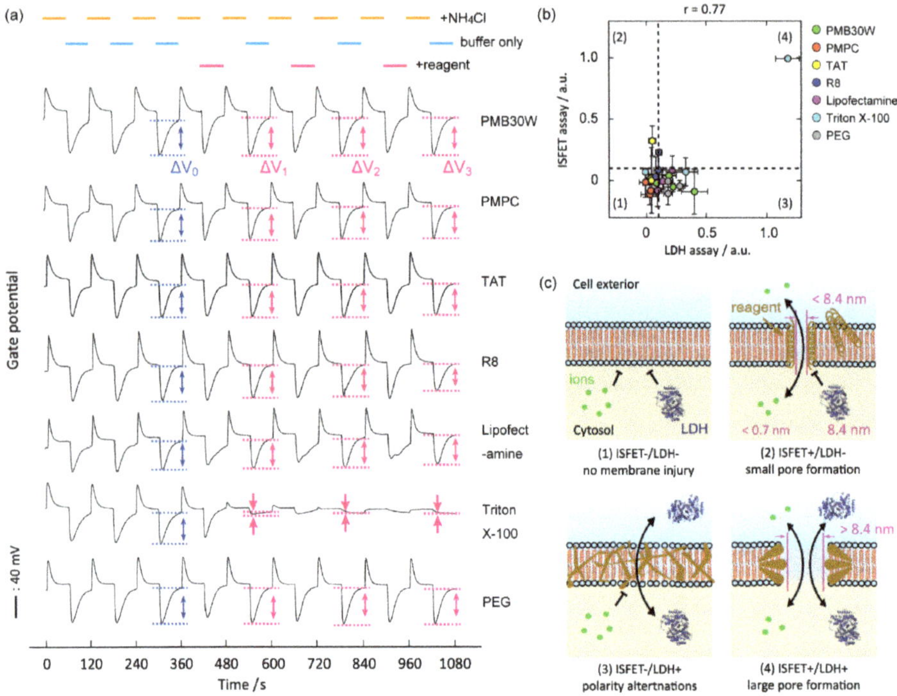

Figure 5. Cluster analysis between the ISFET and LDH assays for identifying the energy-independent permeation mechanism of PMB30W. (**a**) The pH perturbations during repeated exposures of a panel of reagents to cells using a superfusion system. (**b**) Scatter plot between the ISFET and LDH assays. Data points identify the two signals for the samples at set concentrations with mean ± SD ($n = 5$). Dashed lines represent the thresholds defined at 0.1. (**c**) Schematic illustrations showing the types of biomembrane injuries based on the assignment of the four regimes in the correlation diagram. Reproduced with modification from [76] with permission by the American Chemical Society.

8. Detection of the Breach of the Tight Junction on the Epithelial Cell Layer

In addition to cell membranes, TJs are another form of biological barriers found in intercellular gaps of endothelial and epithelial cell layers, including the BBB. TJs, in addition to other form of cell–cell adhesions, are essential in multicellular organisms for partitioning the interior and external world. The TJ barriers enable nutritional reabsorption in the small intestine and the formation of ion gradients in sensory organs. Therefore, breaching epithelial barriers causes various diseases and infection. TJ cancellations are also involved in cancer metastasis mediated by EMT. In DDS, delivering the nanocarrier-payload complexes through epithelial barriers is a challenge. Temporal breaches of TJ barriers are efficient for translocation; however, they create cytotoxicity issues. The drug delivery via an energy-dependent transcytosis mechanism requires no TJ breakdown; however, it has limitations in the design and application of nanocarrier-payload combination. Therefore, the development of a new transport system that can bypass the epithelial barriers is essential for safe and efficient DDS.

To develop a new transport method in DDS, the evaluation of a nanocarrier–epithelial barrier interaction is necessary. Conventional methods for evaluating epithelial barriers in vitro are trans-epithelial electrical resistance (TEER) and permeability tests. TEER provides information about cell adhesion, migration, and proliferation by measuring AC impedance [82,83]. The resistance and capacitance components are determined from ionic currents through the epithelial monolayers. TJs can be interpreted as resistance by modeling the data with an appropriate equivalent circuit. Alternatively, permeability tests use a low molecular weight, cell-impermeable dye for measuring the permeation through paracellular pathways [84,85]. Although this is simple and common, permeability tests cannot detect minor TJ breakdowns that cause leakage of molecules smaller than the indicator. Moreover, permeability tests suffer from low spatiotemporal resolution because of the monitoring of macroscopic process of mass transport. Therefore, an alternative method for sensing TJ breakdown with high sensitivity is required.

Prompted by the blocking features of TJs for small ions, the ammonia-induced pH perturbation technique was used for the non-invasive evaluation of TJ barriers in Madin–Darby canine kidney (MDCK) cells. As a proof-of-concept, the ISFET signal was measured following calcium-chelator or cytotoxin exposure to model epithelial monolayers with TJs on the gate insulator [86]. In contrast to decreases in the pH overshoot by biomembrane lysis using a nonionic detergent Triton™ X-100, the degree of pH perturbation was enhanced by specifically breaching TJ barriers using a calcium-chelator ethylene glycol tetraacetic acid (EGTA) (Figure 6). Numerical analysis revealed that the increased permeability of ammonium ions at the paracellular pathway by TJ breaches enhanced the pH overshoot. Therefore, TJ breakdown can be discriminated from biomembrane damage on the epithelial monolayers by monitoring the amplifying or damping trend in pH perturbation. This is a unique feature of the ISFET assay because the conventional TEER and permeability assays cannot differentiate between the two phenomena. Moreover, a small proton ($R_H < 0.33$ nm) was used to detect the ion barrier breakdown with high sensitivity. The ISFET signal responded to the addition of CPE with a limit of detection (LOD) of 0.03 µg/mL, which was 13-fold less than the LOD of TEER (0.4 µg/mL). Moreover, the effects of the extracellular matrix and a TJ potentiator on the TJ formation process of MDCK cells were successfully evaluated by the ISFET assay [87]. The advanced sensitivity and specificity for examining TJ barriers may create applications including the development of transepithelial nanocarrier, quality control of engineered epithelial tissue, and screening of TJ-targeting drugs.

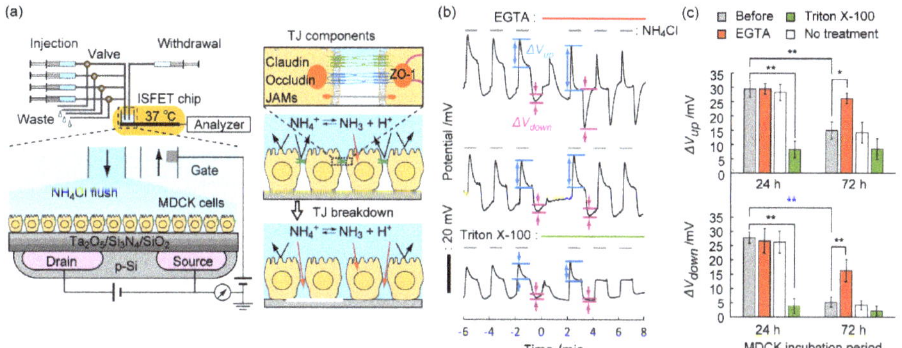

Figure 6. Analysis of barrier functions of epithelial cells with TJs by a pH perturbation assay. (**a**) Schematic diagram showing a system combining ISFET and automated fluidics. Ions permeate cell gaps as a result of TJ breakdown. TJ is mainly composed of claudin, occludin, JAMs, and ZO-1. (**b**) Time course of the gate potential in MDCK cell cultures at 72 h during the NH$_4$Cl-induced pH perturbation assay with/without exposure to 1 mM EGTA or 1 mg/mL Triton X-100. (**c**) ΔV_{up} and ΔV_{down} of MDCK cells cultured for 24 h or 72 h following treatment with EGTA or Triton X-100. Mean ± SD ($n = 3$). * $p < 0.05$, ** $p < 0.01$. Reproduced with modification from [86] with permission by the American Chemical Society.

9. Conclusions

The active pH sensing method was developed to overcome the time course of signaling drift in conventional ISFET-based cell assays. The phenomena of pH perturbation induced by flushing weak acid/base in the cell microenvironment was used for the evaluation of biological barriers, such as cell membranes and TJs, on the gate insulator of ISFET. The high sensitivity and specificity to leakages through these barriers originated from the sensing of the smallest proton indicator. Unlike other techniques, the observation of proton leakage provides evidence of the nanopore formation on biomembranes and TJ breaches in the cell gaps, elucidating the permeation mechanisms of drug nanocarriers through the barriers at molecular levels. Moreover, a combination with conventional assays helps identify the type of biomembrane damage and the cause of cell death. As a CMOS-compatible fabrication process, ISFET sensors can be simply integrated into miniaturized high-density sensor arrays in microfluidic chips. Future applications of multi-parallel and single-cell analysis are expected.

Funding: This research was financed in part by the Grant-in-Aid for "Nanomedicine Molecular Science" (#26107705) from MEXT of Japan, the Nakatani Foundation of Electronic Measuring Technology Advancement, and the Tateisi Science and Technology Foundation.

Conflicts of Interest: The author declares no conflict of interest.

References

1. Gatenby, R.A.; Gawlinski, E.T.; Gmitro, A.F.; Kaylor, B.; Gillies, R.J. Acid-mediated tumor invasion: A multidisciplinary study. *Cancer Res.* **2006**, *66*, 5216–5223. [CrossRef]
2. Curley, G.; Contreras, M.M.; Nichol, A.D.; Higgins, B.D.; Laffey, J.G. Hypercapnia and acidosis in sepsis: A double-edged sword? *Anesthesiology* **2010**, *112*, 462–472. [CrossRef]
3. Ratanaporncharoen, C.; Tabata, M.; Kitasako, Y.; Ikeda, M.; Goda, T.; Matsumoto, A.; Tagami, J.; Miyahara, Y. pH Mapping on Tooth Surfaces for Quantitative Caries Diagnosis Using Micro Ir/IrOx pH Sensor. *Anal. Chem.* **2018**, *90*, 4925–4931. [CrossRef] [PubMed]
4. Ashby, B.S. pH studies in human malignant tumours. *Lancet* **1966**, *2*, 312–315. [CrossRef]
5. Huang, G.; Zhao, T.; Wang, C.; Nham, K.; Xiong, Y.; Gao, X.; Wang, Y.; Hao, G.; Ge, W.P.; Sun, X.; et al. PET imaging of occult tumours by temporal integration of tumour-acidosis signals from pH-sensitive (64)Cu-labelled polymers. *Nat. Biomed. Eng.* **2020**, *4*, 314–324. [CrossRef]
6. Yoshitomi, T.; Suzuki, R.; Mamiya, T.; Matsui, H.; Hirayama, A.; Nagasaki, Y. pH-sensitive radical-containing-nanoparticle (RNP) for the L-band-EPR imaging of low pH circumstances. *Bioconj. Chem.* **2009**, *20*, 1792–1798. [CrossRef] [PubMed]
7. Garcia-Martin, M.L.; Herigault, G.; Remy, C.; Farion, R.; Ballesteros, P.; Coles, J.A.; Cerdan, S.; Ziegler, A. Mapping extracellular pH in rat brain gliomas in vivo by 1H magnetic resonance spectroscopic imaging: Comparison with maps of metabolites. *Cancer Res.* **2001**, *61*, 6524–6531.
8. Cao, H.; Landge, V.; Tata, U.; Seo, Y.S.; Rao, S.; Tang, S.J.; Tibbals, H.F.; Spechler, S.; Chiao, J.C. An Implantable, Batteryless, and Wireless Capsule with Integrated Impedance and pH Sensors for Gastroesophageal Reflux Monitoring. *IEEE Trans. Biomed. Eng.* **2012**, *59*, 3131–3139. [PubMed]
9. Chen, L.Q.; Pagel, M.D. Evaluating pH in the Extracellular Tumor Microenvironment Using CEST MRI and Other Imaging Methods. *Adv. Radiol.* **2015**, *2015*, 206405. [CrossRef]
10. Rothberg, J.M.; Hinz, W.; Rearick, T.M.; Schultz, J.; Mileski, W.; Davey, M.; Leamon, J.H.; Johnson, K.; Milgrew, M.J.; Edwards, M.; et al. An integrated semiconductor device enabling non-optical genome sequencing. *Nature* **2011**, *475*, 348–352. [CrossRef] [PubMed]
11. Toumazou, C.; Shepherd, L.M.; Reed, S.C.; Chen, G.I.; Patel, A.; Garner, D.M.; Wang, C.J.A.; Ou, C.P.; Amin-Desai, K.; Athanasiou, P.; et al. Simultaneous DNA amplification and detection using a pH-sensing semiconductor system. *Nat. Methods* **2013**, *10*, 641–646. [CrossRef] [PubMed]
12. Jensen, L.J.; Sorensen, J.N.; Larsen, E.H.; Willumsen, N.J. Proton pump activity of mitochondria-rich cells. The interpretation of external proton-concentration gradients. *J. Gen. Physiol.* **1997**, *109*, 73–91. [CrossRef]
13. Lehmann, M.; Baumann, W.; Brischwein, M.; Ehret, R.; Kraus, M.; Schwinde, A.; Bitzenhofer, M.; Freund, I.; Wolf, B. Non-invasive measurement of cell membrane associated proton gradients by ion-sensitive field effect transistor arrays for microphysiological and bioelectronical applications. *Biosens. Bioelectron.* **2000**, *15*, 117–124. [CrossRef]
14. Yoshinobu, T.; Schoning, M.J. Light-addressable potentiometric sensors for cell monitoring and biosensing. *Curr. Opin. Electrochem.* **2021**, *28*, 100727. [CrossRef]
15. Yang, C.M.; Yen, T.H.; Liu, H.L.; Lin, Y.J.; Lin, P.Y.; Tsui, L.S.; Chen, C.H.; Chen, Y.P.; Hsu, Y.C.; Lo, C.H.; et al. A real-time mirror-LAPS mini system for dynamic chemical imaging and cell acidification monitoring. *Sens. Actuators B Chem.* **2021**, *341*, 130003. [CrossRef]
16. Schoning, M.J.; Poghossian, A. Recent advances in biologically sensitive field-effect transistors (BioFETs). *Analyst* **2002**, *127*, 1137–1151. [CrossRef]
17. Bergveld, P. Thirty years of ISFETOLOGY—What happened in the past 30 years and what may happen in the next 30 years. *Sens. Actuators B Chem.* **2003**, *88*, 1–20. [CrossRef]

18. Lee, C.S.; Kim, S.K.; Kim, M. Ion-sensitive field-effect transistor for biological sensing. *Sensors* **2009**, *9*, 7111–7131. [CrossRef] [PubMed]
19. Poghossian, A.; Schoning, M.J. Capacitive Field-Effect EIS Chemical Sensors and Biosensors: A Status Report. *Sensors* **2020**, *20*, 5639. [CrossRef]
20. Lehmann, M.; Baumann, W.; Brischwein, M.; Gahle, H.J.; Freund, I.; Ehret, R.; Drechsler, S.; Palzer, H.; Kleintges, M.; Sieben, U.; et al. Simultaneous measurement of cellular respiration and acidification with a single CMOS ISFET. *Biosens. Bioelectron.* **2001**, *16*, 195–203. [CrossRef]
21. Chiang, J.L.; Jan, S.S.; Chou, J.C.; Chen, Y.C. Study on the temperature effect, hysteresis and drift of pH-ISFET devices based on amorphous tungsten oxide. *Sens. Actuators B Chem.* **2001**, *76*, 624–628. [CrossRef]
22. Poghossian, A.; Schoning, M.J. Recent progress in silicon-based biologically sensitive field-effect devices. *Curr. Opin. Electrochem.* **2021**, *29*, 100811. [CrossRef]
23. Yang, Y.; Cuartero, M.; Goncales, V.R.; Gooding, J.J.; Bakker, E. Light-addressable ion sensing for real-time monitoring of extracellular potassium. *Angew. Chem. Int. Ed.* **2018**, *57*, 16801–16805. [CrossRef]
24. Bratov, A.; Abramova, N.; Ipatov, A. Recent trends in potentiometric sensor arrays-A review. *Anal. Chim. Acta* **2010**, *678*, 149–159. [CrossRef]
25. Poghossian, A.; Ingebrandt, S.; Offenhausser, A.; Schoning, M.J. Field-effect devices for detecting cellular signals. *Semin. Cell Dev. Biol.* **2009**, *20*, 41–48. [CrossRef]
26. Walsh, K.B.; DeRoller, N.; Zhu, Y.H.; Koley, G. Application of ion-sensitive field effect transistors for ion channel screening. *Biosens. Bioelectron.* **2014**, *54*, 448–454. [CrossRef]
27. Koppenhofer, D.; Susloparova, A.; Docter, D.; Stauber, R.H.; Ingebrandt, S. Monitoring nanoparticle induced cell death in H441 cells using field-effect transistors. *Biosens. Bioelectron.* **2013**, *40*, 89–95. [CrossRef] [PubMed]
28. Sun, T.M.; Zhang, Y.S.; Pang, B.; Hyun, D.C.; Yang, M.X.; Xia, Y.N. Engineered Nanoparticles for Drug Delivery in Cancer Therapy. *Angew. Chem. Int. Ed.* **2014**, *53*, 12320–12364. [CrossRef] [PubMed]
29. Chauhan, G.; Madou, M.J.; Kalra, S.; Chopra, V.; Ghosh, D.; Martinez-Chapa, S.O. Nanotechnology for COVID-19: Therapeutics and Vaccine Research. *ACS Nano* **2020**, *14*, 7760–7782. [CrossRef]
30. Elsabahy, M.; Wooley, K.L. Design of polymeric nanoparticles for biomedical delivery applications. *Chem. Soc. Rev.* **2012**, *41*, 2545–2561. [CrossRef]
31. Patel, S.; Kim, J.; Herrera, M.; Mukherjee, A.; Kabanov, A.V.; Sahay, G. Brief update on endocytosis of nanomedicines. *Adv. Drug Deliv. Rev.* **2019**, *144*, 90–111. [CrossRef] [PubMed]
32. Cheng, X.L.; Smith, J.C. Biological Membrane Organization and Cellular Signaling. *Chem. Rev.* **2019**, *119*, 5849–5880. [CrossRef]
33. Hillaireau, H.; Couvreur, P. Nanocarriers' entry into the cell: Relevance to drug delivery. *Cell. Mol. Life Sci.* **2009**, *66*, 2873–2896. [CrossRef] [PubMed]
34. Stewart, M.P.; Langer, R.; Jensen, K.F. Intracellular Delivery by Membrane Disruption: Mechanisms, Strategies, and Concepts. *Chem. Rev.* **2018**, *118*, 7409–7531. [CrossRef]
35. Tsukita, S.; Furuse, M. Claudin-based barrier in simple and stratified cellular sheets. *Curr. Opin. Cell Biol.* **2002**, *14*, 531–536. [CrossRef]
36. Yamada, K.M.; Geiger, B. Molecular interactions in cell adhesion complexes. *Curr. Opin. Cell Biol.* **1997**, *9*, 76–85. [CrossRef]
37. Tsukita, S.; Furuse, M.; Itoh, M. Multifunctional strands in tight junctions. *Nat. Rev. Mol. Cell Biol.* **2001**, *2*, 285–293. [CrossRef]
38. Kondoh, M.; Yoshida, T.; Kakutani, H.; Yagi, K. Targeting tight junction proteins-significance for drug development. *Drug Discov. Today* **2008**, *13*, 180–186. [CrossRef]
39. Lee, J.H.; Sahu, A.; Choi, W.I.; Lee, J.Y.; Tae, G. ZOT-derived peptide and chitosan functionalized nanocarrier for oral delivery of protein drug. *Biomaterials* **2016**, *103*, 160–169. [CrossRef] [PubMed]
40. Tang, W.; Fan, W.P.; Lau, J.; Deng, L.M.; Shen, Z.Y.; Chen, X.Y. Emerging blood-brain-barrier-crossing nanotechnology for brain cancer theranostics. *Chem. Soc. Rev.* **2019**, *48*, 2967–3014. [CrossRef] [PubMed]
41. Ribatti, D.; Tamma, R.; Annese, T. Epithelial-Mesenchymal Transition in Cancer: A Historical Overview. *Transl. Oncol.* **2020**, *13*, 100773. [CrossRef] [PubMed]
42. Korzeniewski, C.; Callewaert, D.M. An enzyme-release assay for natural cytotoxicity. *J. Immunol. Methods* **1983**, *64*, 313–320. [CrossRef]
43. Papadopoulos, N.G.; Dedoussis, G.V.; Spanakos, G.; Gritzapis, A.D.; Baxevanis, C.N.; Papamichail, M. An improved fluorescence assay for the determination of lymphocyte-mediated cytotoxicity using flow cytometry. *J. Immunol. Methods* **1994**, *177*, 101–111. [CrossRef]
44. Lichtenberg, D.; Ahyayauch, H.; Alonso, A.; Goni, F.M. Detergent solubilization of lipid bilayers: A balance of driving forces. *Trends Biochem. Sci.* **2013**, *38*, 85–93. [CrossRef]
45. Vaidyanathan, S.; Orr, B.G.; Holl, M.M.B. Detergent Induction of HEK 293A Cell Membrane Permeability Measured under Quiescent and Superfusion Conditions Using Whole Cell Patch Clamp. *J. Phys. Chem. B* **2014**, *118*, 2112–2123. [CrossRef] [PubMed]
46. Imaizumi, Y.; Goda, T.; Schaffhauser, D.F.; Okada, J.; Matsumoto, A.; Miyahara, Y. Proton-sensing transistor systems for detecting ion leakage from plasma membranes under chemical stimuli. *Acta Biomater.* **2017**, *50*, 502–509. [CrossRef]

47. Dantism, S.; Rohlen, D.; Wagner, T.; Wagner, P.; Schoning, M.J. A LAPS-Based Differential Sensor for Parallelized Metabolism Monitoring of Various Bacteria. *Sensors* **2019**, *19*, 4692. [CrossRef] [PubMed]

48. Liang, T.; Gu, C.L.; Gan, Y.; Wu, Q.; He, C.J.; Tu, J.W.; Pan, Y.X.; Qiu, Y.; Kong, L.B.; Wan, H.; et al. Microfluidic chip system integrated with light addressable potentiometric sensor (LAPS) for real-time extracellular acidification detection. *Sens. Actuators B Chem.* **2019**, *301*, 127004. [CrossRef]

49. Wang, J.; Tian, Y.L.; Chen, F.M.; Chen, W.; Du, L.P.; He, Z.Y.; Wu, C.S.; Zhang, D.W. Scanning Electrochemical Photometric Sensors for Label-Free Single-Cell Imaging and Quantitative Absorption Analysis. *Anal. Chem.* **2020**, *92*, 9739–9744. [CrossRef]

50. Saito, A.; Sakata, T. Sperm-Cultured Gate Ion-Sensitive Field-Effect Transistor for Non-Optical and Live Monitoring of Sperm Capacitation. *Sensors* **2019**, *19*, 1784. [CrossRef]

51. Sakata, T.; Sugimoto, H.; Saito, A. Live Monitoring of Microenvironmental pH Based on Extracellular Acidosis around Cancer Cells with Cell-Coupled Gate Ion-Sensitive Field-Effect Transistor. *Anal. Chem.* **2018**, *90*, 12731–12736. [CrossRef]

52. Schaffhauser, D.F.; Patti, M.; Goda, T.; Miyahara, Y.; Forster, I.C.; Dittrich, P.S. An Integrated field-effect microdevice for monitoring membrane transport in Xenopus laevis oocytes via lateral proton diffusion. *PLoS ONE* **2012**, *7*, e39238. [CrossRef] [PubMed]

53. Eklund, S.E.; Taylor, D.; Kozlov, E.; Prokop, A.; Cliffel, D.E. A microphysiometer for simultaneous measurement of changes in extracellular glucose, lactate, oxygen, and acidification rate. *Anal. Chem.* **2004**, *76*, 519–527. [CrossRef] [PubMed]

54. Sakata, T.; Miyahara, Y. Noninvasive monitoring of transporter-substrate interaction at cell membrane. *Anal. Chem.* **2008**, *80*, 1493–1496. [CrossRef] [PubMed]

55. Schaffhauser, D.; Fine, M.; Tabata, M.; Goda, T.; Miyahara, Y. Measurement of rapid amiloride-dependent pH changes at the cell surface using a proton-sensitive field-effect transistor. *Biosensors* **2016**, *6*, 11. [CrossRef]

56. Wu, P.; Ray, N.G.; Shuler, M.L. A computer model for intracellular pH regulation in Chinese hamster ovary cells. *Biotechnol. Prog.* **1993**, *9*, 374–384. [CrossRef] [PubMed]

57. Imaizumi, Y.; Goda, T.; Matsumoto, A.; Miyahara, Y. Identification of types of membrane injuries and cell death using whole cell-based proton-sensitive field-effect transistor systems. *Analyst* **2017**, *142*, 3451–3458. [CrossRef] [PubMed]

58. Imaizumi, Y.; Goda, T.; Toya, Y.; Matsumoto, A.; Miyahara, Y. Oleyl group-functionalized insulating gate transistors for measuring extracellular pH of floating cells. *Sci. Technol. Adv. Mater.* **2016**, *17*, 337–345. [CrossRef]

59. Hong, S.P.; Leroueil, P.R.; Janus, E.K.; Peters, J.L.; Kober, M.M.; Islam, M.T.; Orr, B.G.; Baker, J.R.; Holl, M.M.B. Interaction of polycationic polymers with supported lipid bilayers and cells: Nanoscale hole formation and enhanced membrane permeability. *Bioconj. Chem.* **2006**, *17*, 728–734. [CrossRef] [PubMed]

60. Sikor, M.; Sabin, J.; Keyvanloo, A.; Schneider, M.F.; Thewalt, J.L.; Bailey, A.E.; Frisken, B.J. Interaction of a Charged Polymer with Zwitterionic Lipid Vesicles. *Langmuir* **2010**, *26*, 4095–4102. [CrossRef]

61. Rothe, G.M.; Purkhanbaba, H. Determination of molecular weights and Stokes' radii of non-denatured proteins by polyacrylamide gradient gel electrophoresis. 1. An equation relating total polymer concentration, the molecular weight of proteins in the range of 10^4–10^6, and duration of electrophoresis. *Electrophoresis* **1982**, *3*, 33–42.

62. Tamba, Y.; Ariyama, H.; Levadny, V.; Yamazaki, M. Kinetic Pathway of Antimicrobial Peptide Magainin 2-Induced Pore Formation in Lipid Membranes. *J. Phys. Chem. B* **2010**, *114*, 12018–12026. [CrossRef] [PubMed]

63. Hong, S.P.; Bielinska, A.U.; Mecke, A.; Keszler, B.; Beals, J.L.; Shi, X.Y.; Balogh, L.; Orr, B.G.; Baker, J.R.; Holl, M.M.B. Interaction of poly(amidoamine) dendrimers with supported lipid bilayers and cells: Hole formation and the relation to transport. *Bioconj. Chem.* **2004**, *15*, 774–782. [CrossRef]

64. Tan, A.M.; Ziegler, A.; Steinbauer, B.; Seelig, J. Thermodynamics of sodium dodecyl sulfate partitioning into lipid membranes. *Biophys. J.* **2002**, *83*, 1547–1556. [CrossRef]

65. Deo, N.; Somasundaran, P. Electron spin resonance study of phosphatidyl choline vesicles using 5-doxyl stearic acid. *Colloids Surf. B* **2002**, *25*, 225–232. [CrossRef]

66. Majno, G.; Joris, I. Apoptosis, oncosis, and necrosis. An overview of cell death. *Am. J. Pathol.* **1995**, *146*, 3–15.

67. Hengartner, M.O. The biochemistry of apoptosis. *Nature* **2000**, *407*, 770–776. [CrossRef]

68. Verma, A.; Uzun, O.; Hu, Y.; Hu, Y.; Han, H.S.; Watson, N.; Chen, S.; Irvine, D.J.; Stellacci, F. Surface-structure-regulated cell-membrane penetration by monolayer-protected nanoparticles. *Nat. Mater.* **2008**, *7*, 588–595. [CrossRef]

69. Cronican, J.J.; Beier, K.T.; Davis, T.N.; Tseng, J.C.; Li, W.D.; Thompson, D.B.; Shih, A.F.; May, E.M.; Cepko, C.L.; Kung, A.L.; et al. A Class of Human Proteins that Deliver Functional Proteins into Mammalian Cells In Vitro and In Vivo. *Chem. Biol.* **2011**, *18*, 833–838. [CrossRef]

70. Heitz, F.; Morris, M.C.; Divita, G. Twenty years of cell-penetrating peptides: From molecular mechanisms to therapeutics. *Br. J. Pharmacol.* **2009**, *157*, 195–206. [CrossRef] [PubMed]

71. Goda, T.; Goto, Y.; Ishihara, K. Cell-penetrating macromolecules: Direct penetration of amphipathic phospholipid polymers across plasma membrane of living cells. *Biomaterials* **2010**, *31*, 2380–2387. [CrossRef]

72. Goda, T.; Miyahara, Y.; Ishihara, K. Phospholipid-mimicking cell-penetrating polymers: Principles and applications. *J. Mater. Chem. B* **2020**, *8*, 7633–7641. [CrossRef] [PubMed]

73. Goda, T.; Ishihara, K.; Miyahara, Y. Critical update on 2-methacryloyloxyethyl phosphorylcholine (MPC) polymer science. *J. Appl. Polym. Sci.* **2015**, *132*, 41766. [CrossRef]

74. Du, S.; Liew, S.S.; Li, L.; Yao, S.Q. Bypassing Endocytosis: Direct Cytosolic Delivery of Proteins. *J. Am. Chem. Soc.* **2018**, *140*, 15986–15996. [CrossRef] [PubMed]

75. Marie, E.; Sagan, S.; Cribier, S.; Tribet, C. Amphiphilic Macromolecules on Cell Membranes: From Protective Layers to Controlled Permeabilization. *J. Membr. Biol.* **2014**, *247*, 861–881. [CrossRef] [PubMed]
76. Goda, T.; Imaizumi, Y.; Hatano, H.; Matsumoto, A.; Ishihara, K.; Miyahara, Y. Translocation Mechanisms of Cell-Penetrating Polymers Identified by Induced Proton Dynamics. *Langmuir* **2019**, *35*, 8167–8173. [CrossRef]
77. Werner, M.; Sommer, J.U. Translocation and Induced Permeability of Random Amphiphilic Copolymers Interacting with Lipid Bilayer Membranes. *Biomacromolecules* **2015**, *16*, 125–135. [CrossRef]
78. Goda, T.; Hatano, H.; Yamamoto, M.; Miyahara, Y.; Morimoto, N. Internalization Mechanisms of Pyridinium Sulfobetaine Polymers Evaluated by Induced Protic Perturbations on Cell Surfaces. *Langmuir* **2020**, *36*, 9977–9984. [CrossRef]
79. Boonstra, E.; Hatano, H.; Miyahara, Y.; Uchida, S.; Goda, T.; Cabral, H. A proton/macromolecule-sensing approach distinguishes changes in biological membrane permeability during polymer/lipid-based nucleic acid delivery. *J. Mater. Chem. B* **2021**, *9*, 4298–4302. [CrossRef]
80. Miyata, K.; Oba, M.; Nakanishi, M.; Fukushima, S.; Yamasaki, Y.; Koyama, H.; Nishiyama, N.; Kataoka, K. Polyplexes from Poly(aspartamide) Bearing 1,2-Diaminoethane Side Chains Induce pH-Selective, Endosomal Membrane Destabilization with Amplified Transfection and Negligible Cytotoxicity. *J. Am. Chem. Soc.* **2008**, *130*, 16287–16294. [CrossRef]
81. Funhoff, A.M.; van Nostrum, C.F.; Koning, G.A.; Schuurmans-Nieuwenbroek, N.M.E.; Crommelin, D.J.A.; Hennink, W.E. Endosomal escape of polymeric gene delivery complexes is not always enhanced by polymers buffering at low pH. *Biomacromolecules* **2004**, *5*, 32–39. [CrossRef]
82. Srinivasan, B.; Kolli, A.R.; Esch, M.B.; Abaci, H.E.; Shuler, M.L.; Hickman, J.J. TEER Measurement Techniques for In Vitro Barrier Model Systems. *Jala-J. Lab. Autom.* **2015**, *20*, 107–126. [CrossRef]
83. Benson, K.; Cramer, S.; Galla, H.J. Impedance-based cell monitoring: Barrier properties and beyond. *Fluids Barriers CNS* **2013**, *10*, 5. [CrossRef] [PubMed]
84. Oltra-Noguera, D.; Mangas-Sanjuan, V.; Centelles-Sanguesa, A.; Gonzalez-Garcia, I.; Sanchez-Castano, G.; Gonzalez-Alvarez, M.; Casabo, V.G.; Merino, V.; Gonzalez-Alvarez, I.; Bermejo, M. Variability of permeability estimation from different protocols of subculture and transport experiments in cell monolayers. *J. Pharmacol. Toxicol. Methods* **2015**, *71*, 21–32. [CrossRef]
85. Ujhelyi, Z.; Fenyvesi, F.; Varadi, J.; Feher, P.; Kiss, T.; Veszelka, S.; Deli, M.; Vecsernyes, M.; Bacskay, I. Evaluation of cytotoxicity of surfactants used in self-micro emulsifying drug delivery systems and their effects on paracellular transport in Caco-2 cell monolayer. *Eur. J. Pharm. Sci.* **2012**, *47*, 564–573. [CrossRef] [PubMed]
86. Hatano, H.; Goda, T.; Matsumoto, A.; Miyahara, Y. Induced Proton Perturbation for Sensitive and Selective Detection of Tight Junction Breakdown. *Anal. Chem.* **2019**, *91*, 3525–3532. [CrossRef]
87. Hatano, H.; Goda, T.; Matsumoto, A.; Miyahara, Y. Induced Proton Dynamics on Semiconductor Surfaces for Sensing Tight Junction Formation Enhanced by an Extracellular Matrix and Drug. *ACS Sens.* **2019**, *4*, 3195–3202. [CrossRef] [PubMed]

Review

Field-Effect Sensors Using Biomaterials for Chemical Sensing

Chunsheng Wu [1], Ping Zhu [1], Yage Liu [1], Liping Du [1] and Ping Wang [2,3,*]

[1] Institute of Medical Engineering, Department of Biophysics, School of Basic Medical Science, Health Science Center, Xi'an Jiaotong University, Xi'an 710061, China; wuchunsheng@xjtu.edu.cn (C.W.); jewel121@stu.xjtu.edu.cn (P.Z.); yageliu@xjtu.edu.cn (Y.L.); duliping@xjtu.edu.cn (L.D.)

[2] Biosensor National Special Laboratory, Department of Biomedical Engineering, Zhejiang University, Hangzhou 310027, China

[3] Key Laboratory for Biomedical Engineering of Ministry of Education, Department of Biomedical Engineering, Zhejiang University, Hangzhou 310027, China

*   Correspondence: cnpwang@zju.edu.cn

Abstract: After millions of years of evolution, biological chemical sensing systems (i.e., olfactory and taste systems) have become very powerful natural systems which show extreme high performances in detecting and discriminating various chemical substances. Creating field-effect sensors using biomaterials that are able to detect specific target chemical substances with high sensitivity would have broad applications in many areas, ranging from biomedicine and environments to the food industry, but this has proved extremely challenging. Over decades of intense research, field-effect sensors using biomaterials for chemical sensing have achieved significant progress and have shown promising prospects and potential applications. This review will summarize the most recent advances in the development of field-effect sensors using biomaterials for chemical sensing with an emphasis on those using functional biomaterials as sensing elements such as olfactory and taste cells and receptors. Firstly, unique principles and approaches for the development of these field-effect sensors using biomaterials will be introduced. Then, the major types of field-effect sensors using biomaterials will be presented, which includes field-effect transistor (FET), light-addressable potentiometric sensor (LAPS), and capacitive electrolyte–insulator–semiconductor (EIS) sensors. Finally, the current limitations, main challenges and future trends of field-effect sensors using biomaterials for chemical sensing will be proposed and discussed.

Keywords: field-effect sensors; chemical sensors; biosensors; olfactory; taste; biomaterials

Citation: Wu, C.; Zhu, P.; Liu, Y.; Du, L.; Wang, P. Field-Effect Sensors Using Biomaterials for Chemical Sensing. *Sensors* **2021**, *21*, 7874. https://doi.org/10.3390/s21237874

Academic Editor: Marco Consales

Received: 29 October 2021
Accepted: 25 November 2021
Published: 26 November 2021

Publisher's Note: MDPI stays neutral with regard to jurisdictional claims in published maps and institutional affiliations.

1. Introduction

Biological olfactory and taste systems are two main categories of natural chemical sensing systems, which play crucial roles for almost all the creatures in survival, feeding, and breeding [1–5]. After millions of years of evolution, these biological chemical sensing systems have become very powerful natural systems which show extreme high performances in detecting and discriminating various chemical substances [2,6–8]. For instance, biological olfactory systems are able to detect specific chemical signals presented by the odorant molecules, even at the trace level [9,10]. Similarly, biological taste systems show unique performance and versatility for the detection of chemical signals transmitted by various tastants [2,11]. Creatures are able to obtain essential chemical information about their surroundings from biological chemical sensing systems in order to find food, to communicate with partners, and to avoid predators [12–15]. The key components of biological chemical sensing systems include functional biomaterials that are able to recognize specific chemical substances and transduce the sensed chemical signals into cellular and molecular responses [2,6,7]. These functional biomaterials, which are chemical sensitive cells and molecules, mainly include olfactory sensory neurons, olfactory receptors, taste cells, and taste receptors [16,17]. They have been considered the primary source of high performances of biological chemical sensing systems [7,18,19].

Creating field-effect sensors using biomaterials that are able to detect specific target chemical substances with high sensitivity would have broad applications in many areas, ranging from biomedicine and environments to the food industry, but this has proved extremely challenging [20–22]. The excellent performances of functional biomaterials from biological chemical sensing systems are ideal candidates of sensitive elements for the development of field-effect sensors using biomaterials towards chemical sensing in complex environments [23,24]. For this reason, these biomaterials have been employed for chemical sensing to mimic the mechanisms of biological chemical sensing systems. In recent decades, with rapid advancements in molecular biology and microfabrication process, inspirations from natural chemical sensing systems have led to the development of various field-effect sensors using biomaterials that rely on the combination of functional biomaterials with various field-effect devices [25–28]. Over decades of intense research, field-effect sensors using biomaterials for chemical sensing have achieved significant progress and shown promising prospects and potential applications.

The development of chemical sensors has been inspired by utilizing the biological sensitive materials or mimicking natural porous structures [29–33]. For the latter situation, many chemical sensors were developed to improve the sensing performance [34]. A typical example is the architecture hierarchy of butterfly wings, which can be synthesized chemically via specific approaches. For example, well-organized porous hierarchical SnO_2 was fabricated with connective hollow interiors and thin mesoporous walls for the sensing of chemical vapors [35,36]; the photonic structures from Morpho butterfly wings were prepared for the sensitive optical sensing of ethanol [37–40]. In addition, other biological templates have also been mimicked to fabricate sensitive materials with hierarchical micro/nanostructures, such as the eggshell membrane [41] and the bristles on the wings of the Alpine Black Swallowtail butterfly (*Papilio maackii*) [42]. Considering that the assembly of biological micro/nanostructures mainly belong to the category of material chemistry and has been summarized in other reviews [29,30,43], here we would like to focus on how to utilize the biological sensitive materials with secondary transducers for chemical sensing. Among various chemical sensors, field-effect sensors using biomaterials could retain the biological chemical sensing mechanisms to some extent and could achieve a performance comparable to biological chemical sensing systems by the using of functional biomaterials as sensitive elements for chemical sensing, which are characterized with high sensitivity, high specificity, and low detection limit [16,44].

Despite the rapid advancements and growing interests in the research and development of field-effect sensors using biomaterials for chemical sensing, limited literature is available that outlines recent advances in this field. This review will summarize the state of the art in field-effect sensors using biomaterials for chemical sensing with an emphasis on those using functional biomaterials as sensing elements, such as olfactory and taste cells and receptors. Firstly, unique principles and approaches for the development of these field-effect sensors using biomaterials will be introduced. Then, the major types of field-effect sensors using biomaterials will be presented, which includes field-effect transistor (FET), light-addressable potentiometric sensor (LAPS), and capacitive electrolyte–insulator–semiconductor (EIS) sensors. Finally, the current limitations, main challenges and future trends of field-effect sensors using biomaterials for chemical sensing will be proposed and discussed.

2. Fundamental of Field-Effect Sensors Using Biomaterials

In biological chemical sensing systems, the process of chemical signal detection is initialized by the special interactions between molecular detectors and specific chemical substances, which can trigger a cascade of intracellular biochemical reactions to convert the chemical signals into cellular responses such as cell membrane potential changes [45–47]. These cellular responses are transmitted to the central neural system for the further processing of chemical signals, which allows for the perception of specific chemical substances. Biological chemical sensing systems are the most powerful system for the detection of specific chemical substances with very high performances that cannot be matched by most existing artificial devices. Therefore, it is worthwhile to develop biosensors using biomaterials in order to obtain artificial chemical sensing devices with performances comparable to biological chemical sensing systems.

The main components of biosensors using biomaterials for chemical sensing include sensitive elements and transducers, which are combined to mimic the functions of biological chemical sensing systems to realize the conversion of chemical signals into measurable signals by existing devices such as electrical signals and optical signals. As shown in Figure 1, the basic idea of biosensors using biomaterials is to employ the extreme high capability of functional biomaterials originating from biological systems for the detection of specific chemical substances. The coupling of highly specialized biomaterials with a transducer could lead to the generation of potential devices and instruments with a performance comparable to that of biological chemical sensing systems for the detection of chemical signals in a trace level within complex environmental conditions.

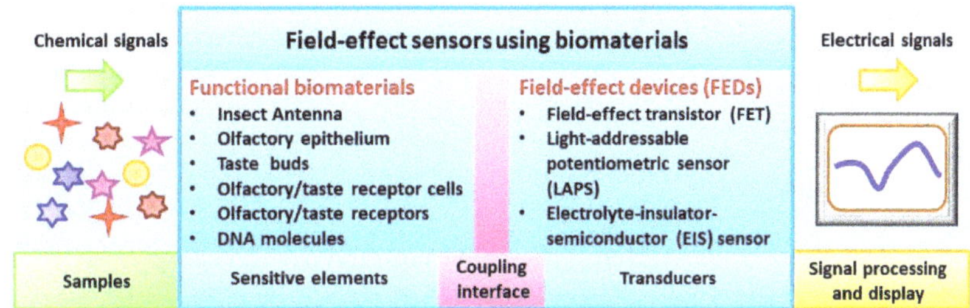

Figure 1. Schematic diagram of configurations of field-effect sensors using biomaterials.

2.1. Preparation of Functional Biomaterials

For the development of biosensors using biomaterials, it is required to obtain functional biomaterials, which maintain their unique capability of chemical sensing and are suitable to be used as sensitive elements to couple with transducers [16]. Because the activity of biomaterials has a direct influence on the performances of biosensors with regard to sensitivity, specificity, and stability, it is of great importance to obtain functional biomaterials for chemical sensing. In addition to maintaining the natural structures and native functions of biomaterials, it is also desirable to produce them in a cost-effective manner and store them in a convenient manner. At present, several methods have been applied in the preparation of functional biomaterials for chemical sensing, which can be divided into two main categories: one is direct isolation from natural biological chemical sensing systems, the other one is preparation based on biotechnology.

Direct isolation from natural biological chemical sensing systems is the most convenient approach to achieving functional biomaterials for the development of biosensors for chemical sensing. It is widely used in the early stage of biosensors, which has the advantages of maintaining the natural structure and functions of biomaterials allowing for the recognition of their natural ligands with high performances. In addition, the powerful capability of biological chemical sensing systems could be preserved to some extent, which helps to enhance the performance of biosensors. Different types of functional biomaterials have been isolated from biological chemical sensing systems and successfully utilized as sensitive elements for the development of biosensors. For instance, olfactory sensory neurons and olfactory receptors have been isolated from animals or insects and have served as sensitive elements in biosensors for odor detection [26,48–51]. Similarly, taste bud cells and taste receptors have also been isolated from animals and applied in the biosensors for taste substance detection [27]. However, this approach has some limitations that hamper further development. The main problem is related to the purification of desired biomaterials, which have crucial influences in the specificity of the biosensors. It is usually time-consuming and expensive to achieve sufficient functional biomaterials for biosensors. In addition, it is also challenging to maintain their native function during the preparation and measurement process of biosensors. All these limitations make it difficult to develop a practical applicable or commercially available biosensors, especially for those in-field applications.

Fast advances in biotechnology provide an alternative approach for the preparation of functional biomaterials for biosensors. This approach can be used to achieve functional biomaterials by the expression of desired type of olfactory or taste receptors either in a heterologous cell system or a cell-free protein synthesis system. This allows for the preparation of functional biomaterials with desired types of olfactory or taste receptors. In addition, this approach makes it easy to graft tags in the prepared receptors, which could greatly facilitate the purification and immobilization of functional biomaterials to improve the performance of biosensors. For example, desired types of olfactory receptors have been expressed in human embryonic kidney (HEK) cells [52,53] and yeast [54–57] and utilized as sensitive elements for biosensors towards odorant detection. Taste receptors have also been expressed based on biotechnology to prepare functional biomaterials for the development of biosensors for taste substance detection [24,58–60]. However, this approach still suffered from the labor-intensive and complex purification process of functional biomaterials. In addition, the expression of receptors in a heterologous cell system usually led to cellular toxic effects that are mainly induced by the membrane incorporation and incompatibility of heterologous expressed olfactory or taste receptors. This results in low expression efficiency, which makes it difficult to improve the preparation efficiency of functional biomaterials. Therefore, cell-free protein synthesis is introduced as an alternative method to prepare functional biomaterials to address this limitation. The synthesis system provides all the necessary components for receptor synthesis such as amino acids, nucleotides, salts and energy-generating factors [61,62]. This cell-free system can not only avoid the cell toxic effect induced by receptor expression, but also could make the preparation process faster, which could mean that the whole expression process could finish within a few hours. Recently, olfactory and taste receptors (Figure 2) have been prepared by a cell-free protein synthesis method and coupled with different transducers for the development of biosensors towards chemical sensing [63,64]. This method could also help the right receptor protein folding via the modification of synthesis reaction conditions. However, it is still a big challenge to produce olfactory receptors in a highly efficient and convenient manner due to their hydrophobicity and dependence on a lipid bilayer environment [26].

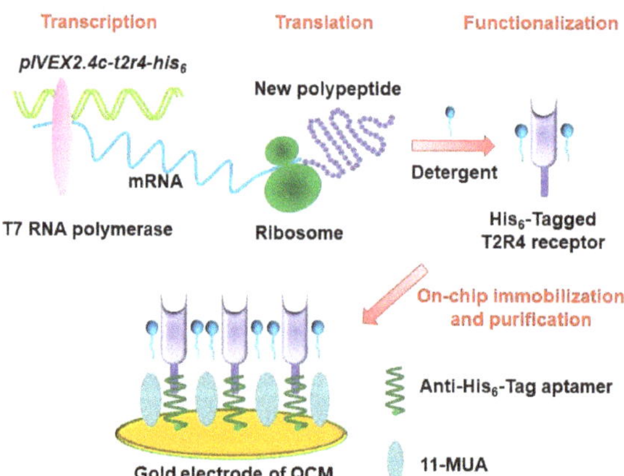

Figure 2. Schematics of preparation of a bitter receptor from cell-free protein expression system for chemical sensing. (Reprinted with permission from ref. [64]. Copyright 2020 Elsevier).

2.2. Fabrication of Field-Effect Devices

Another key component of biosensors is the transducers. Appropriate transducers are also highly essential in order to convert the chemical signals sensed by the functional biomaterials into the measurable signals. For the development of biosensors using biomaterials for chemical sensing, mass-sensitive devices (e.g., quartz crystal microbalances, QCM, and surface acoustic wave, SAW) and field-effect devices (FEDs) are the most commonly used transducers [65–67]. Both of them can record the responsive signals from functional biomaterials upon exposure of chemical substances. Basically, FEDs function as transducers to detect the chemical signals sensed by functional biomaterials and transmit the responsive signals to the peripheral circuits for further signal processing [68]. Therefore, it is crucial to achieve very good and stable coupling between field-effect devices and functional biomaterials in order to develop biosensors with high performances. These biosensors are usually configured with corresponding measurement setup and peripheral circuits in order to readout, collect, and process the detected chemical signals. In this review, we will focus on the biosensors using FEDs as transducers, which mainly include field-effect transistor (FET), light-addressable potentiometric sensor (LAPS), and capacitive electrolyte–insulator–semiconductor (EIS) sensors (Figure 3).

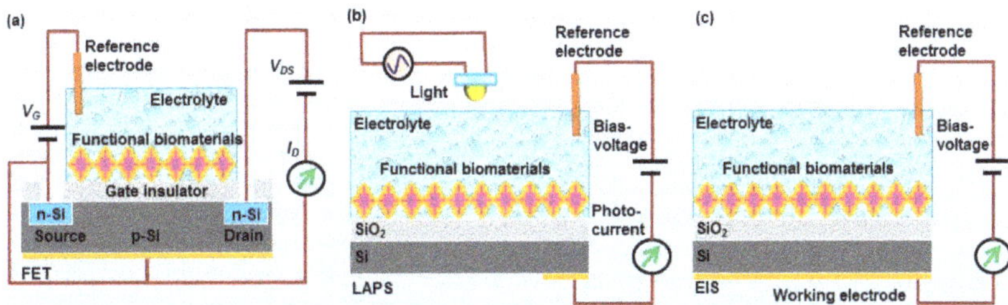

Figure 3. Schematics of field-effect devices utilkized for the development of field-effect sensors using biomaterials for chemical sensing, including (**a**) field-effect transistor (FET), (**b**) light-addressable potentiometric sensor (LAPS), and (**c**) electrolyte-insulator-semiconductor (EIS) sensor.

Fast advances in the micro-fabrication process have greatly facilitated the design and fabrication of various specialized field-effect devices, which could be used as transducers for the development of biosensors towards chemical sensing. For example, FET can be fabricated via a standard micro-fabrication process on silicon wafer [20,69]. The mechanisms and structure of FET are schematically shown in Figure 3a. Usually, an insulator layer is first grown on the surface of silicon wafer via thermal oxidation, which can be used as the gate of FET devices. In some cases, the insulator layer was further deposited with a Si$_3$N$_4$ layer to improve the performance of FET devices. By the following, polyimide is often utilized to form a passivation layer in order to fix with a printed circuit board. Then, the source and drain electrodes are usually fabricated based on photolithography process. Finally, epoxy resin could be used to encapsulate FET devices, which is then fixed with a detection chamber allowing for the exposure of gate surface to the measurement solution inside the detection chamber. With this configuration, the chemical signals sensed by functional biomaterials can be coupled to the gate electrodes of FET, which are then transmitted to the peripheral circuit via the source and drain electrodes of FET.

The LAPS devices and EIS devices are also silicon-based FEDs, as shown in Figure 3b,c [70–72]. Both of them have the same structure of electrolyte–insulator–semiconductor. The difference between them is the measurement configuration. LAPS usually require a moveable focused light to realize addressable measurement on the desired points, while EIS do not require any light illumination during measurement. The structures of LAPS and EIS devices are much simpler than that of FET devices, which greatly facilitated the fabrication process. They are often fabricated based on silicon wafer, which is first thermal oxide with a layer of SiO_2 on its surface to service as insulator layer. In most cases, the insulation layer surface was further grown with a layer of Ta_2O_5 or Si_3N_4 to improve their performance. Then, the oxide layer was removed from the rear side of the wafer, which is then deposited with a metal layer (e.g., Al or Au) to be utilized as Ohmic contact. Finally, the wafer was cut into separate small chips and fixed with a detection chamber. They can thus be applied to the development of biosensors by the immobilization of functional biomaterials onto the gate surface of FEDs exposed to the detection chamber.

2.3. Coupling of Functional Biomaterials with Field-Effect Devices

The coupling of functional biomaterials with FEDs has a significant influence on the performance of biosensors. It is thus highly essential to achieve highly efficient coupling between functional biomaterials and FEDs [23,73]. Highly efficient coupling means not only maintaining the structure and functions of functional biomaterials to make them suitable to serve as the sensitive elements for chemical sensing, but also to transduce the responsive signals into the output signals via FEDs. The output signals will then be further processed by the peripheral circuits [74,75]. Therefore, biosensors usually require the related peripheral circuits and measurement setup to realize the detection of chemical signals.

Functional biomaterials used for the development of biosensors are mainly divided into two categories, i.e., cellular/tissue biomaterials [28] and biomolecules [26,75]. As shown in Figure 4, for cellular/tissue biomaterials, it is ideal to provide a surface that is similar to the cell culture dish, which can provide good surface hydrophilicity and proper surface charges for cell or tissue culture and attachment. However, the surface of FEDs usually consists of silicon dioxide or metal oxide, which shows poor biocompatibility and makes it unsuitable for direct cell or tissue attachment and culture. To improve the biocompatibility of FEDs, a surface modification process is usually required before cell or tissue attachment as reported in some cases [28]. For example, poly-l-ornithine and laminin mixture with a proper rate have been utilized to treat the surface of FEDs to achieve better coupling between cells and FEDs [71]. However, at present, it is still a huge challenge to obtain ideal coupling between cell membrane and the surface of FEDs for the development of biosensors towards chemical sensing.

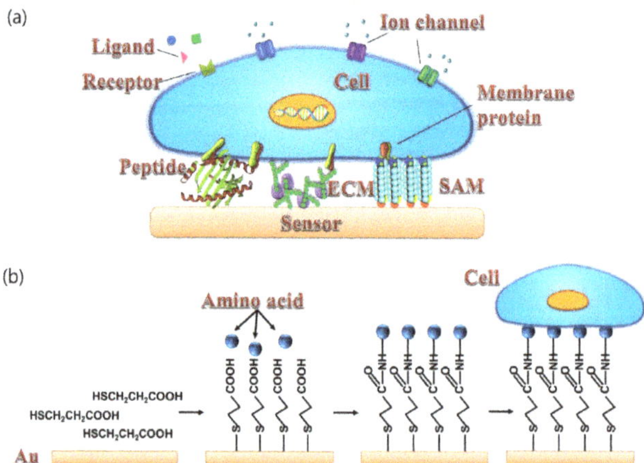

Figure 4. (**a**) Schematics of different surface modification of transducers for cell coupling with sensor including peptide, ECM, and SAM. (**b**) Schematic diagram of cells coupled with gold surface via SAM. (Reprinted with permission from ref. [28]. Copyright 2014 American Chemical Society).

For biomolecules, highly efficient coupling with FEDs usually requires capturing functional biomolecules and avoiding the non-specific adsorption of unrelated molecules to improve the specificity of the biosensors. Current available immobilization approaches mainly include physical adsorption, covalent attachment via chemical reactions, and specific binding via couple molecular pairs, such as a biotin–avidin system. It is crucial to choose the optimal approach to develop biosensors according to the properties of functional biomaterials and surface characters of transducers, since each approach has its intrinsic advantages and disadvantages. For example, physical adsorption has the advantage of being simple, label-free, and reproducible, but it often suffers from the instability of coupling since it can be easily disrupted by minor changes in the microenvironment such as salt density. On the other hand, covalent attachment is much more stable and robust than physical adsorption. In addition, it provides an approach to regulate the surface density of biomolecules, which is very important for achieving optical performances of biosensors [63]. However, the process of covalent attachment is complex and usually require the modification of biomaterials or sensitive surface of transducers, which hamper their applications to some extent. Similarly, the biotin–avidin system can provide strong and robust noncovalent binding between biomolecules and the gate surface of FEDs, which shows very high affinity due to the specific strong interactions between avidin and streptavidin. The biotin–(strept)avidin complex is very strong and robust even in complex environments, which contribute greatly to the repeatability and reproducibility of biosensors. However, the biotin–avidin system also suffers from the complex labelling and reaction process. In general, to obtain the best performances of biosensors, the key point is to specifically couple the functional biomaterials with transducers with high specificity and high stability, which could help to avoid the nonspecific adsorption and generate stable and highly sensitive responsive signals. In addition, it is also very important to maintain the natural sensing functions of biomolecules, especially for those membrane receptors such as olfactory and taste receptors. A hydrophobic environment often needs to be provided, which is crucial to maintaining the chemical sensing function of membrane receptors [66].

3. Development of Field-Effect Sensors Using Biomaterials

Significant progress has been achieved in the field of field-effect sensors using biomaterials as sensitive elements and FEDs as transducers. There are three main types of FEDs that have been applied in the development of biosensors using biomaterials for chemical sensing, which include FET, LAPS, and EIS sensors. Each of them has shown promising prospects in various applications.

3.1. FET-Based Biosensors Using Biomaterials

The most commonly used FEDs is FET, in which the gate surface can be modified with various charge-sensitive layers for the sake of detecting charged biomolecules as well as potential changes induced by excitable cells such as neurons [76]. The obvious advantages of FET come from its innate signal amplification capability, which has shown promising prospects for the detection of weak biological electrical signals. The earliest study of applying FET to biosensors using biomaterials was reported in 2000, in which FET was utilized to couple with the antenna of Colorado potato beetles to form a bioelectronic interface for the detection of a volatile marker (i.e., (Z)-3-hexen-1-ol) of plant damage, as shown in Figure 5 [20]. This biosensor was able to detect the beetle-damaged plants with high performance in the field, which represents a powerful tool for plant protection and food safety. FET can provide a particular reliable joining between an insect antenna and transducer, which makes it ideal for recording the responsive electrical signals from antennae of beetle in response to (Z)-3-hexen-1-ol [77,78]. In short, the potential changes in insect antenna induced by the exposure of specific volatile compounds were recorded by monitoring the changes in the drain current from the FET source and drain electrodes, which are dependent on the concentration of specific chemical volatile compounds. In addition, the small size of biosensors based on FET devices makes it possible to develop portable instruments for the in-field applications such as the detection of explosive compounds in the field of public safety, plant damage detection in the field of plant protection, and smoke detection for building safety.

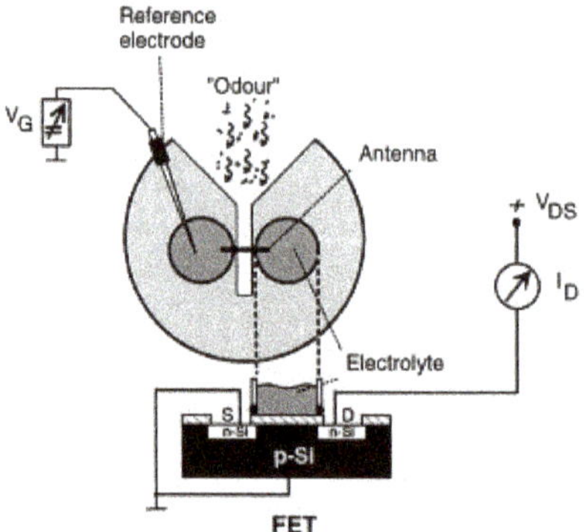

Figure 5. Schematics of field-effect sensors using insect antenna as sensitive element and FET as transducer for volatile marker detection towards plant protection. (Reprinted with permission from ref. [20]. Copyright 2000 Elsevier).

At the molecule level, single wall carbon nanotube (swCNT) has been used to modify the gate surface of FET to generate swCNT-FET, which has been used as transducer and combined with DNA molecules to develop biosensors for chemical sensing [79]. It is indicated that this biosensor with hybrid nanostructure is capable of detecting specific volatile compounds with high sensitivity and specificity. It has been proven that distinct responsive signals can be recorded from swCNT-FET coupled with different bases of DNA molecules. The DNA base sequence-dependent responses suggested that this biosensor is suitable to be utilized to construct gas sensor array towards electronic noses since the responsive signals to the compounds are mainly dependent on the specific base sequences of ssDNA molecules. Similarly, olfactory receptor protein (i.e., hOR2AG1) has been immobilized onto the gate surface of FET that had been previously modified with swCNT [52] or carboxylated-polypyrrole nanotubes (CPNT) [80] for the development of biosensors in order to detect specific odorants with high sensitivity. Similarly, taste receptors have also been attached onto the gate surface of swCNT-FET or CPNT-FET to develop biosensors for the detection of bitter compounds [24,59]. The basic mechanism of these biosensors using receptors rely on the specific interactions between receptors and their ligands, which can often be measured by monitoring the changes in the drain-source current of FETs (Figure 6). These biosensors using human taste receptor protein for bitter compound detection have been applied for the detection of real food samples, which provide valuable tool and show promising prospects in the field of food safety.

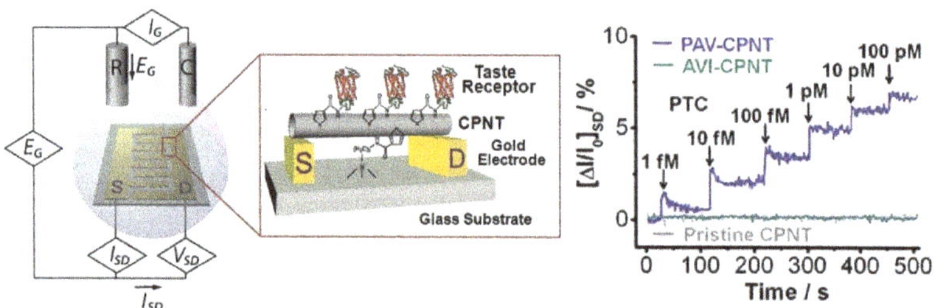

Figure 6. Schematic diagram of a field-effect sensor using taste receptors and its responses to different concentrations of bitter substances. (Reprinted with permission from ref. [59]. Copyright 2013 American Chemical Society).

3.2. LAPS-Based Biosensors Using Biomaterials

LAPS is a surface potential detector, which is suitable for use as a transducer for the development of cell-based biosensors [71,81,82]. It is able to record the changes in extracellular potentials of chemical sensitive cells such as olfactory sensory neurons and taste receptor cells. LAPS has the advantage of having a flat surface and light addressability, which make it ideal for cell measurement. Cells can be cultured randomly on the LAPS surface and a focused light is used to choose the desirable cells for measurement. This overcomes the limitations of FETs, which usually require cells to be cultured precisely on the gate area of the devices.

LAPS has been used to develop various biosensors by the combination with different types of chemical sensitive cells. For instance, olfactory sensory neurons isolated from rat epithelium have been cultured on the LAPS surface to develop a biosensor towards the detection of odorants or neurotransmitters, such as acetic acid and glutamic acid [50,51]. In addition, LAPS has also been reported to be able to record the responsive signals from an intact rat olfactory epithelium induced by different odorants [83,84]. However, the utilization of biomaterials directly originating from animals usually limited by their intrinsic properties such as unknown types of olfactory receptors existing in the biomaterials. To address this issue, bioengineered olfactory receptor neurons expressed with well-defined olfactory receptors were employed to serve as sensitive elements for biosensors towards odorant detection [85]. It is reported that this biosensor based on bioengineered olfactory receptor neurons can be used to detect the specific target odorant in a dose-dependent manner. Furthermore, HEK-293 cells expressed with a specific olfactory receptor, ODR-10, were utilized to couple with LAPS to develop a biosensor for the detection of specific odorant, diacetyl. It has been proven that the measurement of the cell acidification signals recorded by LAPS from single cells can also be used as the responsive signals for the detection of specific odorant stimulation [60].

Similarly, taste cells isolated from rat tongue have been cultured on the LAPS surface to develop biosensors for various taste signals such as bitter [86] and acid [87]. For example, the LAPS surface has been modified with a thin serotonin-sensitive polyvinyl chloride (PVC) membrane, which has been applied in the research of taste cell-to-cell communications via the monitoring of serotonin released from single taste cells [88]. It is reported that this biosensor was able to record the cell membrane potential changes as well as serotonin release from single taste cells in response to acid stimulation and taste mixture (bitter and sweet). In addition, the LAPS surface was modified with a layer of ATP-sensitive aptamers and applied in the detection of ATP release as well as membrane potential changes from single taste bud cells under taste mixture stimulations (Figure 7) [89,90]. It has been proven that this biosensor was able to detect local ATP secretion from a single taste cells in a dose-dependent manner. Biosensors based on LAPS provide a novel and powerful approach to researching taste sensation, which could potentially contribute to the understanding of taste signal transduction mechanisms and cell-to-cell communication.

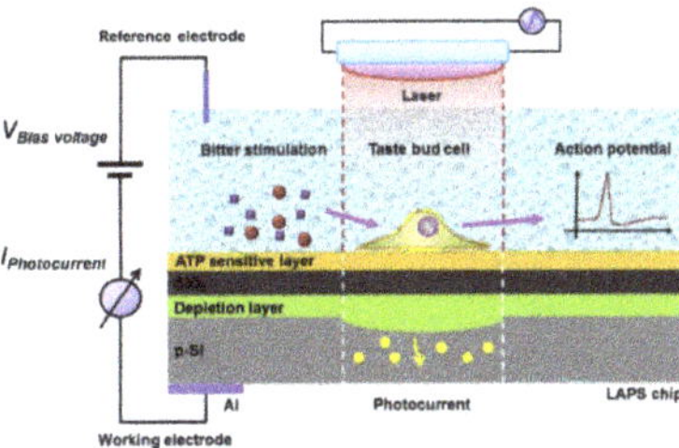

Figure 7. Schematics of a field-effect sensors using LAPS as transducer for the detection of ATP release and membrane potential changes from single taste bud cell. (Reprinted with permission from ref. [90]. Copyright 2018 Elsevier).

3.3. EIS-Based Biosensors Using Biomaterials

Similar to LAPS, capacitive EIS sensors belong to the FEDs category and are a kind of charge-sensitive devices. EIS sensors are able to detect surface charge changes induced by the attachment of charged molecules onto the sensor surface. The most common applications of EIS sensors are related to the label-free detection of pH changes [91,92], ion concentrations [93–98], charged molecules [99], and charged nanoparticles [100,101]. In principle, the attachment of charged molecules or the binding of receptor and ligand occurring on the gate surface of an EIS sensor will lead to the redistribution of surface charge, which will, in turn, result in changes in the space–charge distribution in the semiconductor layer of the EIS sensors. These changes can be reflected by the changes in the output signals of the ES sensors. The decisive advantages of EIS sensors are their simple structure and low cost, which can be fabricated in a convenient and low-cost manner due to the unnecessary involvement of photolithographic process steps or complicated encapsulation procedures. In addition, the capability of surface charge detection makes them suitable for use as transducers for the development of biosensors towards label-free chemical sensing. For instance, EIS sensors have been combined with an olfactory receptor, ODR-10, to develop a biosensor for the detection of a specific odorant, diacetyl [63]. The mechanism of this biosensor was schematically shown in Figure 8a. To improve the coupling efficiency of olfactory receptors with the EIS sensor, the olfactory receptors were prepared using a cell-free protein expression system and fused with a His_6-tag to realize the on-chip purification of sensitive elements based on EIS sensor modified with anti-His_6-tag aptamers. The responsive signals induced by the specific binding between olfactory receptor and its ligand were measured by the monitoring the capacitance changes in the EIS sensor, which is performed by the capacitance $-$ voltage $(C - V)$ and constant-capacitance (ConCap) measurements. It has been proven that this biosensor is able to detect diacetyl in a linear concentration-dependent manner at concentrations ranging from 0.01 nM to 1 nM with a detection limit of 0.01 nM (Figure 8b). This biosensor has great potential to be applied in various fields related to chemical sensing such as biomedicine, food safety, and environmental protection.

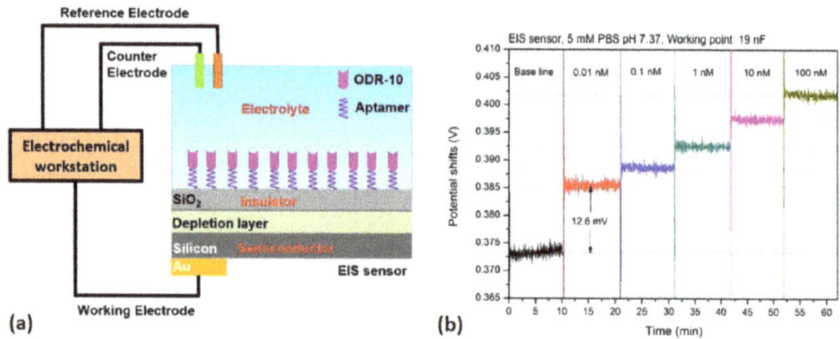

Figure 8. (**a**) Schematics of an EIS sensor using olfactory receptors for the detection of specific odorant. (**b**) Responses of this EIS sensor to different concentrations of odorant. (Reprinted with permission from ref. [63]. Copyright 2019 Elsevier).

4. Conclusions and Prospects

With fast advancement in the microfabrication process, more and more FEDs have been designed and fabricated for various applications. The increasing utilization of FEDs as transducers and functional biomaterials as sensitive elements as part of the development of biosensors for chemical sensing has become a recent trend, which is attracting more and more attention. Biosensors based on FEDs have also shown promising prospects and potential applications in a wide range of fields such as biomedicine, food safety, and environmental protection. However, there are also some limitations that hamper the further development and applications of field-effect sensors using biomaterials. At present, the challenges faced regarding the further development of field-effect sensors using biomaterials mainly include: (1) how to obtain sufficient functional biomaterials that are suitable to serve as sensitive elements, (2) how to fabricate microscale/nanoscale FEDs with sizes that are comparable to the sizes of functional biomaterials, and (3) how to improve the coupling efficiency of biomaterials and transducers as well as the responsive signal transduction efficiency. In the near future, the development of biosensors based on FEDs will probably part of the method of addressing the challenges mentioned above.

Field-effect sensors using biomaterials have shown a powerful capability for chemical sensing, and can not only be used as a novel approach to chemical sensing, but can also be applied in the research for mechanisms of chemical sensations. The final goal of future research and development on field-effect sensors using biomaterials is to improve their performances for chemical sensing in complex environments, which includes improvements on sensitivity, specificity, repeatability, and stability. This usually requires the incorporation of multiple technique advancements in different fields such as biotechnology, nanotechnology, and microfabrication processes. For instance, progress in nanotechnology and microfabrication processes allows for the micro/nano FEDs that could facilitate the coupling with biomaterials. Similarly, advancement in biotechnology could provide novel approaches for the preparation of functional biomaterials that are more suitable to being used as sensitive elements in the development of field-effect sensors using biomaterials for chemical sensing. It is expected that these advancements will greatly contribute to the further development and applications of field-effect sensors using biomaterials.

Author Contributions: C.W.: Conceptualization, Writing—Original draft preparation, Supervision, Funding acquisition; P.Z.: Writing —Review and Editing, Visualization; Y.L.: Writing—Review and Editing; L.D.: Writing—Review and Editing, Supervision, Funding acquisition; P.W.: Conceptualization, Writing—Review and Editing; Supervision, Funding acquisition. All authors have read and agreed to the published version of the manuscript.

Funding: This research was funded by the National Natural Science Foundation of China (Grant No. 32071370, 51861145307, and 31700859).

Institutional Review Board Statement: Not applicable.

Informed Consent Statement: Not applicable.

Data Availability Statement: Not applicable.

Conflicts of Interest: The authors declare no conflict of interest.

References

1. Ache, B.W.; Young, J.M. Olfaction: Diverse species, conserved principles. *Neuron* **2005**, *48*, 417–430. [CrossRef]
2. Chandrashekar, J.; Hoon, M.A.; Ryba, N.J.P.; Zuker, C.S. The receptors and cells for mammalian taste. *Nature* **2006**, *444*, 288–294. [CrossRef] [PubMed]
3. de March, C.A.; Titlow, W.B.; Sengoku, T.; Breheny, P.; Matsunami, H.; McClintock, T.S. Modulation of the combinatorial code of odorant receptor response patterns in odorant mixtures. *Mol. Cell. Neurosci.* **2020**, *104*, 103469. [CrossRef]
4. Efeyan, A.; Comb, W.C.; Sabatini, D.M. Nutrient-sensing mechanisms and pathways. *Nature* **2015**, *517*, 302–310. [CrossRef]
5. Damasio, A.; Carvalho, G.B. The nature of feelings: Evolutionary and neurobiological origins. *Nat. Rev. Neurosci.* **2013**, *14*, 143–152. [CrossRef]
6. Firestein, S. How the olfactory system makes sense of scents. *Nature* **2001**, *413*, 211–218. [CrossRef]
7. Glatz, R.; Bailey-Hill, K. Mimicking nature's noses: From receptor deorphaning to olfactory biosensing. *Prog. Neurobiol.* **2011**, *93*, 270–296. [CrossRef] [PubMed]
8. Pelosi, P.; Zhou, J.J.; Ban, L.P.; Calvello, M. Soluble proteins in insect chemical communication. *Cell. Mol. Life Sci.* **2006**, *63*, 1658–1676. [CrossRef]
9. Buck, L.; Axel, R. A Novel Multigene Family May Encode Odorant Receptors—A Molecular-Basis for Odor Recognition. *Cell* **1991**, *65*, 175–187. [CrossRef]
10. Gire, D.H.; Restrepo, D.; Sejnowski, T.J.; Greer, C.; DeCarlos, J.A.; Lopez-Mascaraque, L. Temporal Processing in the Olfactory System: Can We See a Smell? *Neuron* **2013**, *78*, 416–432. [CrossRef]

11. Matsunami, H.; Montmayeur, J.P.; Buck, L.B. A family of candidate taste receptors in human and mouse. *Nature* **2000**, *404*, 601–604. [CrossRef]
12. Barretto, R.P.J.; Gillis-Smith, S.; Chandrashekar, J.; Yarmolinsky, D.A.; Schnitzer, M.J.; Ryba, N.J.P.; Zuker, C.S. The neural representation of taste quality at the periphery. *Nature* **2015**, *517*, 373–376. [CrossRef] [PubMed]
13. Behrens, M.; Meyerhof, W. Bitter taste receptors and human bitter taste perception. *Cell. Mol. Life Sci.* **2006**, *63*, 1501–1509. [CrossRef] [PubMed]
14. Sankaran, S.; Khot, L.R.; Panigrahi, S. Biology and applications of olfactory sensing system: A review. *Sens. Actuators B Chem.* **2012**, *171–172*, 1–17. [CrossRef]
15. Bessac, B.F.; Jordt, S.E. Sensory detection and responses to toxic gases: Mechanisms, health effects, and countermeasures. *Proc. Am. Thorac. Soc.* **2010**, *7*, 269–277. [CrossRef] [PubMed]
16. Cheema, J.A.; Carraher, C.; Plank, N.O.V.; Travas-Sejdic, J.; Kralicek, A. Insect odorant receptor-based biosensors: Current status and prospects. *Biotechnol. Adv.* **2021**, *53*, 107840. [CrossRef]
17. Er, S.; Laraib, U.; Arshad, R.; Sargazi, S.; Rahdar, A.; Pandey, S.; Thakur, V.K.; Díez-Pascual, A.M. Amino acids, peptides, and proteins: Implications for nanotechnological applications in biosensing and drug/gene delivery. *Nanomaterials* **2021**, *11*, 3002. [CrossRef]
18. Ishimaru, Y. Molecular mechanisms of taste transduction in vertebrates. *Odontology* **2009**, *97*, 1–7. [CrossRef] [PubMed]
19. Huang, Y.A.; Maruyama, Y.; Stimac, R.; Roper, S.D. Presynaptic (Type III) cells in mouse taste buds sense sour (acid) taste. *J. Physiol.* **2008**, *586*, 2903–2912. [CrossRef]
20. Schutz, S.; Schoning, M.J.; Schroth, P.; Malkoc, U.; Weissbecker, B.; Kordos, P.; Luth, H.; Hummel, H.E. An insect-based BioFET as a bioelectronic nose. *Sens. Actuat. B Chem.* **2000**, *65*, 291–295. [CrossRef]
21. Pohlmann, C.; Wang, Y.R.; Humenik, M.; Heidenreich, B.; Gareis, M.; Sprinzl, M. Rapid, specific and sensitive electrochemical detection of foodborne bacteria. *Biosens. Bioelectron.* **2009**, *24*, 2766–2771. [CrossRef]
22. Tahara, Y.; Toko, K. Electronic Tongues-A Review. *IEEE Sens. J.* **2013**, *13*, 3001–3011. [CrossRef]
23. Son, M.; Kim, D.; Ko, H.J.; Hong, S.; Park, T.H. A portable and multiplexed bioelectronic sensor using human olfactory and taste receptors. *Biosens. Bioelectron.* **2017**, *87*, 901–907. [CrossRef] [PubMed]
24. Kim, T.H.; Song, H.S.; Jin, H.J.; Lee, S.H.; Namgung, S.; Kim, U.K.; Park, T.H.; Hong, S. "Bioelectronic super-taster" device based on taste receptor-carbon nanotube hybrid structures. *Lab Chip* **2011**, *11*, 2262–2267. [CrossRef] [PubMed]
25. Lee, S.H.; Park, T.H. Recent Advances in the Development of Bioelectronic Nose. *Biotechnol. Bioproc. Eng.* **2010**, *15*, 22–29. [CrossRef]
26. Du, L.P.; Wu, C.S.; Liu, Q.J.; Huang, L.Q.; Wang, P. Recent advances in olfactory receptor-based biosensors. *Biosens. Bioelectron.* **2013**, *42*, 570–580. [CrossRef]
27. Wu, C.S.; Du, L.P.; Zou, L.; Zhao, L.H.; Huang, L.Q.; Wang, P. Recent advances in taste cell- and receptor-based biosensors. *Sens. Actuators B Chem.* **2014**, *201*, 75–85. [CrossRef]
28. Liu, Q.J.; Wu, C.S.; Cai, H.; Hu, N.; Zhou, J.; Wang, P. Cell-Based Biosensors and Their Application in Biomedicine. *Chem. Rev.* **2014**, *114*, 6423–6461. [CrossRef]
29. Laucirica, G.; Terrones, Y.T.; Cayon, V.; Cortez, M.L.; Toimil-Molares, M.E.; Trautmann, C.; Marmisolle, W.; Azzaroni, O. Biomimetic solid-state nanochannels for chemical and biological sensing applications. *TrAC Trends Anal. Chem.* **2021**, *144*, 116425. [CrossRef]
30. Yaraghi, N.A.; Kisailus, D. Biomimetic Structural Materials: Inspiration from Design and Assembly. *Annu. Rev. Phys. Chem.* **2018**, *69*, 23–57. [CrossRef]
31. Liu, K.; Jiang, L. Bio-inspired design of multiscale structures for function integration. *Nano Today* **2011**, *6*, 155–175. [CrossRef]
32. Huang, Y.F.; Chattopadhyay, S.; Jen, Y.J.; Peng, C.Y.; Liu, T.A.; Hsu, Y.K.; Pan, C.L.; Lo, H.C.; Hsu, C.H.; Chang, Y.H.; et al. Improved broadband and quasi-omnidirectional anti-reflection properties with biomimetic silicon nanostructures. *Nat. Nanotechnol.* **2007**, *2*, 770–774. [CrossRef]
33. Zhang, C.; McAdams, D.A., II; Grunlan, J.C. Nano/micro-manufacturing of bioinspired materials: A review of methods to mimic natural structures. *Adv. Mater.* **2016**, *28*, 6292–6321. [CrossRef] [PubMed]
34. Tripathy, A.; Nine, M.J.; Losic, D.; Silva, F.S. Nature inspired emerging sensing technology: Recent progress and perspectives. *Mater. Sci. Eng. R Rep.* **2021**, *146*, 100647. [CrossRef]
35. Song, F.; Su, H.L.; Han, J.; Zhang, D.; Chen, Z.X. Fabrication and good ethanol sensing of biomorphic SnO2 with architecture hierarchy of butterfly wings. *Nanotechnology* **2009**, *20*, 495502. [CrossRef] [PubMed]
36. Song, F.; Su, H.L.; Han, J.; Xu, J.Q.; Zhang, D. Controllable synthesis and gas response of biomorphic SnO2 with architecture hierarchy of butterfly wings. *Sens. Actuators B Chem.* **2010**, *145*, 39–45. [CrossRef]
37. Zhu, S.M.; Zhang, D.; Chen, Z.X.; Gu, J.J.; Li, W.F.; Jiang, H.B.; Zhou, G. A simple and effective approach towards biomimetic replication of photonic structures from butterfly wings. *Nanotechnology* **2009**, *20*, 315303. [CrossRef]
38. Potyrailo, R.A.; Bonam, R.K.; Hartley, J.G.; Starkey, T.A.; Vukusic, P.; Vasudev, M.; Bunning, T.; Naik, R.R.; Tang, Z.X.; Palacios, M.A.; et al. Towards outperforming conventional sensor arrays with fabricated individual photonic vapour sensors inspired by Morpho butterflies. *Nat. Commun.* **2015**, *6*, 7959. [CrossRef] [PubMed]
39. Poncelet, O.; Tallier, G.; Mouchet, S.R.; Crahay, A.; Rasson, J.; Kotipalli, R.; Deparis, O.; Francis, L.A. Vapour sensitivity of an ALD hierarchical photonic structure inspired by Morpho. *Bioinspir. Biomim.* **2016**, *11*, 036011. [CrossRef]

40. Rasson, J.; Poncelet, O.; Mouchet, S.R.; Deparis, O.; Francis, L.A. Vapor sensing using a bio-inspired porous silicon photonic crystal. *Mater. Today Proc.* **2017**, *4*, 5006–5012. [CrossRef]

41. Dong, Q.; Su, H.L.; Zhang, D.; Zhang, F.Y. Fabrication and gas sensitivity of SnO2 hierarchical films with interwoven tubular conformation by a biotemplate-directed sol-gel technique. *Nanotechnology* **2006**, *17*, 3968–3972. [CrossRef]

42. Tian, J.L.; Pan, F.; Xue, R.Y.; Zhang, W.; Fang, X.T.; Liu, Q.L.; Wang, Y.H.; Zhang, Z.J.; Zhang, D. A highly sensitive room temperature H2S gas sensor based on SnO_2 multi-tube arrays bio-templated from insect bristles. *Dalton Trans.* **2015**, *44*, 7911–7916. [CrossRef]

43. Biswas, A.; Bayer, I.S.; Biris, A.S.; Wang, T.; Dervishi, E.; Faupel, F. Advances in top-down and bottom-up surface nanofabrication: Techniques, applications & future prospects. *Adv. Colloid Interface Sci.* **2012**, *170*, 2–27. [CrossRef] [PubMed]

44. Poghossian, A.; Schoning, M.J. Recent progress in silicon-based biologically sensitive field-effect devices. *Curr. Opin. Electrochem.* **2021**, *29*, 100811. [CrossRef]

45. Bushdid, C.; Magnasco, M.O.; Vosshall, L.B.; Keller, A. Humans Can Discriminate More than 1 Trillion Olfactory Stimuli. *Science* **2014**, *343*, 1370–1372. [CrossRef] [PubMed]

46. Maffei, A.; Haley, M.; Fontanini, A. Neural processing of gustatory information in insular circuits. *Curr. Opin. Neurobiol.* **2012**, *22*, 709–716. [CrossRef] [PubMed]

47. Leal, W.S. Odorant Reception in Insects: Roles of Receptors, Binding Proteins, and Degrading Enzymes. *Annu. Rev. Entomol.* **2013**, *58*, 373–391. [CrossRef] [PubMed]

48. Wu, T.Z. A piezoelectric biosensor as an olfactory receptor for odour detection: Electronic nose. *Biosens. Bioelectron.* **1999**, *14*, 9–18. [CrossRef]

49. Huotari, M.J. Biosensing by insect olfactory receptor neurons. *Sens. Actuators B Chem.* **2000**, *71*, 212–222. [CrossRef]

50. Liu, Q.J.; Cai, H.; Xu, Y.; Li, Y.; Li, R.; Wang, P. Olfactory cell-based biosensor: A first step towards a neurochip of bioelectronic nose. *Biosens. Bioelectron.* **2006**, *22*, 318–322. [CrossRef]

51. Wu, C.S.; Chen, P.H.; Yu, H.; Liu, Q.J.; Zong, X.L.; Cai, H.; Wang, P. A novel biomimetic olfactory-based biosensor for single olfactory sensory neuron monitoring. *Biosens. Bioelectron.* **2009**, *24*, 1498–1502. [CrossRef]

52. Kim, T.H.; Lee, S.H.; Lee, J.; Song, H.S.; Oh, E.H.; Park, T.H.; Hong, S. Single-Carbon-Atomic-Resolution Detection of Odorant Molecules using a Human Olfactory Receptor-based Bioelectronic Nose. *Adv. Mater.* **2009**, *21*, 91–94. [CrossRef]

53. Wu, C.S.; Du, L.P.; Wang, D.; Wang, L.; Zhao, L.H.; Wang, P. A novel surface acoustic wave-based biosensor for highly sensitive functional assays of olfactory receptors. *Biochem. Bioph. Res. Commun.* **2011**, *407*, 18–22. [CrossRef] [PubMed]

54. Benilova, I.V.; Vidic, J.M.; Pajot-Augy, E.; Soldatkin, A.P.; Martelet, C.; Jafftezic-Renault, N. Electrochemical study of human olfactory receptor OR17-40 stimulation by odorants in solution. *Mat. Sci. Eng. C* **2008**, *28*, 633–639. [CrossRef]

55. Hou, Y.X.; Jaffrezic-Renault, N.; Martelet, C.; Zhang, A.D.; Minic-Vidic, J.; Gorojankina, T.; Persuy, M.A.; Pajot-Augy, E.; Salesse, R.; Akimov, V.; et al. A novel detection strategy for odorant molecules based on controlled bioengineering of rat olfactory receptor 17. *Biosens. Bioelectron.* **2007**, *22*, 1550–1555. [CrossRef] [PubMed]

56. Marrakchi, M.; Vidic, J.; Jaffrezic-Renault, N.; Martelet, C.; Pajot-Augy, E. A new concept of olfactory biosensor based on interdigitated microelectrodes and immobilized yeasts expressing the human receptor OR17-40. *Eur. Biophys. J.* **2007**, *36*, 1015–1018. [CrossRef] [PubMed]

57. Vidic, J.; Pla-Roca, M.; Grosclaude, J.; Persuy, M.A.; Monnerie, R.; Caballero, D.; Errachid, A.; Hou, Y.X.; Jaffrezic-Renault, N.; Salesse, R.; et al. Gold surface functionalization and patterning for specific immobilization of olfactory receptors carried by nanosomes. *Anal. Chem.* **2007**, *79*, 3280–3290. [CrossRef] [PubMed]

58. Wu, C.S.; Du, L.P.; Zou, L.; Huang, L.Q.; Wang, P. A biomimetic bitter receptor-based biosensor with high efficiency immobilization and purification using self-assembled aptamers. *Analyst* **2013**, *138*, 5989–5994. [CrossRef]

59. Song, H.S.; Kwon, O.S.; Lee, S.H.; Park, S.J.; Kim, U.K.; Jang, J.; Park, T.H. Human Taste Receptor-Functionalized Field Effect Transistor as a Human-Like Nanobioelectronic Tongue. *Nano Lett.* **2013**, *13*, 172–178. [CrossRef]

60. Du, L.P.; Zou, L.; Zhao, L.H.; Huang, L.Q.; Wang, P.; Wu, C.S. Label-free functional assays of chemical receptors using a bioengineered cell-based biosensor with localized extracellular acidification measurement. *Biosens. Bioelectron.* **2014**, *54*, 623–627. [CrossRef]

61. Katzen, F.; Chang, G.; Kudlicki, W. The past, present and future of cell-free protein synthesis. *Trends Biotechnol.* **2005**, *23*, 150–156. [CrossRef]

62. Kaiser, L.; Graveland-Bikker, J.; Steuerwald, D.; Vanberghem, M.; Herlihy, K.; Zhang, S.G. Efficient cell-free production of olfactory receptors: Detergent optimization, structure, and ligand binding analyses. *Proc. Natl. Acad. Sci. USA* **2008**, *105*, 15726–15731. [CrossRef] [PubMed]

63. Chen, F.M.; Wang, J.; Du, L.P.; Zhang, X.; Zhang, F.; Chen, W.; Cai, W.; Wu, C.S.; Wang, P. Functional expression of olfactory receptors using cell-free expression system for biomimetic sensors towards odorant detection. *Biosens. Bioelectron.* **2019**, *130*, 382–388. [CrossRef] [PubMed]

64. Du, L.P.; Chen, W.; Tian, Y.L.; Zhu, P.; Wu, C.S.; Wang, P. A biomimetic taste biosensor based on bitter receptors synthesized and purified on chip from a cell-free expression system. *Sens. Actuators B Chem.* **2020**, *312*, 127949. [CrossRef]

65. Wu, C.S.; Wang, L.J.; Zhou, J.; Zhao, L.H.; Wang, P. The progress of olfactory transduction and biomimetic olfactory-based biosensors. *Chin. Sci. Bull.* **2007**, *52*, 1886–1896. [CrossRef]

66. Wu, C.S.; Lillehoj, P.B.; Wang, P. Bioanalytical and chemical sensors using living taste, olfactory, and neural cells and tissues: A short review. *Analyst* **2015**, *140*, 7048–7061. [CrossRef] [PubMed]

67. Wu, C.S.; Du, Y.W.; Huang, L.Q.; Galeczki, Y.B.; Dagan-Wiener, A.; Naim, M.; Niv, M.Y.; Wang, P. Biomimetic Sensors for the Senses: Towards Better Understanding of Taste and Odor Sensation. *Sensors* **2017**, *17*, 2881. [CrossRef] [PubMed]

68. Kwon, D.; Jung, G.; Shin, W.; Jeong, Y.; Hong, S.; Oh, S.; Kim, J.; Bae, J.H.; Park, B.G.; Lee, J.H. Efficient fusion of spiking neural networks and FET-type gas sensors for a fast and reliable artificial olfactory system. *Sens. Actuators B Chem.* **2021**, *345*, 130419. [CrossRef]

69. Schoning, M.J.; Schroth, P.; Schutz, S. The use of insect chemoreceptors for the assembly of biosensors based on semiconductor field-effect transistors. *Electroanalysis* **2000**, *12*, 645–652.

70. Hafeman, D.G.; Parce, J.W.; Mcconnell, H.M. Light-Addressable Potentiometric Sensor for Biochemical Systems. *Science* **1988**, *240*, 1182–1185. [CrossRef]

71. Ismail, A.B.M.; Yoshinobu, T.; Iwasaki, H.; Sugihara, H.; Yukimasa, T.; Hirata, I.; Iwata, H. Investigation on light-addressable potentiometric sensor as a possible cell-semiconductor hybrid. *Biosens. Bioelectron.* **2003**, *18*, 1509–1514. [CrossRef]

72. Bronder, T.S.; Poghossian, A.; Scheja, S.; Wu, C.S.; Keusgen, M.; Mewes, D.; Schoning, M.J. DNA Immobilization and Hybridization Detection by the Intrinsic Molecular Charge Using Capacitive Field-Effect Sensors Modified with a Charged Weak Polyelectrolyte Layer. *ACS Appl. Mater. Inter.* **2015**, *7*, 20068–20075. [CrossRef]

73. Yang, H.; Lee, M.; Kim, D.; Hong, S.; Park, T.H. Bioelectronic nose using olfactory receptor-embedded nanodiscs. *Methods Mol. Biol.* **2018**, *1820*, 239–249.

74. Full, J.; Baumgarten, Y.; Delbrück, L.; Sauer, A.; Miehe, R. Market perspectives and future fields of application of odor detection biosensors within the biological transformation—A systematic analysis. *Biosensors* **2021**, *11*, 93. [CrossRef]

75. Barbosa, A.J.M.; Oliveira, A.R.; Roque, A.C.A. Protein- and Peptide-Based Biosensors in Artificial Olfaction. *Trends Biotechnol.* **2018**, *36*, 1244–1258. [CrossRef] [PubMed]

76. Schoning, M.J.; Poghossian, A. Bio FEDs (Field-Effect devices): State-of-the-art and new directions. *Electroanalysis* **2006**, *18*, 1893–1900. [CrossRef]

77. Schoning, M.J.; Schutz, S.; Schroth, P.; Weissbecker, B.; Steffen, A.; Kordos, P.; Hummel, H.E.; Luth, H. A BioFET on the basis of intact insect antennae. *Sens. Actuators B Chem.* **1998**, *47*, 235–238. [CrossRef]

78. Schroth, P.; Schoning, M.J.; Luth, H.; Weissbecker, B.; Hummel, H.E.; Schutz, S. Extending the capabilities of an antenna/chip biosensor by employing various insect species. *Sens. Actuators B Chem.* **2001**, *78*, 1–5. [CrossRef]

79. Staii, C.; Johnson, A.T. DNA-decorated carbon nanotubes for chemical sensing. *Nano Lett.* **2005**, *5*, 1774–1778. [CrossRef]

80. Yoon, H.; Lee, S.H.; Kwon, O.S.; Song, H.S.; Oh, E.H.; Park, T.H.; Jang, J. Polypyrrole Nanotubes Conjugated with Human Olfactory Receptors: High-Performance Transducers for FET-Type Bioelectronic Noses. *Angew. Chem.* **2009**, *48*, 2755–2758. [CrossRef]

81. Stein, B.; George, M.; Gaub, H.E.; Parak, W.J. Extracellular measurements of averaged ionic currents with the light-addressable potentiometric sensor (LAPS). *Sens. Actuators B Chem.* **2004**, *98*, 299–304. [CrossRef]

82. Xu, G.X.; Ye, X.S.; Qin, L.F.; Xu, Y.; Li, Y.; Li, R.; Wang, P. Cell-based biosensors based on light-addressable potentiometric sensors for single cell monitoring. *Biosens. Bioelectron.* **2005**, *20*, 1757–1763. [CrossRef] [PubMed]

83. Liu, Q.J.; Ye, W.W.; Hu, N.; Cai, H.; Yu, H.; Wang, P. Olfactory receptor cells respond to odors in a tissue and semiconductor hybrid neuron chip. *Biosens. Bioelectron.* **2010**, *26*, 1672–1678. [CrossRef] [PubMed]

84. Liu, Q.J.; Ye, W.W.; Yu, H.; Hu, N.; Du, L.P.; Wang, P.; Yang, M. Olfactory mucosa tissue-based biosensor: A bioelectronic nose with receptor cells in intact olfactory epithelium. *Sens. Actuators B Chem.* **2010**, *146*, 527–533. [CrossRef]

85. Du, L.P.; Wu, C.S.; Peng, H.; Zhao, L.H.; Huang, L.Q.; Wang, P. Bioengineered olfactory sensory neuron-based biosensor for specific odorant detection. *Biosens. Bioelectron.* **2013**, *40*, 401–406. [CrossRef]

86. Wu, C.S.; Du, L.P.; Mao, L.H.; Wang, P. A Novel Bitter Detection Biosensor Based on Light Addressable Potentiometric Sensor. *J. Innov. Opt. Health Sci.* **2012**, *5*, 1250008. [CrossRef]

87. Chen, P.H.; Liu, X.D.; Wang, B.Q.; Cheng, G.; Wang, P. A biomimetic taste receptor cell-based biosensor for electrophysiology recording and acidic sensation. *Sens. Actuators B Chem.* **2009**, *139*, 576–583. [CrossRef]

88. Chen, P.H.; Zhang, W.; Chen, P.; Zhou, Z.Y.; Chen, C.; Hu, J.S.; Wang, P. A serotonin-sensitive sensor for investigation of taste cell-to-cell communication. *Biosens. Bioelectron.* **2011**, *26*, 3054–3058. [CrossRef]

89. Wu, C.S.; Du, L.P.; Zou, L.; Zhao, L.H.; Wang, P. An ATP sensitive light addressable biosensor for extracellular monitoring of single taste receptor cell. *Biomed. Microdevices* **2012**, *14*, 1047–1053. [CrossRef]

90. Du, L.P.; Wang, J.; Chen, W.; Zhao, L.H.; Wu, C.S.; Wang, P. Dual functional extracellular recording using a light-addressable potentiometric sensor for bitter signal transduction. *Anal. Chim. Acta* **2018**, *1022*, 106–112. [CrossRef]

91. Poghossian, A.S. The Super-Nernstian Ph Sensitivity of Ta_2O_5-Gate Isfets. *Sens. Actuators B Chem.* **1992**, *7*, 367–370. [CrossRef]

92. Poghossian, A.; Baade, A.; Emons, H.; Schoning, M.J. Application of ISFETs for pH measurement in rain droplets. *Sens. Actuators B Chem.* **2001**, *76*, 634–638. [CrossRef]

93. Poghossian, A.; Mai, D.T.; Mourzina, Y.; Schoning, M.J. Impedance effect of an ion-sensitive membrane: Characterisation of an EMIS sensor by impedance spectroscopy, capacitance-voltage and constant-capacitance method. *Sens. Actuators B Chem.* **2004**, *103*, 423–428. [CrossRef]

94. Jimenez-Jorquera, C.; Orozco, J.; Baldi, A. ISFET Based Microsensors for Environmental Monitoring. *Sensors* **2010**, *10*, 61–83. [CrossRef] [PubMed]
95. Gun, J.; Schoning, M.J.; Abouzar, M.H.; Poghossian, A.; Katz, E. Field-effect nanoparticle-based glucose sensor on a chip: Amplification effect of coimmobilized redox species. *Electroanalysis* **2008**, *20*, 1748–1753. [CrossRef]
96. Lee, C.S.; Kim, S.K.; Kim, M. Ion-Sensitive Field-Effect Transistor for Biological Sensing. *Sensors* **2009**, *9*, 7111–7131. [CrossRef]
97. Siqueira, J.R.; Abouzar, M.H.; Backer, M.; Zucolotto, V.; Poghossian, A.; Oliveira, O.N.; Schoning, M.J. Carbon nanotubes in nanostructured films: Potential application as amperometric and potentiometric field-effect (bio-)chemical sensors. *Phys. Status Solidi A* **2009**, *206*, 462–467. [CrossRef]
98. Nakazato, K. An Integrated ISFET Sensor Array. *Sensors* **2009**, *9*, 8831–8851. [CrossRef]
99. Stern, E.; Vacic, A.; Reed, M.A. Semiconducting Nanowire Field-Effect Transistor Biomolecular Sensors. *IEEE Trans. Electron Devices* **2008**, *55*, 3119–3130. [CrossRef]
100. Gun, J.; Gutkin, V.; Lev, O.; Boyen, H.G.; Saitner, M.; Wagner, P.; D'Olieslaeger, M.; Abouzar, M.H.; Poghossian, A.; Schoning, M.J. Tracing Gold Nanoparticle Charge by Electrolyte Insulator Semiconductor Devices. *J. Phys. Chem. C* **2011**, *115*, 4439–4445. [CrossRef]
101. Poghossian, A.; Backer, M.; Mayer, D.; Schoning, M.J. Gating capacitive field-effect sensors by the charge of nanoparticle/molecule hybrids. *Nanoscale* **2015**, *7*, 1023–1031. [CrossRef] [PubMed]

sensors

Review

Process Variability in Top-Down Fabrication of Silicon Nanowire-Based Biosensor Arrays

Marcel Tintelott, Vivek Pachauri, Sven Ingebrandt and Xuan Thang Vu *

Institute of Materials in Electrical Engineering 1, RWTH Aachen University, Sommerfeldstr. 24, 52074 Aachen, Germany; tintelott@iwe1.rwth-aachen.de (M.T.); pachauri@iwe1.rwth-aachen.de (V.P.); ingebrandt@iwe1.rwth-aachen.de (S.I.)
* Correspondence: vu@iwe1.rwth-aachen.de; Tel.: +49-241-80-27816

Abstract: Silicon nanowire field-effect transistors (SiNW-FET) have been studied as ultra-high sensitive sensors for the detection of biomolecules, metal ions, gas molecules and as an interface for biological systems due to their remarkable electronic properties. "Bottom-up" or "top-down" approaches that are used for the fabrication of SiNW-FET sensors have their respective limitations in terms of technology development. The "bottom-up" approach allows the synthesis of silicon nanowires (SiNW) in the range from a few nm to hundreds of nm in diameter. However, it is technologically challenging to realize reproducible bottom-up devices on a large scale for clinical biosensing applications. The top-down approach involves state-of-the-art lithography and nanofabrication techniques to cast SiNW down to a few 10s of nanometers in diameter out of high-quality Silicon-on-Insulator (SOI) wafers in a controlled environment, enabling the large-scale fabrication of sensors for a myriad of applications. The possibility of their wafer-scale integration in standard semiconductor processes makes SiNW-FETs one of the most promising candidates for the next generation of biosensor platforms for applications in healthcare and medicine. Although advanced fabrication techniques are employed for fabricating SiNW, the sensor-to-sensor variation in the fabrication processes is one of the limiting factors for a large-scale production towards commercial applications. To provide a detailed overview of the technical aspects responsible for this sensor-to-sensor variation, we critically review and discuss the fundamental aspects that could lead to such a sensor-to-sensor variation, focusing on fabrication parameters and processes described in the state-of-the-art literature. Furthermore, we discuss the impact of functionalization aspects, surface modification, and system integration of the SiNW-FET biosensors on post-fabrication-induced sensor-to-sensor variations for biosensing experiments.

Keywords: silicon nanowire field-effect transistor; device-to-device variation; biosensor; top-down fabrication; surface modification

Citation: Tintelott, M.; Pachauri, V.; Ingebrandt, S.; Vu, X.T. Process Variability in Top-Down Fabrication of Silicon Nanowire-Based Biosensor Arrays. *Sensors* **2021**, *21*, 5153. https://doi.org/10.3390/s21155153

Academic Editor: Goutam Koley

Received: 9 July 2021
Accepted: 27 July 2021
Published: 29 July 2021

Publisher's Note: MDPI stays neutral with regard to jurisdictional claims in published maps and institutional affiliations.

1. Introduction

Devices for point-of-care testing (POCT) gained attention in recent years due to the societal need for on-demand analysis and a rising market for such devices. New technologies and device miniaturization foster this ever-increasing growth in the development of POCT devices. The sensor needs to provide a clear signal with low false-positive and low false-negative rates for point-of-care applications. More importantly, it should be easy to use and disposable [1]. Biosensors based on silicon nanowire field-effect transistors (SiNW-FET) are amongst the most promising candidates for future clinical POCT diagnostic technology due to their low limit-of-detection (LoD), the possibility for multiplexing, and label-free sensing [2–4]. As illustrated in Figure 1, the SiNW-FET is used for versatile applications ranging from sensing of ions and biomolecular detection, action potential recording. SiNW-FETs show ultra-high sensitivity to detect different biomolecules such as DNA, proteins, or antibody-antigens [5–8]. Furthermore, SiNW-FETs have been utilized to study not only the action potential of cardiac muscle cells or neurons [9,10] but also the action potential propagation along the axon of a neuron [11]. Compared to their planar and microscale counterpart, SiNW-FETs show an increased signal-to-noise (S/N) ratio during the recording of action potentials [9]. By modifying the surface of the SiNW with an ion-specific aptamer enables local monitoring of K^+ efflux during neurotransmission [12].

Nevertheless, a commercial breakthrough of this remarkable biosensor is still pending [13]. One of the hurdles for the applications is the sensor-to-sensor variation, which is caused by the complexity of the sensor preparations. The sensor-to-sensor variation induces the variation in the electrical performance of the sensors and thus creates the need for recalibration for the response of different devices [14]. The need for calibration increases the chance of user errors, leading to an incorrect response of the sensor and limiting the applicability of label-free SiNW-FET biosensors in general.

Several factors are involved in the sensor-to-sensor variation of the SiNW-FETs, including sensor design, sensor fabrication, surface chemistry, and readout methods. These aspects need to be optimized for final products using the SiNW-FETs to meet the standard requirements of point-of-care diagnostic tools. A reliable and reproducible sensor design and fabrication processes are the first and most crucial steps in the SiNW-FET biosensor fabrication blockchain. It is important to identify aspects in the design and fabrication process that may cause the variations.

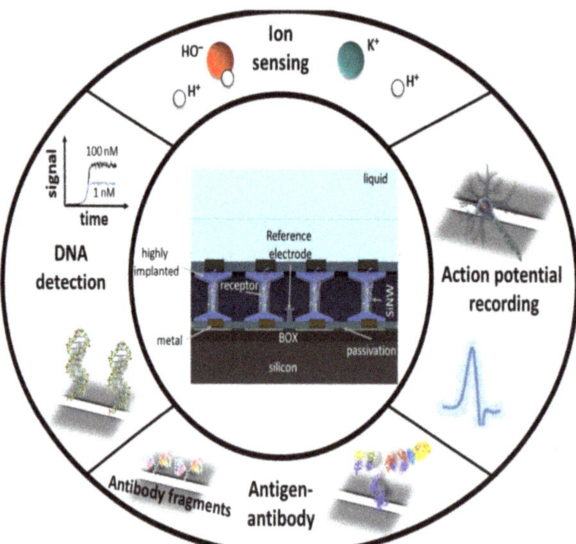

Figure 1. Schematic overview of different applications of SiNW-FETs. The inner ring shows a schematic illustration of a SiNW-FET and a sensing setup. The outer ring illustrates different applications of SiNW-FETs.

SiNW-FET sensors are fabricated by either "top-down" or "bottom-up" approaches [15,16]. In the "bottom-up" approach, firstly, SiNWs are vertically grown on a silicon substrate using Vapor-Liquid-Solid (VLS) technique or oxide assisted growth (OAG) technique [17–19]. Secondly, the SiNWs are transferred and laid down to another substrate using different methods, such as polydimethylsiloxane (PDMS) transfer or Langmuir–Blodgett transfer techniques [4,15,20]. Finally, electrical contacts to the SiNWs by electron beam lithography and lift-off techniques using noble metals are created. A precise arrangement of the SiNWs on a wafer-scale level is challenging with the current transfer techniques, and thus, the "bottom-up" is limited in the device integration and large-scale production, a key factor for POCT application. Due to its intrinsic limitations, the "bottom-up" approach is less favorable for large-scale biosensors fabrication [16].

The "top-down" approach is based on the well-established complementary metal-oxide semiconductors (CMOS) industrial processes allowing very-large-scale integration and thus enabling low-cost fabrication [4,21]. Hence, this approach is much more attractive in large-scale production and system integration. Starting from a Silicon-on-Insulator (SOI) wafer, the structure of the SiNW sensor is firstly defined at desire positions on top of the wafer by advanced lithographic methods such as electron-beam lithography (EBL), nanoimprint lithography (NIL), or sidewall transfer lithography (STL) [22–24]. Subsequent etching techniques, either by reactive ion etching (RIE) or wet chemical etching using tetramethylammonium hydroxide (TMAH) or a combination of both techniques, are used to transfer the structure to the top silicon layer of the SOI wafer. Afterward, microfabrication techniques are used to finalize the devices. Ion-implantation was used to create the source and the drain as well as to create the ohmic contact for the device. An ultra-thin layer of oxide was grown on top of the SiNW to create the gate dielectric layer. A thick passivation layer was deposited on the source and the drain contact to enable the device to work reliably when interfacing with the liquid environment [6,13,22,25–30]. Each fabrication step induces variations that may alter the electronic characteristic from device to device. Even though variations will always occur during fabrication, they can be minimized by the layout of the sensor and the choice of the process. The patterning and etching of the top silicon layer can induce geometrical variations, influencing the electrical parameters such as the threshold voltage, the subthreshold slope, or the transconductance (and thus the sensitivity) of single devices. Furthermore, the formation of high-quality ohmic contacts is crucial for the reliable readout of the SiNW-FET devices. Variations in feed line resistance will alter the sensitivity from device to device. The sensing layer —the gate dielectric—of SiNW-FET devices affects many characteristics of the sensor and thus needs to be controlled to reduce variations. However, insufficient reproducibility is not only limited by the fabrication process itself but can also occur during packaging or surface chemistry processes.

In literature reviews on the usage of SiNWs in cancer detection [31,32], biologically sensitive field-effect transistors [33], nanowires bioelectric interfaces [34], the detection principles of biological field-effect transistors [35], and the overall application and functionality of (hybrid) nanowires as (bio)sensors [36–38] have been already discussed. This review will summarize the technological "top-down" approaches of SiNWs-based biosensor fabrication to obtain highly sensitive nanoscale SiNW-FETs and analyze aspects that may lead to sensor-to-sensor variation. Chronologically, a short introduction to the SiNW biosensor and its detection principles for sensing applications following by the discussion for the design and fabrication considerations, the state-of-the art fabrication techniques, the effects of microfluidic integration and surface chemistry concerning the variation between different devices. Finally, we will discuss how to decrease the sensor-to-sensor variation and improve the fabrication processes.

2. SiNW-FET Biosensor

2.1. Structure of SiNW FET-Based Biosensors

Label-free biosensors are analytical devices that transduce the binding of target molecules to their biologically sensitive layer into an electrical signal (Figure 2a). Biological sensitive SiNW-FETs have a similar structure to the traditional metal-oxide-semiconductor field-effect transistor (MOSFET) except from the metal gate electrode. As shown in Figure 2b, the gate dielectric is in direct contact with a liquid, and a reference electrode that is submerged in the liquid provides the gate voltage for the SiNW-FET sensor. Other voltage sources are connected to the source and the drain contacts during the device operation. Varying the gate voltage will lead to the electrical current change between the source and drain of the SiNW-FET. A bio(receptor) layer is introduced on the gate dielectric layer using a surface chemistry process. A binding event of target molecules to the bio(receptor) layer causes a change in the electrical response of the SiNW-FET (transducer). A SiNW-FET sensor consists of small wires, with a width in the nanometer regime and a length of a few micrometers (Figure 2c). The wires are contacted via extended feed line contacts to source and drain, which have a typical length of a few millimeters (Figure 2d). Ohmic contact to the SiNW is formed either by ion implantation, silicidation, or using a metal or combination of all techniques mentioned earlier [16,39]. The feed line contacts are passivated to avoid the electrical contacts shortcutting with the liquid (Figure 2c).

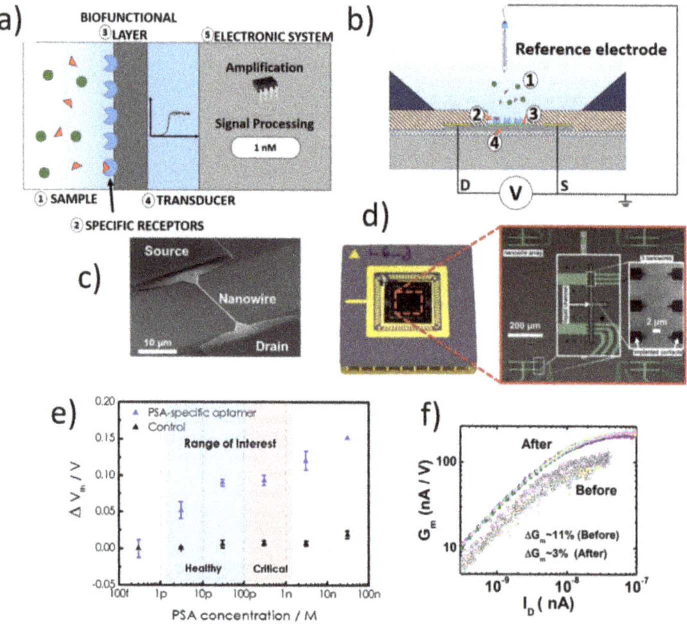

Figure 2. (**a**) Schematic illustration of an electrical biosensor: The analyte of interest (1) interacts with the specific receptor layer (2), which will be recognized by the biofunctional layer (3). The transducer (4) alters its electrical characteristic, which is read by the electronic system (5). (**b**) Schematic setup of a biosensor based on SiNW-FETs. (**c**) scanning electron microscopy (SEM) image of a SiNW and its contacts in the micrometer regime. [Reprinted with permission from [8]. Copyright (2018), Wiley]. (**d**) Encapsulated SiNW chip with microfluidic structures. [Reprinted with permission from [40]. Copyright (2014), Elsevier]. (**e**) Dose-response curve of a SiNW-FET to detect PSA using PSA-specific aptamers. [Reprinted with permission from [6]]. (**f**) Variations of the g_m value before and after optimizing the fabrication process to reduce the sensor-to-sensor variations. [Reprinted with permission from [13]. Copyright (2018) American Chemical Society].

2.2. Readout Methods of SiNW-FETs

There are two principles to read out the electrical signal of the SiNW-FET upon the binding of target molecules to the bioreceptor layer on the functionalized gate oxide, namely potentiometric and impedimetric readouts [41,42]. The potentiometric readout is based on the change of the surface potential caused by the binding of charged molecules. As shown in Figure 2e, the change in the surface potential results in a shift of the threshold voltage (V_{th}) or a change of the drain-source current (I_{ds}) at a fixed working point (V_{gs} = constant and V_{ds} = constant). A difference in the sensitivity (the transconductance g_m value), the subthreshold slope (when measuring in the subthreshold regime), or thickness of the functional layer (e.g., silanes) from device-to-device causes the sensor-to-sensor variation on their electrical signal [43–45]. Figure 2f visualizes the fabrication-induced variation of the g_m value, which results in varying sensitivity from device to device.

The impedimetric readout is based on a change in input impedance due to a biomolecule binding onto the nanowire surface [8]. The SiNW-FET is set at a fixed working point, and a small sinusoidal signal, 5–10 mV, is added to its gate electrode. The binding of biomolecules on the gate oxide causes a change in its effective gate capacitance and resistance of the SiNW-FET [8,41,46,47]. The change of the input impedance results in a change in its frequency response. Variations in the capacitance and serial resistance of the feed lines, the thickness of the functional layer, the gate oxide capacitance, and the reference electrode will cause the sensor-to-sensor variation [8,46].

3. Design and Fabrication Considerations of SiNW-FET Biosensors

3.1. Nanowire: Dimensions and Pattering Method

The nanowire determines the electrical properties, LoD, and signal-to-noise (S/N) ratio of the biosensor. It is well-known that the sensitivity of Si NWs-based biosensors increases with a higher surface-to-volume (S/V) ratio [16,48]. The conductance change of an NW defines the sensitivity parameter S of such devices due to binding events occurring on their surface. According to Park et al. [49], the sensitivity of SiNW as the change of the conductance can be expressed as the following equation for a nanoscale p-type SiNW-FET:

$$S = \frac{\Delta G}{G_o} \approx -\frac{(w + 2h)}{w \times h} \frac{N_S}{N_A} \tag{1}$$

where ΔG is the change in conductance, h is the NW height, w is the width of the NW, N_S is the surface charge density, and N_A represents the doping concentration of the NW channel [49,50]. From Equation (1), it is clear that the sensitivity increases with decreasing the cross-section of the nanowire (smaller height and width). However, downscaling of NWs have a high impact on the sensor-to-sensor variation as well, since the width of the NW becomes more dominant in the regime of a few tens of nanometer and thus leads to higher variations from device-to-device. Here, it should also be noted that shorter nanowires show a higher sensitivity compared to longer ones [16,48]. As shown in Equation (1), the sensitivity of a SiNW-FET increases with decreasing doping concentration (N_A) in the SiNW. Nair et al. showed that a low doping concentration of dopant in the SiNW is required to be smaller than 10^{17} cm^{-3} to ensure a highly sensitive biological sensing performance of the biosensor [48].

Top-down fabricated SiNW-FETs are usually fabricated on SOI wafers with a low doping concentration [3,6,22,24,51]. However, the choice of the starting material (in general, the SOI wafer) has an extreme impact on the electrical properties of the device. In most cases, the top silicon layer needs to be thinned down to define the height of the resulting NW. Therefore, SOI wafers with low top Si layer thicknesses (<90 nm) are favored to avoid thickness variations induced by the thinning processes [13,27]. Thinning of the top Si layer can be performed by either thermal oxidation combined with an HF-dip or by wet etching using the standard cleaning one (SC1) solution ($NH_4OH:H_2O_2:H_2O$) [13,26,52]. Thermal oxidation of the top Si layer leads to thickness variations. A process with very low thickness variation down to ±0.9 nm has been demonstrated by Zafar et al. [13]. Due to the low etching rate (between 0.32–0.66 nm/min) of Si in SC1 solution, a thinning of the Si layer by wet etching can be precisely controlled, with thickness variations of less than ±0.3 nm [26,52]. The lower the thickness variation of the Si-layer on SOI wafers is, is the lower the variation in the resultant SiNW height, and this is expected to reduce the difference in sensitivity of different devices and, therefore, reduces the sensor-to-sensor variations.

Sensor variations can occur due to random dopant fluctuations within the nanowire channel. For instance, a 10 µm long SiNW with a 10 nm diameter having a doping density of 10^{17} cm^{-3} would contain only about 80 dopant atoms in the active channel, and shorter wires have even less dopant [48]. For such small devices, random fluctuation of the channel doping concentration N_A will induce sensitivity variations between different devices.

The variation in the threshold voltage σV_{th} due to random doping fluctuation can be estimated by the following equation

$$\sigma V_{th} = \left(\frac{\sqrt[4]{q^3 \varepsilon_{Si} \phi_B}}{2} \right) \frac{T_{OX}}{\varepsilon_{OX}} \left(\frac{\sqrt[4]{N_A}}{\sqrt{W_{eff} L_{eff}}} \right) \tag{2}$$

where q is the electron charge, ε_{Si} and ε_{OX} are the permittivity of silicon and the dielectric material, T_{OX} is the thickness of the dielectric layer, ϕ_B is the built-in potential of the drain/source-to-channel pn junction, and W_{eff} and L_{eff} are the effective width and length of the SiNW, respectively [53]. Thus, with a large and a long SiNW, the impact of random doping fluctuation decreases, and so does the sensor-to-sensor variation [48]. However, it will decrease the sensitivity of the sensors, as shown in Equation (1). A trade-off between the sensitivity and the doping fluctuation needs to be taken into account to decrease the sensor-to-sensor variation. A higher sensitivity of the SiNW-FET sensor can be achieved by operating the sensor in the subthreshold regime [45].

In addition, Zafar et al. have shown the dependency of V_{th} on the SiNW width as a basis for sensor-to-sensor variation for long channel devices. As depicted in Figure 3, V_{th} shows a high dependency on the SiNW widths below 25 nm [13]. Lithographic processes such as EBL, NIL, or STL are typically used to define the geometry of the SiNW. The line edge roughness (LER) of the lithography processes is a major source of device-to-device variation since LER is becoming a larger fraction of the width of downscaled SiNW sensors. By considering this effect, the SiNW width should not be too small to achieve a low sensor-to-sensor variation. Besides, Regonda et al. have shown that devices consisting of more than one SiNW (e.g., a SiNW-FET consisting of 100 SiNWs in parallel) would reduce the variation in threshold voltage and subthreshold slope to a minimum of 1.8% and 4.73%, respectively [54].

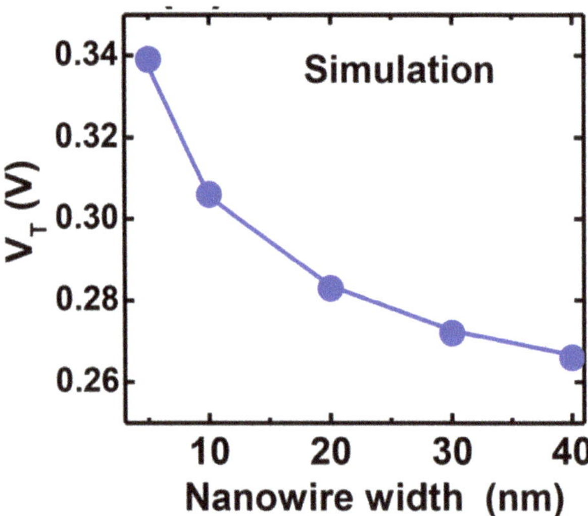

Figure 3. The simulation result shows the dependency of the threshold voltage (V_{th}) on the nanowire width. [Reprinted with permission from [13]. Copyright (2018) American Chemical Society].

Furthermore, the structuring of the SiNW needs to be controlled to reduce geometrical variations. The structuring of the silicon is conducted by either wet or dry etching [13,25]. Anisotropic wet etching of Si can be realized by using TMAH [25], resulting in a trapezoidal shape of the SiNW. The patterning of the SiNW with RIE would result in vertical sidewalls with (110) orientation. Figure 4 shows the resulting structure of dry and wet etched NWs. It has been reported that wet-etched SiNW-FETs have a lower subthreshold swing and a higher S/N ratio than that of the dry-etched NWs [55]. As shown in Table 1, it should be considered that dry-etched NWs have a low S/N ratio due to plasma-induced defects on the SiNW surface [55,56]. The $1/f$ noise of SiNW-FETs is proportional to the Hooge Constant α_H. The low-frequency noise S_I is defined as

$$S_I = \frac{\alpha_H I_d^2}{f^\beta N} \tag{3}$$

where N is the number of carriers, f is the frequency. The exponential factor β is usually found in a range $0.8 < \beta < 1.2$ [55]. Therefore, a lower α_H indicates a higher S/N ratio. The defects of dry-etched SiNW-FETs can be reduced by reducing the ion energy during the etching process or by additional dry oxidation, followed by an HF-dip to remove the damaged silicon [13]. A wet etch has the advantage of being highly controllable due to the slow etching of the (111) plane. However, changes in the etching rate of Si in TMAH solution due to a change in TMAH concentration caused by water evaporation need to be considered [57]. This is of high importance when it comes to the large-scale fabrication of SiNW-FETs.

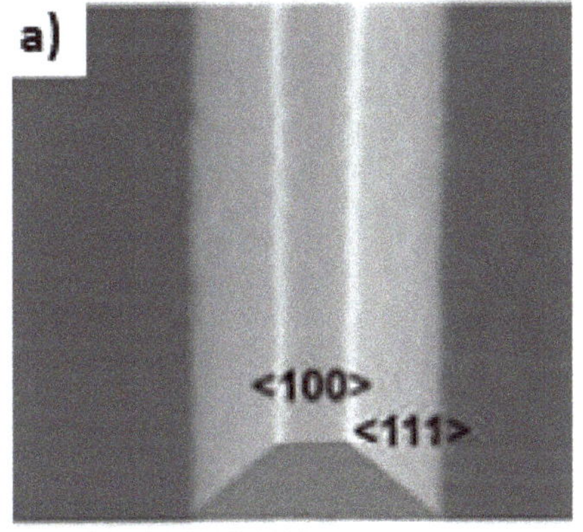

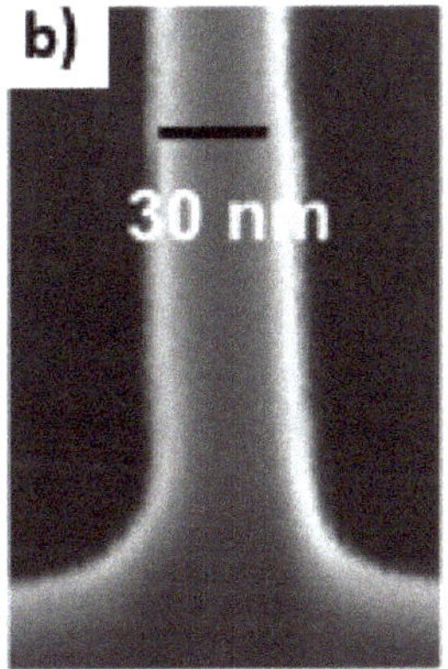

Figure 4. SEM images of wet etched (**a**) [Reprinted with permission from [22]. Copyright (2009), Wiley] and dry etched (**b**) [Reprinted with permission from [13]. Copyright (2018) American Chemical Society] SiNWs. The wet etched SiNW has a trapezoid structure due to sidewalls with (111) orientation compared to the dry-etched having vertical sidewalls with a (110) orientation.

Table 1. Comparison of device characteristics of SiNW-FETs fabricated by different etching processes. The low-frequency noise is proportional to Hooge constant.

Etching Process	Hooge Constant α_H	Subthreshold Swing	Reference
TMAH	0.0021	1.0 V/decade	[55]
Cl_2 (ICP)	0.015	2.6 V/decade	[55]
CF_4 (RIE)	0.017	3.0 V/decade	[55]

3.2. The Drain and Source Contacts

The electrical contacts, known as the *drain* and the *source* contacts or feed lines, play a crucial role in the sensor-to-sensor variations. Since the electrical performance (e.g., transconductance [43], high-frequency behavior, low-frequency noise, and power consumption) of SiNW-FETs is based on the electrical resistance of the drain and the source contacts, low-resistance feed lines are important [58]. The drain-source current I_d, in the unsaturated region, through the NW channel can be expressed as

$$I_d = \beta\left[\left(V_{gs} - V_{th}\right)V_{ds} - \frac{1}{2}V_{ds}^2\right] \tag{4}$$

where $\beta = \mu C_{ox}W/L$ is a geometry constant, V_{gs} and V_{ds} are the gate-source and drain-source voltages, and V_{th} is the threshold voltage. This approximation of I_d, however, does not consider the resistance of the feed lines. With the incorporation of the drain resistance (R_d) and source resistance (R_s), the drain current I_d of real NW devices is given by

$$I_d = \beta V_d \frac{V_{gs} - \frac{1}{2}V_{ds} - V_{th}}{1 + \beta(R_d + R_s)\left(V_{gs} - \frac{1}{2}V_{ds} - V_{th}\right)} \tag{5}$$

Equation (5) implies that the drain current I_d of the transistor is influenced by the drain and source resistance [43]. Figure 5c,d illustrates the impact of the drain and the source feed line resistance on the resulting $I_d - V_{gs}$ characteristic. Higher serial resistance will decrease the current. Consequently, a higher resistance of the drain and the source contacts has an impact on the transconductance of the device and thus affects the sensitivity. Variations of drain and source feed lines also cause sensor-to-sensor variation. Therefore, the resistance of the feed lines of different devices needs to be identical to obtain identical sensitivity of the devices and thus eliminate the effect of the feed line contact resistance to the sensor-to-sensor variation. The feed line resistances of source and drain contact can be optimized in the layout design of the sensor by taking into account the sheet resistance value of the feed lines and controlling the homogeneity of the thickness or doping level of the feed lines in the fabrication. As shown in Figure 5a,b two different approaches are used to create ohmic contacts. The metal contacts can be created close to the NWs (Figure 5a) or at a certain distance (Figure 5b). A sensor design with an intermediate highly doped silicon feed line allows the passivation by high-quality thermal oxide [27], while sensors with metal feed lines next to the NWs need to be passivated by CVD processes [22] or polyimide [59].

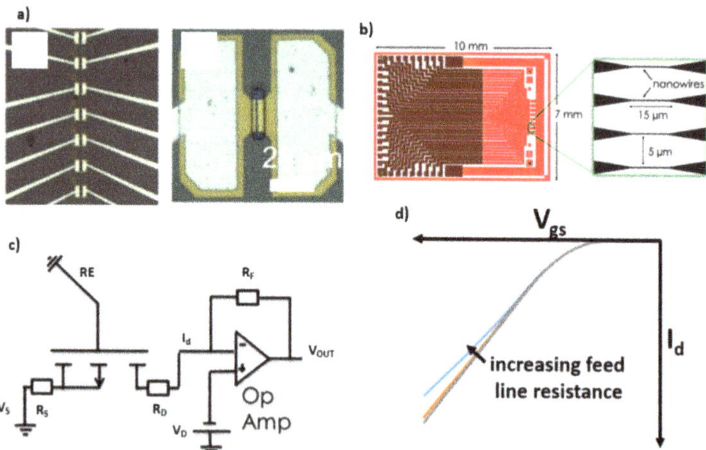

Figure 5. Illustration of two possible methods to form ohmic feed line contacts. Formation of ohmic contacts close to the NW (**a**) [Reprinted with permission from [2]] and formation of ohmic contacts on top of silicon feed lines (**b**) [Reprinted with permission from [27]]. Electrical readout configuration for DC readout of liquid gated FETs (**c**). Schematic illustration of the impact of drain and source feed line resistance R_D and R_S on the resulting drain current I_d (**d**).

As discussed above, the feed lines affect the device sensitivity of the sensors in a DC readout method, and it also affects the frequency response of SiNW-FETs in an impedimetric readout method. Here, variations in the feed lines resistance of the drain and the source contacts cause a minor impact on the frequency response of the device [60]. Indeed, the parasitic capacitance of the drain and the source feed lines influences the frequency response of the SiNW sensor. A dependency of the cut-off frequency and the amplitude of a SiNW-FET transfer function was intensively discussed by Abhiroop et al. and Nguyen et al. [46,60]. As shown in Figure 6, the frequency response of a SiNW-FET depends on the solution resistance (R_{sol}), the capacitance (C_{Bio}) and resistance (R_{Bio}) of the biological layer, and the parasitic capacitance (C_{CLS} and C_{CLD}) of the feed lines. Therefore, sensor-to-sensor variations can be compensated by reducing variations between the feed line resistance and by minimizing area variations of feed lines.

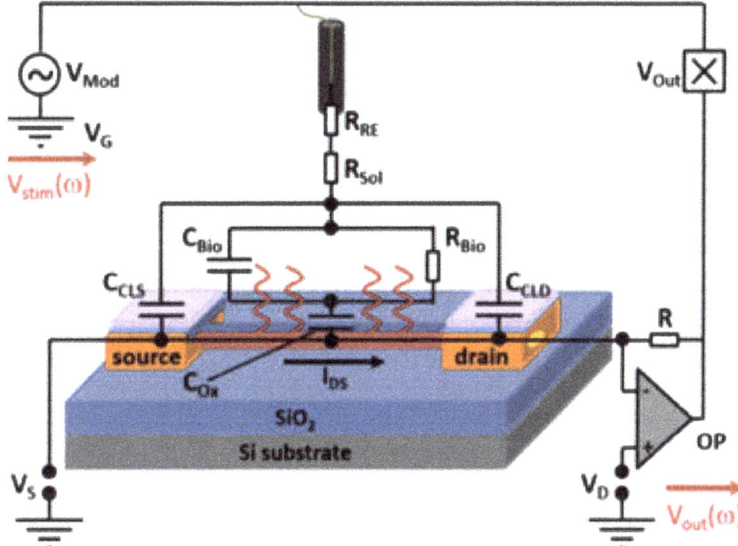

Figure 6. Schematic view of the electrical equivalent circuit of the SiNW FET in AC-mode. Variation in drain and source capacitance will lead to variations in the output signal. [Reprinted with permission from [8]. Copyright (2018), Wiley].

Since SiNW-FETs are often fabricated on an ultra-thin top Si-layer of the SOI wafer, a further modification of the feed lines to lower their resistance is required. A heavy ions implantation in combination with a metal or a stack of metals is most commonly used in the fabrication of the SiNW–FET, as presented in Table 2 [3,13,22]. Due to the skinny top silicon layer on the SOI wafer, the ion implantation needs to be carried out in a low energy process to obtain a homogenous distribution of the dopant in the feed line. Due to the required heavy ion- implantation, the implantation cost is higher when the doping energy is lower, thus increasing the fabrication costs per wafer. Al is used to form an Ohmic contact with the heavily doped Si [25–27,30], and a protective metal layer is used to prevent reactions of the Al with the surrounding environment since Al is a highly reactive metal. These processes are highly controllable, and thus resulting in a low device-to-device variation. A second approach to create low-resistance contacts is the use of silicide contacts [40]. Here, metals (e.g., Ti [61] or Ni [62]) are sintered on undoped silicon to form a metal-silicon alloy. However, the uncontrollable consumption of silicon during annealing can lead to higher sensor-to-sensor variations compared to the ion-implantation method [16,62].

Table 2. Overview of different processes to form ohmic feed line contacts.

Approach	Doping Process Parameters	Doping Concentration	Metal	References
Ion implantation and silicide formation	(B) 2.5 keV, 4×10^{15} ions/cm^2	~8×10^{19} atoms/cm^{-3}	NiPt (10% Pt)/TiN	[13]
Ion implantation and Al contacts	(B) 7 keV, 1×10^{14} ions/cm^2	N/A	Al/Ti/Au	[22]
Ion implantation and Ti/Al contacts	(BF$_2^+$) 8 keV, 5×10^{15} ions/cm^2	N/A	Ti/Al	[3]

3.3. The Gate Oxide

Since the gate oxide affects many characteristics of SiNW-FET devices, such as threshold voltage, hysteresis, and subthreshold swing, a high-quality gate dielectric is needed [13,50]. One of the most important parameters of SiNW-based biosensors is the threshold voltage V_{th} since the shift in V_{th} is a measure for the detection of biomolecules. Generally, the V_{th} of a SiNW-FET is given by

$$V_{th} = E_{ref} - \Psi_s + \chi_{sol} - \frac{\Psi_{Si}}{q} - \frac{Q_{ox} + Q_{ss}}{C_{ox}} - \frac{Q_B}{C_{ox}} + 2\phi_F \qquad (6)$$

here, E_{ref} is the potential of the reference electrode, Ψ_s the surface potential, χ_{sol} the surface dipole potential, Ψ_{Si} the work function of silicon, q the elementary charge, ϕ_F is the difference between the Fermi level of intrinsic silicon and the actual Fermi level of the device, C_{ox} the capacitance of the gate oxide, Q_{ox}, Q_{ss} and Q_B are the fixed charges in the oxide, the surface state density, and the depletion charge, respectively. Derived from Equation (5), the V_{th} is dependent on the gate capacitance, the fixed charges, and the surface state density, which is influenced by the thickness and quality of the dielectric material and the interface between the dielectric and silicon. On the one hand, thickness variations along the wafer result in a variation of the gate capacitance and, thereby, varying V_{th}. On the other hand, variations in dielectric thickness along a single NW induce changes in the subthreshold slope [13]. Figure 7 shows a comparison of the cross-section of an NW with homogeneous and nonhomogeneous SiO$_2$ layers and the simulation results showing the changes in the subthreshold slope. Furthermore, alignment variation of the gate area is known to induce sensor-to-sensor variations leading to changes in V_{th} [27]. An additional oxide growth during plasma-enhanced chemical vapor deposition (PECVD) processes to passivate the drain and source feed lines should be compensated in order to reduce oxide thickness variations (compare Figure 7c) [13]. It has been shown that the formation of the gate oxide after the feed line passivation in a fabrication protocol leads to a minimum variation in oxide thickness resulting in only a low variation of V_{th} [25,27]. In the following, we will summarize state-of-the-art processes to reduce these variations during gate oxide fabrication.

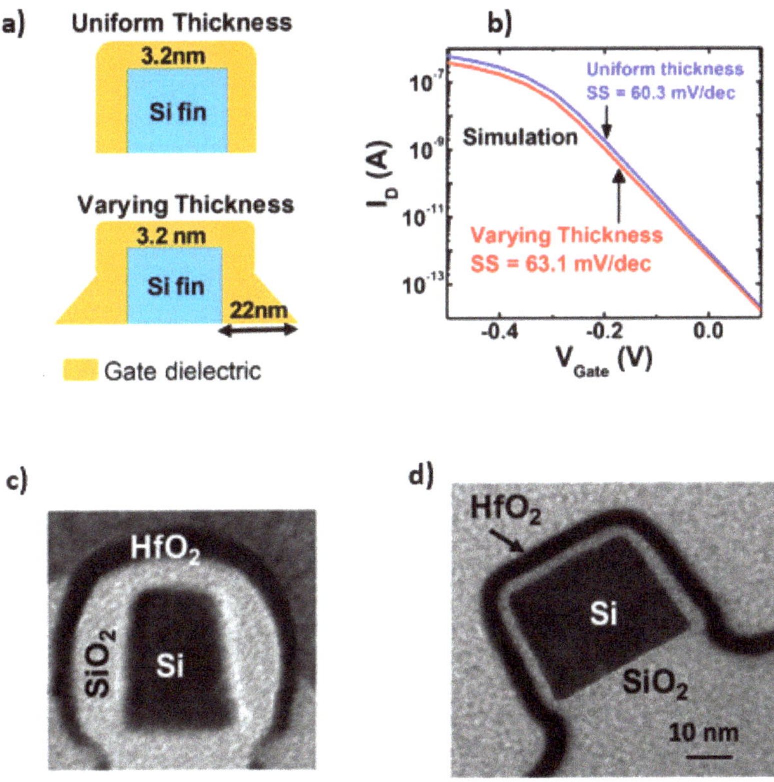

Figure 7. Illustration of uniform and varying thicknesses of the gate dielectric (**a**) and simulation results of how the varying thickness influences the subthreshold slope (**b**). SEM images of varying and uniform gate oxide thickness (**c,d**). [Reprinted with permission from [13]. Copyright (2018) American Chemical Society].

Silicon dioxide (SiO_2) is the most common gate material in the semiconductor industry due to its dielectric properties and CMOS compatibility. The growth of SiO_2 is a well-controlled process leading to a high-quality Si/SiO_2 interface with minimal variation in oxide thickness [13,27,63]. To create a high-quality Si/SiO_2 interface, a standard RCA cleaning protocol prior to the gate oxidation is of high importance. Differences in the cleaning procedure can create differences in the Si/SiO_2 interface quality and thus lead to V_{th} variations and hysteresis of the device characteristics. In addition, SiO_2 has drawbacks, such as uncontrollable drifting behavior, low pH buffer capacity, and incorporation of charged ions present in the analyte sample [35,50,51,64,65]. Materials with a high dielectric constant, so-called high-k materials, such as aluminum oxide and hafnium oxide, can overcome these issues. Higher gate capacitances achievable from such high-k dielectrics allow an increase in the thickness of the gate dielectric resulting in favorable conditions such as reduction in gate leakage current [36]. Even so, the use of high-k materials adds more complexity to the fabrication process. These materials are often deposited using atomic layer deposition (ALD), which can create defects at the Si/high-k material interface [13,16,66–69]. Furthermore, it has been reported that the carrier mobility of FET devices with a high-k material in contact with silicon is usually less than that of FETs with SiO_2 as gate oxide dielectric [67]. A stack of SiO_2 and high-k materials as gate dielectrics combines the advantages of both materials. Thermal oxidation leads to a high-quality Si/SiO_2 interface with a low interfacial trap density. The additional high-k material offers nearly

Nernstian pH sensitivity, an effective ion diffusion barrier, a low leakage current, and low leakage voltage operation [13,16,66]. Bae et al. reported a drift rate of only 0.25 mV/h for a dielectric layer stack made of SiO_2/Al_2O_3 while a SiNW-FET made of SiO_2 had a drift rate of 45.24 mV/h [50]. Besides, Table 3 provides a performance overview of different gate material combinations of SiO_2 and other high-k materials. A combination of SiO_2/Al_2O_3 leads to the lowest drifting rate and lowest hysteresis with an increased pH-sensitivity compared to the SiO_2 layer.

Table 3. An overview of the performance of different combinations of gate dielectrics. Data adapted from [50].

Gate Material	pH Sensitivity (mV/pH)	Drift Rate (mV/h)	Hysteresis (mV)
SiO_2	38.7	45.24	173
SiO_2/Si_3N_4	49.7	3.86	20.9
SiO_2/HfO_2	55.3	1.88	6.9
SiO_2/Ta_2O_5	52.6	0.61	13.9
SiO_2/ZrO_2	53.9	0.44	22.1
SiO_2/Al_2O_3	53.1	0.25	0.6

4. Fabrication Methods for SiNW Based Biosensors

4.1. Electron Beam Lithography (EBL)

EBL is one of the most common, advanced lithographic processes involved in the fabrication of SiNW based biosensors. A typical fabrication process of SiNW-FET using EBL is presented in Figure 8 (top). EBL has demonstrated its ability to process high-resolution nanostructures with high flexibility due to maskless patterning. However, EBL is a time-consuming and high-cost fabrication process. To reduce the cost and to increase the high throughput of the fabrication, a combination between EBL using negative tone resists such as hydrogen silsesquioxane (HSQ) and optical lithography was used and thus far have been able to achieve large scale fabrication with variations in V_{th} down to ± 28 mV [13,70]. To achieve such low variations, practical factors such as stage tilt, inhomogeneous resist thickness, write field alignment, and thermal drift during long-term writing need to be compensated to reduce variations in the nanowire width and position of the nanowire on wafers. During long-term exposures, the thermal drifting effect can be reduced by minimizing the writing time and changing the carrier material [13].

Geometrical variations are one of the most relevant factors that lead to sensor-to-sensor variation. Therefore, line edge roughness (LER) is a crucial parameter that needs to be investigated during the fabrication of SiNW-FETs. Since lithographic features are not perfectly smooth, LER defines the deviation of a real photoresist edge from an expected one. The effect of LER concerning sensor-to-sensor variations has been investigated for MOSFETs as well as for SiNW-FETs [13,71]. The reduction of LER leads to a lower sensor-to-sensor variation. The LER depends on the resist thickness and the electron beam dose. A higher electron beam dose results in a lower LER but increases the nanowire width. The resist thickness has to be as thin as possible to reduce the LER since the LER increases with the resist thickness. Figure 8 (bottom) shows the results of wet etched nanowires using EBL processes with HSQ resist for patterning.

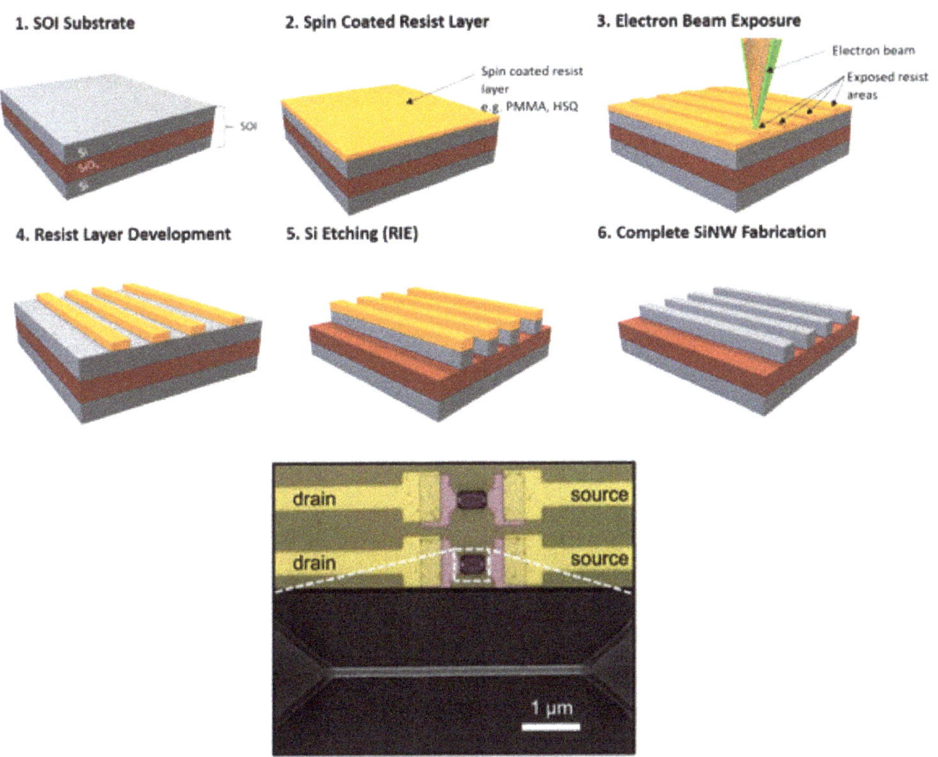

Figure 8. Schematic process flow to fabricate SiNW-FETs using EBL (**top**). [Reprinted with permission from [31]. Copyright (2020), American Chemical Society]. SEM image of top-down fabricated SiNW-FETs using EBL (**bottom**). [Reprinted with permission from [72]. Copyright (2011), AIP].

Table 4 provides an overview of the fabrication results and the variation in threshold voltage. Zafar et al. have shown that the variation can be reduced (e.g., the variation in g_m was reduced from 11% to 3%) by considering the design of the SiNW and by optimizing other steps in the fabrication process [13].

Table 4. Overview of SiNW-based biosensors fabricated in different EBL processes. Note that the fabrication process described in Ref. [3] does not include EBL.

Fabrication Approach	NW Size in Width and Length	V_{th} and Its Variation	CMOS Integration	References
Top-down fabrication on SOI wafer, EBL process using HSQ combined with optical lithography	30 nm, 5 µm	0.28 ± 0.028 V	No	[13]
Top-down fabrication on SOI wafer, EBL process using HSQ combined with optical lithography	50 nm, 20 µm	1.15 ± 0.16 V	No	[54]
Top-down fabrication on SOI wafer, EBL process using HSQ combined with optical lithography	55 nm, N/A	N/A	Yes	[73]
Top-down fabrication on SOI wafer, optical lithography	Nanoribbon	-2.3 ± 0.15 V	No	[3]

4.2. Sidewall Transfer Lithography (STL)

STL is a low-cost and high-throughput patterning technique to transfer nanoscale structures using standard lithography processes. As shown in Figure 9, an STL process involves the deposition of a dielectric material and a sacrificial support material [23]. The support material is deposited and structured to define the position of the resulting NWs. A hard mask material (e.g., Si_3N_4) is deposited by plasma-enhanced chemical vapor deposition (PECVD) and structured using RIE. The reliability and reproducibility and thus the sensor-to-sensor variation of STL fabricated nanowires depend on the control of the thickness of the deposited material, the conformal deposition of the sidewall layer, the selective etching of the sacrificial material, and the anisotropy of the RIE process.

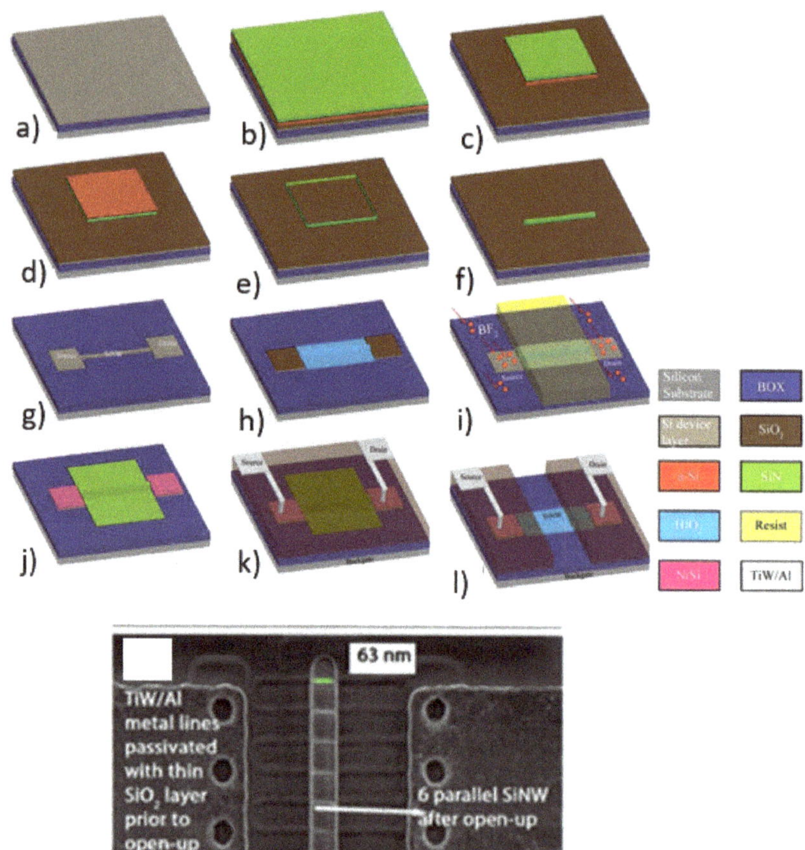

Figure 9. Process flow of top-down fabrication of SiNWs using STL (top): SOI is used as a starting material (**a**). Deposition of a tri-layer stack of SiO_2, amorphous silicon (a-Si), and silicon nitride (SiN) (**b**). Selective etching of a-Si using SiN as a hard mask (**c**). Deposition of a SiN spacer (**d**). Etching of a-Si using TMAH (**e**). Removal of the spacers (**f**). Patterning of drain/source contacts and SiNW (**g**). Formation of a gate oxide using thermal oxidation of silicon and subsequent HfO_2 ALD deposition (**h**). Ion-implantation to form conductive drain and source regions (**i**). Formation of nickel silicide (NiSi) ohmic contacts (**j**). Passivation of feed lines and contact metallization (**k**). Opening of the gate area (**l**). [Reprinted with permission from [74]]. SEM picture of the resulting device (bottom). [Reprinted with permission from [74]].

4.3. Nanoimprint Lithography (NIL)

NIL is a fully CMOS compatible nanofabrication process, in which a stamp is used to transfer its negative image into a temperature- (T-NIL) or light-sensitive (UV-NIL) resist. As shown in Figure 10, the imprinting technique relies on the mechanical transfer of the pattern into the nanoimprint resist followed by a polymerization process of the resist. Typically, the stamp is coated by a release layer to guarantee the quality of the resist pattern upon release of the stamp after polymerization. After imprinting the pattern into the resist, the residual layer, which is the remaining resist in the imprinted areas of the pattern, is removed using an anisotropic reactive ion etching (RIE) process [75]. As for other lithography techniques, LER is an issue of NIL as well. Yu et al. presented a low-cost and easy implementation method for reduced LER of nanoimprint resists. A thermal treatment above the glass transition temperature reduces the LER of imprint resists drastically [76]. Besides its major advantages, such as high throughput (up to 80 wafers per hour) and low-cost fabrication, NIL also allows the transfer of micro-and nanostructures simultaneously [22,26,27,77]. Since nano- and microstructures are patterned in the same step, variations due to misalignment of micro- and nanostructures are reduced. However, NIL also has some drawbacks, such as inhomogeneous residual layer thickness and alignment problems between nanoimprint mold and the lithography masks, which can induce sensor-to-sensor variation [27].

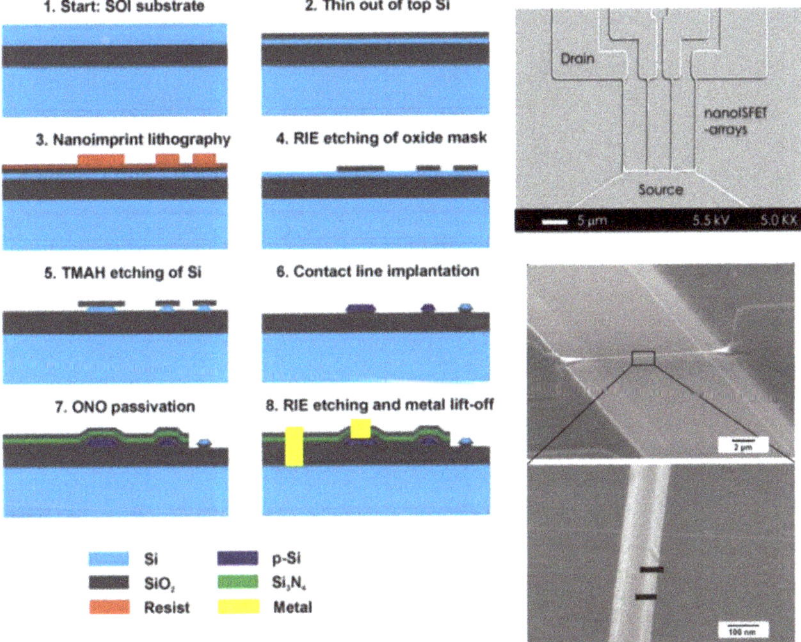

Figure 10. Schematic illustration of the process flow for fabrication of SiNW FETs using NIL (**left**) [Reprinted with permission from [22]. Copyright (2009), Wiley]. SEM images of wet etched SiNW fabricated using NIL (**right**) [Reprinted with permission from [29]. Copyright (2010), Wiley, and reprinted with permission from [27]].

Nevertheless, the fabrication of SiNW biosensors using NIL can result in performance variation of different devices down to 7% [27]. Table 5 presents an overview of sensor-to-sensor variation of wafer-scale NIL processes. The sensor-to-sensor variation is addressed not only to the NIL process itself but also to the quality of the mold and the size variation of the nanowire's template on the mold. Therefore, size variations of structures on the mold need to be reduced. Since EBL is commonly used to fabricate such molds, aspects discussed for the EBL fabrication of nanostructures need to be considered for the fabrication of nanoimprint molds.

Table 5. Comparison of Si NWs-based biosensors fabricated with NIL processes.

Fabrication Approach	NW Size in Width and Length	V_{th} and Its Variation	CMOS Integration	References
Top-down fabrication on SOI wafer, NIL	125 nm × 15 μm	0.384 ± 0.106 V	No	[27]
Top-down fabrication on SOI wafer, NIL	100 nm × 7 μm	0.65 ± 0.3 V	No	[26]

5. System Integration

5.1. Surface Functionalization for Biosensing Applications

Surface functionalization is of significant importance when it comes to label-free biosensing applications. To realize a high sensitivity and specificity, the choice of receptor molecules needs to be considered. The target molecule must bind with high affinity and selectivity to the receptor molecules on the sensing area. Silanization with 3-aminopropyltriethoxysilane (APTES) or Glycidyloxypropyltrimethoxysilane (GPTES) is the most common method for surface modification, used for covalent binding of receptor biomolecules to the gate oxide surface [5,8,44,78,79]. This process can be carried out either in gas-phase or in liquid-phase [6,8,27,44,80]. It applies that the thinner the silane layer, the higher the sensitivity of a SiNW-FET [81]. A monolayer of siloxane resulting from the surface modification process increase sensitivity and reduce sensor-to-sensor variations. It has been reported that gas-phase silanization can lead to APTES layer thickness of 20 ± 2 Å in comparison to a liquid phase silanization, which usually results in a minimum layer thickness of 40 ± 5 Å [44,79]. Therefore, sensor-to-sensor variations can be reduced by favoring gas-phase silanization processes over liquid-phase methods. Munief et al. presented a protocol for gas-phase deposition of different silanes with a low silane thickness and a versatile, uniform, and large-area coating of SiO_2 substrates [80], which can be applied to the surface modification of the SiNW-FET.

After surface modification, the analyte-specific receptor molecules (e.g., aptamers or ssDNA) are immobilized on the SiNW-FET surface via covalent bonding between the receptor and silane-modified oxide surface. A non-uniform immobilization of charged receptor molecules onto the SiNW-FET surface is expected to induce variable surface charges and influence the V_{th} of the sensors. Here, the composition of the charged biofunctional layer determines the sensor characteristics of the SiNW-FET device. In an ideal case, the receptor molecules are located only at the SiNW-FET surface and enable high specific localized binding of analytes exclusively to the NW surface, as presented in Figure 11a,b. As shown in Figure 11c, a selective surface modification (SSM) decreases the LoD compared to that of an all-area modification (AAM) approach [78]. Park et al. have demonstrated a method for selective functionalization of single silicon nanowires via joule heating [82]. Here, a protective polymer layer was used to prevent the functionalization of other areas than the desired NW. The protective polymer (polytetrafluoroethylene (PTFE)) was removed from the NW surface using joule heating. After a cleaning procedure, the NW could be selectively functionalized by linker molecules. The whole process of the functionalization of single NWs is illustrated in Figure 11b.

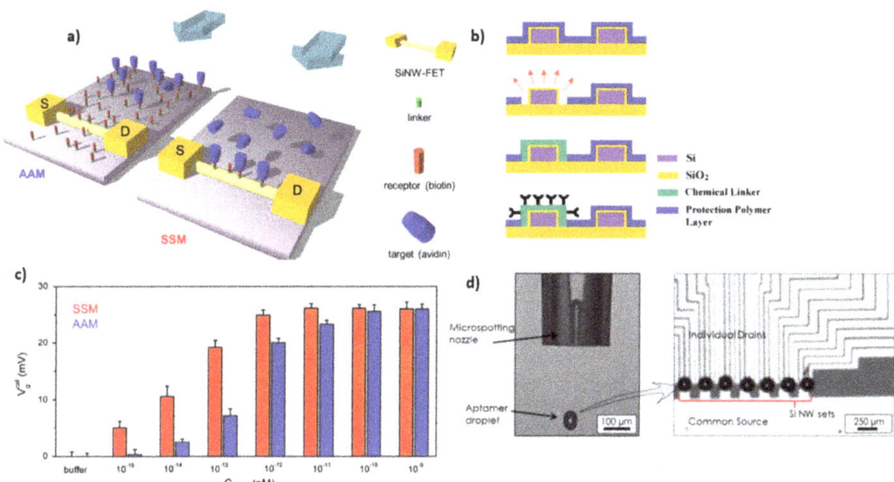

Figure 11. Visualization of AAM and SSM modification of SiNW-FETs (**a**) [Reprinted with permission from [78]. Copyright (2013), Elsevier]. Schematic illustration of a single NW functionalization using a protective polymer layer (**b**). [Reprinted with permission from [82]. Copyright (2007), American Chemical Society]. Comparison of the signal response of AAM and SSM modified SiNW-FETs (**c**) [Reprinted with permission from [78]. Copyright (2013), Elsevier]. Micro spotting technique for localized surface modification (**d**). [Reprinted with permission from [6]].

High-temperature processes such as joule heating of nanowires may be unsuitable for specific applications or sensor structures. Therefore, localized immobilization is carried out using the micro spotting technique, as shown in Figure 11d [6]. Single droplets containing relevant receptor molecules (e.g., aptamers) are spotted onto the desired area with a diameter of about 200 µm. However, differences in capture molecule concentration or misalignment of the droplet lead to sensor-to-sensor variations. However, threshold variations of only 4.9% have been reported for such localized immobilization of capture molecules using micro spotting [83].

The type of receptor molecules influences the sensor performance. To achieve high selectivity and specificity, the chemistry for binding the molecule to the surface needs to be considered [84]. We refer to already exiting reviews for a detailed overview of how to graft recognition elements onto solid surfaces [85–87]. In the following, we briefly discuss the use of different kinds of recognition molecules. Antibodies are often used in biosensing applications due to their high specificity antibody-antigen binding. The use of antibody fragments results in the same specificity as the whole antibody and provides a smaller size, which is of great interest when considering general limitations such as Debye screening [84,88,89]. A loss of biological activity of the antibodies upon immobilization has been noticed due to the random orientation of the asymmetric antibody on the supported surface [90]. Several approaches for achieving oriented coupling of antibodies to the surfaces and the antigen-binding capacity are summarized by Lu et al. [90].

Aptamers (single-stranded DNA or RNA sequences folded into a three-dimensional structure) are often used for the detection of specific target molecules. They show a high affinity and specificity to their targets. Furthermore, they feature an easy coupling to the sensor surface and high reproducibility, which is of great interest to sensor-to-sensor variations [84]. As described above, sensor-to-sensor variations mainly depend on the homogeneity of the silane layer and the density of receptor molecules bound to the SiNW-FET surface. In general, an ideal surface modification of the oxide surface, a choice of the suitable receptor molecules, and controlling the density of the receptor layer will increase the sensor sensitivity and decrease the sensor-to-sensor variation.

5.2. Microfluidic Integration

The microfluidic integration to the SiNW-FETs allows a controlled supply of fluids containing target molecules of interest. Concerning commercial applications of SiNW-FETs, the microfluidic integration of such sensors allows automated fluid handling, which enables high throughput and low-cost analyses [91]. Microfluidic channels of dimensions of several 10 s up to 100 s of micrometers are typically used for fluidic integration of biosensors to handle small quantities of analyte samples allowing for rapid and low-cost analysis. These fluidic channels are often made from polydimethylsiloxane (PDMS) containing an inlet and an outlet (compare Figure 12) [29]. The geometrical variations of the microfluidic channel will alter the transport of species. Especially for diffusion-based sensing approached or investigations of molecular interactions, differences in the geometry will change the sensor response. The need to include a reference electrode without a fluidic leak increases the complexity of the sensor integration and may induce additional sensor-to-sensor variation due to changes in the relative position of the reference electrode to the NW devices [27].

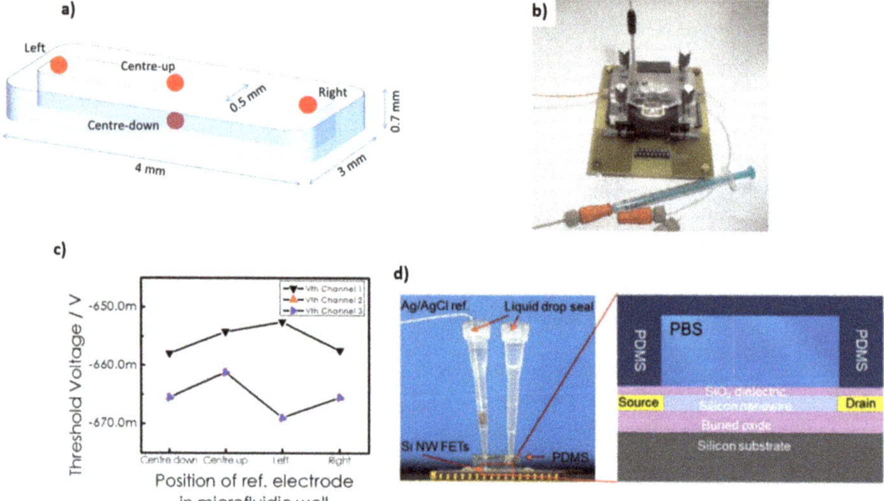

Figure 12. Schematic illustration of a microfluidic well and different positions of the reference electrode (**a**). [Reprinted with permission from [27]] Experimental setups for SiNW-FETs using PDMS-based microfluidic channels (**b**) [Reprinted with permission from [29]. Copyright (2010), Wiley] and (**d**) [Reprinted with permission from [2]]. Threshold voltage dependency on the position of the reference electrode (**c**). [Reprinted with permission from [27]].

As a solution to the fluidic integration of the reference electrode, the realization of an on-chip reference electrode would reduce sensor-to-sensor variations. The reference electrode position is of major importance, particularly for the *AC* readout, since the resistance of the analyte has an impact on the recorded spectra [8,46,47]. Several approaches for on-chip pseudo-reference electrodes have been investigated. Silver-silver chloride (Ag/AgCl) based redox systems are the most accurate ones of the available pseudo-reference electrode types. The fabrication of such solid-state pseudo-reference electrodes has been described [92,93]. To enhance the stability of the Ag/AgCl on-chip pseudo-reference electrodes, KCl membranes were used to prevent corrosion caused by the electrolyte and to provide a constant potential independent of the Cl^- ion concentration [92,93]. Other concepts of on-chip pseudo-reference electrodes are based on the catalytic properties of platinum or iridium oxide. These, however, show high pH sensitivity or low potential stability [94,95]. As an alternative, the mixed electronic-ionic conduction of conductive polymers (e.g., polypyrrole), which provide a stable interface in liquids, can be used as on-chip pseudo-reference electrodes for the applications [96,97].

6. Conclusions and Outlook

We discussed different fabrication and design-induced parameters, including the design of NWs, feed line configuration, and the impact of the gate dielectric, which critically influence sensor-to-sensor variations of NW-based biosensor platforms. The fabrication process of such downscaled NW structures needs to be precisely controlled to reduce geometrical variations between the different devices. It is difficult to find the balance between sensitivity and low sensor-to-sensor variation since the sensitivity increases with smaller dimensions (high S/V ratio) while the variation among individual devices increases. The starting SOI wafer should have a low doping concentration to ensure high sensitivity and a low initial thickness to reduce the height variations of the SiNW-FET. The thinning process of the top Si-layer needs to be controlled to reduce variations in the height of the SiNWs. The wet-etching process using the SC1 solution is a suitable candidate to decrease the height variation and also to decrease the complexity in the overall "top-down" fabrication approach.

Furthermore, the diameter or width of the SiNW-FET has a substantial impact on the sensitivity and the sensor-to-sensor variation. The impact of random doping fluctuation on sensor-to-sensor variation is also reduced with a "larger" width of the SiNW. Small SiNW-FETs have high sensitivity but also have the ability for higher sensor-to sensor variation. Depending on applications (target molecules of interest), an optimized nanowires diameters or nanowire width must be decided to meet the required sensitivity and minimal sensor-to-sensor variation. In addition, devices consisting of multiple NWs result in lower sensor-to-sensor variations.

The drain and the source resistances and capacitances, which affect the sensor sensitivity and the frequency response, are one of the factors affecting the sensor-to sensor variation. A minimal difference in the feed line parameters is required for all SiNW-FETs of a sensor array and on the final product. The feed line parameter can be optimized by combing the sensor design parameters and the selection of the feed line materials.

The quality and thickness of the gate oxide on the NWs, as a dielectric, influences various device characteristics. The formation of a gate dielectric based on SiO_2 results in a low variation in thickness and thus in a lower variation in gate capacitance. In case a passivation layer using a CVD process is employed, the growth of gate oxide is required after the passivation of the feed lines to reduce thickness variations due to eventually additional oxide growth during CVD processes. However, the unstable nature of SiO2 in aqueous solutions makes it less favorable for stable and highly sensitive biosensors. Therefore, a stack of SiO_2 and high-k materials is a promising approach.

To reduce sensor-to-sensor variations in the "top-down" fabrication protocols, reducing the pattern size differences of the nanostructure is required. The line-edge roughness needs to be carefully addressed during the fabrication process. Choosing the right parameters for EBL processes such as the write-field, beam side, beam current, and stage compensation will minimize the size variations SiNW-FET. The LER in EBL processes can be reduced by optimizing the resist thickness and the electron dose. In addition, a precise loading, unloading of the wafer, and self-calibration of the EBL parameter is needed to ensure a minimal variation from wafer to wafer.

NIL has a clear advantage over other fabrication methods as the imprint technique results in less wafer-to-wafer variation, which is of high importance for mass fabrication. During fabrication of the imprint mold, size variations need to be minimized to ensure lower sensor-to-sensor variations. Since EBL processes are involved in the fabrication of imprint molds, aspects such as LER need to be optimized in the EBL process. Thermal treatment can reduce the LER caused by NIL processes.

STL is a low-cost fabrication method for nanoscale devices without the need for expensive tools for nanoscale patterning. However, the homogeneous and conformal deposition of masking materials is a source that caused size variations from device to device. The deposition process and the post-process are quite complex, thus an improvement in the masking layer deposition is needed for large-scale production.

The chemical functionalization of the SiNWs and the bioimmobilization protocol are of major importance when it comes to sensor-to-sensor variations. Uniform deposition of the functional layers leads to a reduced sensor-to-sensor variation. Gas-phase deposition of silanes has shown a reduced thickness variation and an overall lower thickness compared to liquid phase deposition. Furthermore, controlling the receptor density on the SiNW surface and maintaining its biological activity by choosing the right receptor and the immobilization process is crucial to minimize the sensor-to-sensor variation. Gas-phase silanization, using a micro-spotting machine to locally spot the receptor to the SiNW combining with a covalent binding of the receptor to the modified gate oxide surface, would lead to minimal variation.

SiNW-FETs have remarkable electronic properties and offer ultra-high sensitivity to detect biological binding events of target analyte molecules for the next generation of clinical biosensors. Further reduction of the sensor-to-sensor variation in large-scale production will increase the potential of SiNW-FET based biosensors in translational research and boost the likelihood of this technology reaching its full commercial potential at the biomedical diagnostics market.

Author Contributions: M.T., X.T.V. designed the content and the structure of the review; M.T. wrote the draft with the input of all authors; V.P., S.I., and X.T.V. revised the manuscript. All authors have read and agreed to the published version of the manuscript.

Funding: This research was funded by the Deutsche Forschungsgemeinschaft (DFG) grant number No. 391107823.

Institutional Review Board Statement: Not applicable.

Informed Consent Statement: Not applicable.

Acknowledgments: The authors thank the DFG for funding the research project "Molecular Programs for neurodegenerative diseases markers Biosensing" (No. 391107823). The authors would like to thank Linda Wetzel for linguistic editing of the manuscript.

Conflicts of Interest: The authors declare no conflict of interest.

References

1. Dincer, C.; Bruch, R.; Costa-Rama, E.; Fernandez-Abedul, M.T.; Merkoci, A.; Manz, A.; Urban, G.A.; Guder, F. Disposable Sensors in Diagnostics, Food, and Environmental Monitoring. *Adv. Mater.* **2019**, *31*, e1806739. [CrossRef]
2. Li, J.; Kutovyi, Y.; Zadorozhnyi, I.; Boichuk, N.; Vitusevich, S. Monitoring of Dynamic Processes during Detection of Cardiac Biomarkers Using Silicon Nanowire Field-Effect Transistors. *Adv. Mater. Interfaces* **2020**, *7*, 2000508. [CrossRef]
3. Vacic, A.; Criscione, J.M.; Stern, E.; Rajan, N.K.; Fahmy, T.; Reed, M.A. Multiplexed SOI BioFETs. *Biosens Bioelectron* **2011**, *28*, 239–242. [CrossRef] [PubMed]
4. Zheng, G.; Patolsky, F.; Cui, Y.; Wang, W.U.; Lieber, C.M. Multiplexed electrical detection of cancer markers with nanowire sensor arrays. *Nat. Biotechnol.* **2005**, *23*, 1294–1301. [CrossRef]
5. Cui, Y.; Wei, Q.; Park, H.; Lieber, C.M. Nanowire nanosensors for highly sensitive and selective detection of biological and chemical species. *Science* **2001**, *293*, 1289–1292. [CrossRef] [PubMed]
6. Rani, D.; Pachauri, V.; Madaboosi, N.; Jolly, P.; Vu, X.T.; Estrela, P.; Chu, V.; Conde, J.P.; Ingebrandt, S. Top-Down Fabricated Silicon Nanowire Arrays for Field-Effect Detection of Prostate-Specific Antigen. *ACS Omega* **2018**, *3*, 8471–8482. [CrossRef]
7. Gao, Z.; Agarwal, A.; Trigg, A.D.; Singh, N.; Fang, C.; Tung, C.-H.; Fan, Y.; Buddharaju, K.D.; Kong, J.J.A.C. Silicon nanowire arrays for label-free detection of DNA. *Anal. Chem.* **2007**, *79*, 3291–3297. [CrossRef]
8. Schwartz, M.; Nguyen, T.C.; Vu, X.T.; Wagner, P.; Thoelen, R.; Ingebrandt, S. Impedimetric Sensing of DNA with Silicon Nanowire Transistors as Alternative Transducer Principle. *Phys. Status Solidi* **2018**, *215*, 1700740. [CrossRef]
9. Eschermann, J.F.; Stockmann, R.; Hueske, M.; Vu, X.T.; Ingebrandt, S.; Offenhäusser, A. Action potentials of HL-1 cells recorded with silicon nanowire transistors. *Appl. Phys. Lett.* **2009**, *95*, 083703. [CrossRef]
10. Delacour, C.; Veliev, F.; Crozes, T.; Bres, G.; Minet, J.; Ionica, I.; Ernst, T.; Briançon-Marjollet, A.; Albrieux, M.; Villard, C.J.A.E.M. Neuron-Gated Silicon Nanowire Field Effect Transistors to Follow Single Spike Propagation within Neuronal Network. *Adv. Eng. Mater.* **2021**, *23*, 2001226. [CrossRef]

11. Patolsky, F.; Timko, B.P.; Yu, G.; Fang, Y.; Greytak, A.B.; Zheng, G.; Lieber, C.M.J.S. Detection, stimulation, and inhibition of neuronal signals with high-density nanowire transistor arrays. *Science* **2006**, *313*, 1100–1104. [CrossRef] [PubMed]

12. Anand, A.; Liu, C.-R.; Chou, A.-C.; Hsu, W.-H.; Ulaganathan, R.K.; Lin, Y.-C.; Dai, C.-A.; Tseng, F.-G.; Pan, C.-Y.; Chen, Y.-T.J.A.S. Detection of K+ efflux from stimulated cortical neurons by an aptamer-modified silicon nanowire field-effect transistor. *ACS Sens.* **2017**, *2*, 69–79. [CrossRef]

13. Zafar, S.; D'Emic, C.; Jagtiani, A.; Kratschmer, E.; Miao, X.; Zhu, Y.; Mo, R.; Sosa, N.; Hamann, H.; Shahidi, G. Silicon Nanowire Field Effect Transistor Sensors with Minimal Sensor-to-Sensor Variations and Enhanced Sensing Characteristics. *ACS Nano* **2018**, *12*, 6577–6587. [CrossRef]

14. Li, H.; Dauphin-Ducharme, P.; Ortega, G.; Plaxco, K.W. Calibration-Free Electrochemical Biosensors Supporting Accurate Molecular Measurements Directly in Undiluted Whole Blood. *J. Am. Chem. Soc.* **2017**, *139*, 11207–11213. [CrossRef]

15. Rani, D.; Pachauri, V.; Ingebrandt, S. Silicon Nanowire Field-Effect Biosensors. In *Label-Free Biosensing*; Springer: Berlin, Germany, 2018; pp. 27–57.

16. Tran, D.P.; Pham, T.T.T.; Wolfrum, B.; Offenhausser, A.; Thierry, B. CMOS-Compatible Silicon Nanowire Field-Effect Transistor Biosensor: Technology Development toward Commercialization. *Materials* **2018**, *11*, 785. [CrossRef] [PubMed]

17. Namdari, P.; Daraee, H.; Eatemadi, A. Recent Advances in Silicon Nanowire Biosensors: Synthesis Methods, Properties, and Applications. *Nanoscale Res. Lett* **2016**, *11*, 406. [CrossRef] [PubMed]

18. Lee, S.; Wang, N.; Zhang, Y.; Tang, Y.J.M.B. Oxide-assisted semiconductor nanowire growth. *Mrs Bull.* **1999**, *24*, 36–42. [CrossRef]

19. Patolsky, F.; Zheng, G.; Lieber, C.M. Fabrication of silicon nanowire devices for ultrasensitive, label-free, real-time detection of biological and chemical species. *Nat. Protoc.* **2006**, *1*, 1711–1724. [CrossRef] [PubMed]

20. Kim, A.; Ah, C.S.; Yu, H.Y.; Yang, J.-H.; Baek, I.-B.; Ahn, C.-G.; Park, C.W.; Jun, M.S.; Lee, S.J.A.P.L. Ultrasensitive, label-free, and real-time immunodetection using silicon field-effect transistors. *Appl. Phys. Lett.* **2007**, *91*, 103901. [CrossRef]

21. Gao, A.; Lu, N.; Dai, P.; Li, T.; Pei, H.; Gao, X.; Gong, Y.; Wang, Y.; Fan, C. Silicon-nanowire-based CMOS-compatible field-effect transistor nanosensors for ultrasensitive electrical detection of nucleic acids. *Nano Lett.* **2011**, *11*, 3974–3978. [CrossRef]

22. Vu, X.T.; Eschermann, J.F.; Stockmann, R.; GhoshMoulick, R.; Offenhäusser, A.; Ingebrandt, S. Top-down processed silicon nanowire transistor arrays for biosensing. *Phys. Status Solidi* **2009**, *206*, 426–434. [CrossRef]

23. Jayakumar, G.; Garidis, K.; Hellström, P.-E.; Östling, M. Fabrication and characterization of silicon nanowires using STL for biosensing applications. In Proceedings of the 15th International Conference on Ultimate Integration on Silicon (ULIS), Stockholm, Sweden, 7–9 April 2014; pp. 109–112.

24. Park, I.; Li, Z.; Pisano, A.P.; Williams, R.S. Top-down fabricated silicon nanowire sensors for real-time chemical detection. *Nanotechnology* **2010**, *21*, 015501. [CrossRef]

25. Vu, X.; GhoshMoulick, R.; Eschermann, J.; Stockmann, R.; Offenhäusser, A.; Ingebrandt, S. Fabrication and application of silicon nanowire transistor arrays for biomolecular detection. *Sens. Actuators B Chem.* **2010**, *144*, 354–360. [CrossRef]

26. Müller, A.; Vu, X.T.; Pachauri, V.; Francis, L.A.; Flandre, D.; Ingebrandt, S. Wafer-Scale Nanoimprint Lithography Process Towards Complementary Silicon Nanowire Field-Effect Transistors for Biosensor Applications. *Phys. Status Solidi* **2018**, *215*, 1800234. [CrossRef]

27. Rani, D.; Pachauri, V.; Mueller, A.; Vu, X.T.; Nguyen, T.C.; Ingebrandt, S. On the Use of Scalable NanoISFET Arrays of Silicon with Highly Reproducible Sensor Performance for Biosensor Applications. *ACS Omega* **2016**, *1*, 84–92. [CrossRef] [PubMed]

28. Rani, D.; Singh, Y.; Salker, M.; Vu, X.T.; Ingebrandt, S.; Pachauri, V. Point-of-care-ready nanoscale ISFET arrays for sub-picomolar detection of cytokines in cell cultures. *Anal. Bioanal. Chem.* **2020**, 1–12. [CrossRef]

29. Vu, X.T.; Stockmann, R.; Wolfrum, B.; Offenhäusser, A.; Ingebrandt, S. Fabrication and application of a microfluidic-embedded silicon nanowire biosensor chip. *Phys. Status Solidi* **2010**, *207*, 850–857. [CrossRef]

30. Stern, E.; Klemic, J.F.; Routenberg, D.A.; Wyrembak, P.N.; Turner-Evans, D.B.; Hamilton, A.D.; LaVan, D.A.; Fahmy, T.M.; Reed, M.A. Label-free immunodetection with CMOS-compatible semiconducting nanowires. *Nature* **2007**, *445*, 519–522. [CrossRef]

31. Smith, R.; Geary, S.M.; Salem, A.K. Silicon Nanowires and Their Impact on Cancer Detection and Monitoring. *ACS Appl. Nano Mater.* **2020**, *3*, 8522–8536. [CrossRef]

32. Doucey, M.-A.; Carrara, S. Nanowire sensors in cancer. *Trends Biotechnol.* **2019**, *37*, 86–99. [CrossRef] [PubMed]

33. Pachauri, V.; Ingebrandt, S. Biologically sensitive field-effect transistors: From ISFETs to NanoFETs. *Essays Biochem.* **2016**, *60*, 81–90. [CrossRef]

34. Tian, B.; Lieber, C.M. Nanowired Bioelectric Interfaces. *Chem. Rev.* **2019**, *119*, 9136–9152. [CrossRef] [PubMed]

35. Kaisti, M. Detection principles of biological and chemical FET sensors. *Biosens. Bioelectron.* **2017**, *98*, 437–448. [CrossRef]

36. Noor, M.O.; Krull, U.J. Silicon nanowires as field-effect transducers for biosensor development: A review. *Anal. Chim. Acta* **2014**, *825*, 1–25. [CrossRef]

37. Zhang, G.J.; Ning, Y. Silicon nanowire biosensor and its applications in disease diagnostics: A review. *Anal. Chim. Acta* **2012**, *749*, 1–15. [CrossRef]

38. Baraban, L.; Ibarlucea, B.; Baek, E.; Cuniberti, G. Hybrid Silicon Nanowire Devices and Their Functional Diversity. *Adv. Sci.* **2019**, *6*, 1900522. [CrossRef] [PubMed]

39. Leonard, F.; Talin, A.A. Electrical contacts to one- and two-dimensional nanomaterials. *Nat. Nanotechnol.* **2011**, *6*, 773–783. [CrossRef] [PubMed]

40. Livi, P.; Shadmani, A.; Wipf, M.; Stoop, R.L.; Rothe, J.; Chen, Y.; Calame, M.; Schönenberger, C.; Hierlemann, A. Sensor system including silicon nanowire ion sensitive FET arrays and CMOS readout. *Sens. Actuators B Chem.* **2014**, *204*, 568–577. [CrossRef]

41. Tran, D.P.; Winter, M.; Yang, C.T.; Stockmann, R.; Offenhausser, A.; Thierry, B. Silicon Nanowires Field Effect Transistors: A Comparative Sensing Performance between Electrical Impedance and Potentiometric Measurement Paradigms. *Anal. Chem.* **2019**, *91*, 12568–12573. [CrossRef]

42. Duan, X.; Li, Y.; Rajan, N.K.; Routenberg, D.A.; Modis, Y.; Reed, M.A. Quantification of the affinities and kinetics of protein interactions using silicon nanowire biosensors. *Nat. Nanotechnol.* **2012**, *7*, 401–407. [CrossRef]

43. Bergveld, P. The operation of an ISFET as an electronic device. *Sens. Actuators* **1981**, *1*, 17–29. [CrossRef]

44. Han, Y.; Offenhäusser, A.; Ingebrandt, S. Detection of DNA hybridization by a field-effect transistor with covalently attached catcher molecules. *Surf. Interface Anal.* **2006**, *38*, 176–181. [CrossRef]

45. Gao, X.P.; Zheng, G.; Lieber, C.M. Subthreshold regime has the optimal sensitivity for nanowire FET biosensors. *Nano Lett.* **2010**, *10*, 547–552. [CrossRef] [PubMed]

46. Nguyen, T.C.; Vu, X.T.; Freyler, M.; Ingebrandt, S. PSPICE model for silicon nanowire field-effect transistor biosensors in impedimetric measurement mode. *Phys. Status Solidi* **2013**, *210*, 870–876. [CrossRef]

47. Bhattacharjee, A.; Nguyen, T.C.; Pachauri, V.; Ingebrandt, S.; Vu, X.T.J.M. Comprehensive Understanding of Silicon-Nanowire Field-Effect Transistor Impedimetric Readout for Biomolecular Sensing. *Micromachines* **2021**, *12*, 39. [CrossRef]

48. Nair, P.R.; Alam, M.A. Design Considerations of Silicon Nanowire Biosensors. *IEEE Trans. Electron. Devices* **2007**, *54*, 3400–3408. [CrossRef]

49. Park, C.W.; Ahn, C.G.; Yang, J.H.; Baek, I.B.; Ah, C.S.; Kim, A.; Kim, T.Y.; Sung, G.Y. Control of channel doping concentration for enhancing the sensitivity of 'top-down' fabricated Si nanochannel FET biosensors. *Nanotechnology* **2009**, *20*, 475501. [CrossRef]

50. Bae, T.-E.; Jang, H.-J.; Yang, J.-H.; Cho, W.-J. High Performance of Silicon Nanowire-Based Biosensors using a High-k Stacked Sensing Thin Film. *ACS Appl. Mater. Interfaces* **2013**, *5*, 5214–5218. [CrossRef]

51. Choi, S.; Park, I.; Hao, Z.; Holman, H.-Y.N.; Pisano, A.P. Quantitative studies of long-term stable, top-down fabricated silicon nanowire pH sensors. *Appl. Phys. A* **2012**, *107*, 421–428. [CrossRef]

52. Tang, X.; Reckinger, N.; Larrieu, G.; Dubois, E.; Flandre, D.; Raskin, J.P.; Nysten, B.; Jonas, A.M.; Bayot, V. Characterization of ultrathin SOI film and application to short channel MOSFETs. *Nanotechnology* **2008**, *19*, 165703. [CrossRef]

53. Saha, S.K. Modeling process variability in scaled CMOS technology. *IEEE Des. Test. Comput.* **2010**, *27*, 8–16. [CrossRef]

54. Regonda, S.; Tian, R.; Gao, J.; Greene, S.; Ding, J.; Hu, W. Silicon multi-nanochannel FETs to improve device uniformity/stability and femtomolar detection of insulin in serum. *Biosens. Bioelectron.* **2013**, *45*, 245–251. [CrossRef] [PubMed]

55. Rajan, N.K.; Routenberg, D.A.; Jin, C.; Reed, M.A. 1/f Noise of Silicon Nanowire BioFETs. *IEEE Electron. Device Lett.* **2010**, *31*, 615–617. [CrossRef]

56. Moh, T.S.Y.; Pandraud, G.; Sarro, P.M.; Huang, Q.A.; Sudhölter, E.J.R.; Nie, M.; de Smet, L.C.P.M. Effect of silicon nanowire etching on signal-to-noise ratio of SiNW FETs for (bio)sensor applications. *Electron. Lett.* **2013**, *49*, 782–784. [CrossRef]

57. Schnakenberg, U.; Benecke, W.; Lange, P. TMAHW etchants for silicon micromachining. In Proceedings of the TRANSDUCERS'91: International Conference on Solid-State Sensors and Actuators, Digest of Technical Papers, San Francisco, CA, USA, 24–27 June 1991; pp. 815–818.

58. Cui, Y.; Zhong, Z.; Wang, D.; Wang, W.U.; Lieber, C.M. High Performance Silicon Nanowire Field Effect Transistors. *Nano Lett.* **2003**, *3*, 149–152. [CrossRef]

59. Kutovyi, Y.; Zadorozhnyi, I.; Hlukhova, H.; Handziuk, V.; Petrychuk, M.; Ivanchuk, A.; Vitusevich, S.J.N. Origin of noise in liquid-gated Si nanowire troponin biosensors. *Nanotechnology* **2018**, *29*, 175202. [CrossRef]

60. Susloparova, A.; Vu, X.T.; Koppenhöfer, D.; Law, J.K.-Y.; Ingebrandt, S. Investigation of ISFET device parameters to optimize for impedimetric sensing of cellular adhesion. *Phys. Status Solidi* **2014**, *211*, 1395–1403. [CrossRef]

61. Motayed, A.; Bonevich, J.E.; Krylyuk, S.; Davydov, A.V.; Aluri, G.; Rao, M.V. Correlation between the performance and microstructure of Ti/Al/Ti/Au Ohmic contacts to p-type silicon nanowires. *Nanotechnology* **2011**, *22*, 075206. [CrossRef]

62. Lin, Y.C.; Chen, Y.; Xu, D.; Huang, Y. Growth of nickel silicides in Si and Si/SiOx core/shell nanowires. *Nano Lett.* **2010**, *10*, 4721–4726. [CrossRef]

63. Sze, S.M. *Semiconductor Devices: Physics and Technology*; John Wiley & Sons: Hoboken, NJ, USA, 2008.

64. Zhou, W.; Dai, X.; Fu, T.M.; Xie, C.; Liu, J.; Lieber, C.M. Long term stability of nanowire nanoelectronics in physiological environments. *Nano Lett.* **2014**, *14*, 1614–1619. [CrossRef]

65. Kim, S.; Kwon, D.W.; Lee, R.; Kim, D.H.; Park, B.-G. Investigation of drift effect on silicon nanowire field effect transistor based pH sensor. *Jpn. J. Appl. Phys.* **2016**, *55*. [CrossRef]

66. Rigante, S.; Scarbolo, P.; Wipf, M.; Stoop, R.L.; Bedner, K.; Buitrago, E.; Bazigos, A.; Bouvet, D.; Calame, M.; Schonenberger, C.; et al. Sensing with Advanced Computing Technology: Fin Field-Effect Transistors with High-k Gate Stack on Bulk Silicon. *ACS Nano* **2015**, *9*, 4872–4881. [CrossRef]

67. Robertson, J. Interfaces and defects of high-K oxides on silicon. *Solid State Electron.* **2005**, *49*, 283–293. [CrossRef]

68. Zafar, S.; Callegari, A.; Gusev, E.; Fischetti, M.V. Charge trapping related threshold voltage instabilities in high permittivity gate dielectric stacks. *J. Appl. Phys.* **2003**, *93*, 9298–9303. [CrossRef]

69. Zafar, S.; Kumar, A.; Gusev, E.; Cartier, E. Threshold voltage instabilities in high-/spl kappa/ gate dielectric stacks. *IEEE Trans. Device Mater. Reliab.* **2005**, *5*, 45–64. [CrossRef]

70. Guillorn, M.; Chang, J.; Fuller, N.; Patel, J.; Darnon, M.; Pyzyna, A.; Joseph, E.; Engelmann, S.; Ott, J.; Newbury, J.; et al. Hydrogen silsesquioxane-based hybrid electron beam and optical lithography for high density circuit prototyping. *J. Vac. Sci. Technol. B* **2009**, *27*, 2588–2592. [CrossRef]
71. Drennan, P.G.; McAndrew, C.C. Understanding MOSFET mismatch for analog design. *IEEE J. Solid-State Circuits* **2003**, *38*, 450–456. [CrossRef]
72. Rajan, N.K.; Routenberg, D.A.; Reed, M.A. Optimal signal-to-noise ratio for silicon nanowire biochemical sensors. *Appl. Phys. Lett.* **2011**, *98*, 264107–2641073. [CrossRef]
73. Lee, J.; Jang, J.; Choi, B.; Yoon, J.; Kim, J.Y.; Choi, Y.K.; Kim, D.M.; Kim, D.H.; Choi, S.J. A Highly Responsive Silicon Nanowire/Amplifier MOSFET Hybrid Biosensor. *Sci. Rep.* **2015**, *5*, 12286. [CrossRef]
74. Jayakumar, G.; Legallais, M.; Hellstrom, P.E.; Mouis, M.; Pignot-Paintrand, I.; Stambouli, V.; Ternon, C.; Ostling, M. Wafer-scale HfO2 encapsulated silicon nanowire field effect transistor for efficient label-free DNA hybridization detection in dry environment. *Nanotechnology* **2019**, *30*, 184002. [CrossRef]
75. Lim, C.M.; Lee, I.K.; Lee, K.J.; Oh, Y.K.; Shin, Y.B.; Cho, W.J. Improved sensing characteristics of dual-gate transistor sensor using silicon nanowire arrays defined by nanoimprint lithography. *Sci. Technol. Adv. Mater.* **2017**, *18*, 17–25. [CrossRef]
76. Yu, Z.; Chen, L.; Wu, W.; Ge, H.; Chou, S.Y. Fabrication of nanoscale gratings with reduced line edge roughness using nanoimprint lithography. *J. Vac. Sci. Technol. B* **2003**, *21*. [CrossRef]
77. Sreenivasan, S.V. Nanoimprint lithography steppers for volume fabrication of leading-edge semiconductor integrated circuits. *Microsyst. Nanoeng.* **2017**, *3*, 17075. [CrossRef] [PubMed]
78. Li, B.R.; Chen, C.W.; Yang, W.L.; Lin, T.Y.; Pan, C.Y.; Chen, Y.T. Biomolecular recognition with a sensitivity-enhanced nanowire transistor biosensor. *Biosens. Bioelectron.* **2013**, *45*, 252–259. [CrossRef] [PubMed]
79. Han, Y.; Mayer, D.; Offenhäusser, A.; Ingebrandt, S. Surface activation of thin silicon oxides by wet cleaning and silanization. *Thin Solid Film.* **2006**, *510*, 175–180. [CrossRef]
80. Munief, W.M.; Heib, F.; Hempel, F.; Lu, X.L.; Schwartz, M.; Pachauri, V.; Hempelmann, R.; Schmitt, M.; Ingebrandt, S. Silane Deposition via Gas-Phase Evaporation and High-Resolution Surface Characterization of the Ultrathin Siloxane Coatings. *Langmuir* **2018**, *34*, 10217–10229. [CrossRef]
81. Uslu, F.; Ingebrandt, S.; Mayer, D.; Bocker-Meffert, S.; Odenthal, M.; Offenhausser, A. Labelfree fully electronic nucleic acid detection system based on a field-effect transistor device. *Biosens. Bioelectron.* **2004**, *19*, 1723–1731. [CrossRef] [PubMed]
82. Park, I.; Li, Z.; Pisano, A.P.; Williams, R.S. Selective surface functionalization of silicon nanowires via nanoscale Joule heating. *Nano Lett.* **2007**, *7*, 3106–3111. [CrossRef]
83. Rani, D. Label-free Detection of Biomolecules Using Silicon Nanowire Ion-sensitive Field-effect Transistor Devices. Ph.D. Thesis, Justus-Liebig-Universität Gießen, Giessen, Germany, 2017.
84. Reimhult, E.; Höök, F.J.S. Design of surface modifications for nanoscale sensor applications. *Sensors* **2015**, *15*, 1635–1675. [CrossRef] [PubMed]
85. Kalia, J.; Raines, R.T.J. Advances in bioconjugation. *Curr. Org. Chem.* **2010**, *14*, 138–147. [CrossRef]
86. Knopp, D.; Tang, D.; Niessner, R.J.A. Bioanalytical applications of biomolecule-functionalized nanometer-sized doped silica particles. *Anal. Chim. Acta* **2009**, *647*, 14–30. [CrossRef]
87. Camarero, J.A.J. Recent developments in the site-specific immobilization of proteins onto solid supports. *Pept. Sci.* **2008**, *90*, 450–458. [CrossRef]
88. Tzouvadaki, I.; Zapatero-Rodríguez, J.; Naus, S.; de Micheli, G.; O'Kennedy, R.; Carrara, S.J.S.; Chemical, A.B. Memristive biosensors based on full-size antibodies and antibody fragments. *Sens. Actuators B* **2019**, *286*, 346–352. [CrossRef]
89. Ingebrandt, S.J.N. Sensing beyond the Limit. *Nat. Nanotechnol.* **2015**, *10*, 734–735. [CrossRef]
90. Lu, B.; Smyth, M.R.; O'Kennedy, R.J.A. Tutorial review. Oriented immobilization of antibodies and its applications in immunoassays and immunosensors. *Analyst* **1996**, *121*, 29R–32R. [CrossRef]
91. Linshiz, G.; Jensen, E.; Stawski, N.; Bi, C.; Elsbree, N.; Jiao, H.; Kim, J.; Mathies, R.; Keasling, J.D.; Hillson, N.J. End-to-end automated microfluidic platform for synthetic biology: From design to functional analysis. *J. Biol. Eng.* **2016**, *10*, 1–15. [CrossRef] [PubMed]
92. Huang, I.Y.; Huang, R.S. Fabrication and characterization of a new planar solid-state reference electrode for ISFET sensors. *Thin Solid Film.* **2002**, *406*, 255–261. [CrossRef]
93. Shinwari, M.W.; Zhitomirsky, D.; Deen, I.A.; Selvaganapathy, P.R.; Deen, M.J.; Landheer, D. Microfabricated Reference Electrodes and their Biosensing Applications. *Sensors* **2010**, *10*, 1679–1715. [CrossRef] [PubMed]
94. Yang, H.; Kang, S.K.; Choi, C.A.; Kim, H.; Shin, D.-H.; Kim, Y.S.; Kim, Y.T. An iridium oxide reference electrode for use in microfabricated biosensors and biochips. *Lab Chip* **2004**, *4*, 42–46. [CrossRef] [PubMed]
95. Lewenstam, A. *Handbook of Reference Electrodes*; György, I., Andrzej, L., Fritz, S., Eds.; Springer: Berlin, Germany, 2013; pp. 1–344. ISBN 978-3-642-36187-6.
96. Duarte-Guevara, C.; Swaminathan, V.V.; Burgess, M.; Reddy, B.; Salm, E.; Liu, Y.S.; Rodriguez-Lopez, J.; Bashir, R. On-chip metal/polypyrrole quasi-reference electrodes for robust ISFET operation. *Analyst* **2015**, *140*, 3630–3641. [CrossRef] [PubMed]
97. Kremers, T.; Tintelott, M.; Pachauri, V.; Vu, X.T.; Ingebrandt, S.; Schnakenberg, U. Microelectrode Combinations of Gold and Polypyrrole Enable Highly Stable Two-Electrode Electrochemical Impedance Spectroscopy Measurements Under Turbulent Flow Conditions. *Electroanalysis* **2020**, *33*, 197–207. [CrossRef]

sensors

Article

Planar Junctionless Field-Effect Transistor for Detecting Biomolecular Interactions

Rajendra P. Shukla [1,*], J. G. Bomer [1], Daniel Wijnperle [1], Naveen Kumar [2], Vihar P. Georgiev [2], Aruna Chandra Singh [3], Sivashankar Krishnamoorthy [3], César Pascual García [4], Sergii Pud [1,*] and Wouter Olthuis [1]

[1] BIOS Lab-on-a-Chip Group, MESA+ Institute for Nanotechnology, Max Planck Center for Complex Fluid Dynamics, University of Twente, P.O. Box 217, 7500 AE Enschede, The Netherlands; j.g.bomer@utwente.nl (J.G.B.); d.wijnperle@utwente.nl (D.W.); w.olthuis@utwente.nl (W.O.)
[2] Device Modelling Group, School of Engineering, University of Glasgow, Glasgow G12 8LT, UK; naveen.kumar@glasgow.ac.uk (N.K.); vihar.georgiev@glasgow.ac.uk (V.P.G.)
[3] Nano-Enabled Medicine and Cosmetics Group, Materials Research and Technology Department, Luxembourg Institute of Science and Technology (LIST), L-4362 Belvaux, Luxembourg; aruna.singh@list.lu (A.C.S.); sivashankar.krishnamoorthy@list.lu (S.K.)
[4] Nanoscale Engineering for Devices & Bio-Interfaces, Nanotechnology Unit of the Materials Research and Technology Department, Luxembourg Institute of Science and Technology (LIST), L-4422 Belvaux, Luxembourg; cesar.pascual@list.lu
* Correspondence: r.p.shukla@utwente.nl (R.P.S.); s.pud@utwente.nl (S.P.)

Citation: Shukla, R.P.; Bomer, J.G.; Wijnperle, D.; Kumar, N.; Georgiev, V.P.; Singh, A.C.; Krishnamoorthy, S.; Pascual García, C.; Pud, S.; Olthuis, W. Planar Junctionless Field-Effect Transistor for Detecting Biomolecular Interactions. *Sensors* **2022**, *22*, 5783. https://doi.org/10.3390/s22155783

Academic Editors: Michael J. Schöning and Sven Ingebrandt

Received: 11 July 2022
Accepted: 30 July 2022
Published: 2 August 2022

Publisher's Note: MDPI stays neutral with regard to jurisdictional claims in published maps and institutional affiliations.

Abstract: Label-free field-effect transistor-based immunosensors are promising candidates for proteomics and peptidomics-based diagnostics and therapeutics due to their high multiplexing capability, fast response time, and ability to increase the sensor sensitivity due to the short length of peptides. In this work, planar junctionless field-effect transistor sensors (FETs) were fabricated and characterized for pH sensing. The device with SiO_2 gate oxide has shown voltage sensitivity of 41.8 ± 1.4, 39.9 ± 1.4, 39.0 ± 1.1, and 37.6 ± 1.0 mV/pH for constant drain currents of 5, 10, 20, and 50 nA, respectively, with a drain to source voltage of 0.05 V. The drift analysis shows a stability over time of -18 nA/h (pH 7.75), -3.5 nA/h (pH 6.84), -0.5 nA/h (pH 4.91), 0.5 nA/h (pH 3.43), corresponding to a pH drift of -0.45, -0.09, -0.01, and 0.01 per h. Theoretical modeling and simulation resulted in a mean value of the surface states of $3.8 \times 10^{15}/cm^2$ with a standard deviation of $3.6 \times 10^{15}/cm^2$. We have experimentally verified the number of surface sites due to APTES, peptide, and protein immobilization, which is in line with the theoretical calculations for FETs to be used for detecting peptide-protein interactions for future applications.

Keywords: planar junctionless FETs; pH sensor; proteomics; peptidomics; peptide-protein interaction; therapeutics; diagnostics

1. Introduction

Recent developments in peptidomics and proteomics have enabled the rapid progress of novel personalized therapies [1,2]. Peptides are short sequences of amino acids with high specificity and affinity towards binding targets [3,4]. Some of them represent protein epitopes that carry diagnostic and therapeutic information as their interaction with the major histocompatibility complex (MHC proteins) can determine a patient's specific response to a possible vaccine (e.g., for cancer) [5,6]. Screening such sequences for their interaction with antibodies and MHC proteins is of great interest in modeling the response of the immune system. However, the intrinsic variability of these peptide sequences hinders high throughput screening to cover all possible combinations of amino acids [7]. Transducing such interactions into readable signals requires a multiplexed setup of label-free immunosensors that allows detection in the physiological range [5,8,9]. Current sensing technologies have limited multiplexing capabilities and require labelling of the molecules (e.g., ELISA) [9]. Therefore, there is a need for a multiplexed setup with controlled immobilization of these peptide sequences on devices that enable highly sensitive and label-free sensing of the target analytes. Field-effect transistor (FET)-based immunosensors are good candidates for multiplexed label-free sensing due to their high scalability, compatibility with current CMOS technology, fast response time, and label-free sensitivity [10–13]. When functionalized with short sequences of peptides, a FET gate can detect the binding of proteins in close proximity to the sensitive region within the Debye screening length of the protein solution [14]. Nanowire-based FETs are one of the most highly investigated structures among them because of the 3D gating effect and faster mass transport towards the sensing area [15–18]. The surface area-to-volume ratio allows the adsorption of the analytes in 2D as compared to planar adsorption. However, nanowire devices are still facing several challenges in clinical applications due to reliability issues [19]. Here, we propose a simpler design in terms of fabrication point-of-view, called planar junctionless FETs, where the conducting channel acts as a resistor and the carrier density in the channel resistor can be modulated by applying the gate voltage by means of a reference electrode in a given pH of the electrolyte solution [20]. The advantage of this device lies in its relative simplicity; it does not require the fabrication of shallow implanted p-n junctions in the source and drain areas. Moreover, the planar structure of the device allows more robust functionalization of the sensing surface as compared to any other non-planar structure [17]. We have used lightly doped thin device layer SOI wafers to demonstrate their suitability for detecting small changes in charge at the electrolyte-oxide surfaces (i.e., caused by the interaction of the proteins with peptides immobilized on the gate surface, which is the long-term goal of this work). A larger planar surface area allows a better signal-to-noise ratio and also less stringent requirements to counter reliability issues, e.g., from pin-holes, as compared to the nanowire counterparts [21].

Figure 1 shows the schematic of the proposed device design (Figure 1A) and the cross-sectional view of the device layout (Figure 1B). We have overcome several fabrication related challenges during the process. For example, ohmic contact with the lightly doped thin device layer is best achieved by the formation of a thin layer of PtSi alloys at the interface of Ta/Pt and silicon. However, this process is too sensitive to the thickness of the silicide formation and the annealing temperature to provide reproducible results with our device layer thickness [22–29]. We have overcome this issue by optimization of the annealing process to get a reliable planar junctionless FET device, and in the end, we have demonstrated the pH sensing performance of our fabricated device. Theoretical modeling and simulation were done using experimental data to calculate the surface states and charge density present at the oxide layer. Further, we have experimentally calculated the number of surface sites after silanization of the gate oxide surface with APTES, peptide, and protein functionalization. These data will be used for optimizing future devices where the oxide surface of the FETs will be functionalized with different chemistries. We have tested the stability of the device over time (drift analysis) and confirmed its suitability for future application as a label-free sensor of peptide–protein interactions. We anticipate that the proposed rational device design can be an optimal solution for reproducible multiplexed sensing of peptide–protein interactions [30,31].

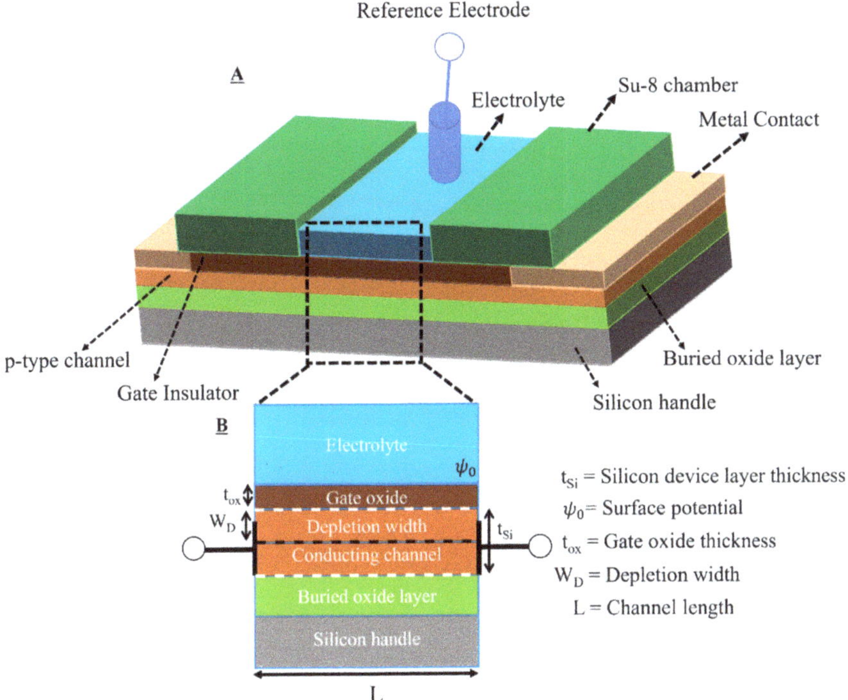

Figure 1. (**A**) Schematic of the proposed planar junctionless FETs and (**B**) cross-sectional view of the proposed device design.

2. Materials and Methods

2.1. pH Test Buffers

Tetrabutylammonium chloride (TBACl), tetrabutylammonium hydroxide solution (TBAOH), acetic acid, boric acid, and orthophosphoric acid were purchased from Merck (Sigma Aldrich). As a background electrolyte, 0.1 M of TBACl was used. First of all, a universal buffer mixture (UBM) was prepared by mixing 0.5 M acetic acid, 0.5 M boric acid, and 0.5 M orthophosphoric acid. To buffer the solution, 200 µL of 0.5 M UBM was mixed with 50 mL of 0.1 M TBACl. The pH at the start was around 2.7 at 25 °C. Titration was performed with 0.1 M TBAOH in 20 steps of 0.4 mL and the pH at the end was found to be around 10.5. Back titration was performed with 0.1 M HCl.

2.2. Design Considerations for Planar Junctionless FETs

2.2.1. Wafer Specifications

SOI wafers were purchased from IceMOS Technology, Ltd. with a diameter of 100.00 ± 0.20 mm, device orientation <100> ± 1.0 degree, silicon device layer thickness of 2.00 ± 0.50 µm, and p-type device layer resistivity of 1–10 Ohm.cm were used for the fabrication of planar junctionless FETs.

2.2.2. Thin and Lightly Doped Device Layer

The thin and lightly doped silicon device layer is required to have a higher sensitivity with a high on/off drain current ratio. For this purpose, the device layer was thinned down using successive wet oxidation and etching of the SiO_2 layer. We have fabricated FET devices with a device layer thickness of 250–300 nm.

2.3. Device Fabrication

The fabrication of the planar junctionless FETs consists of the following steps as depicted in Figure 2.

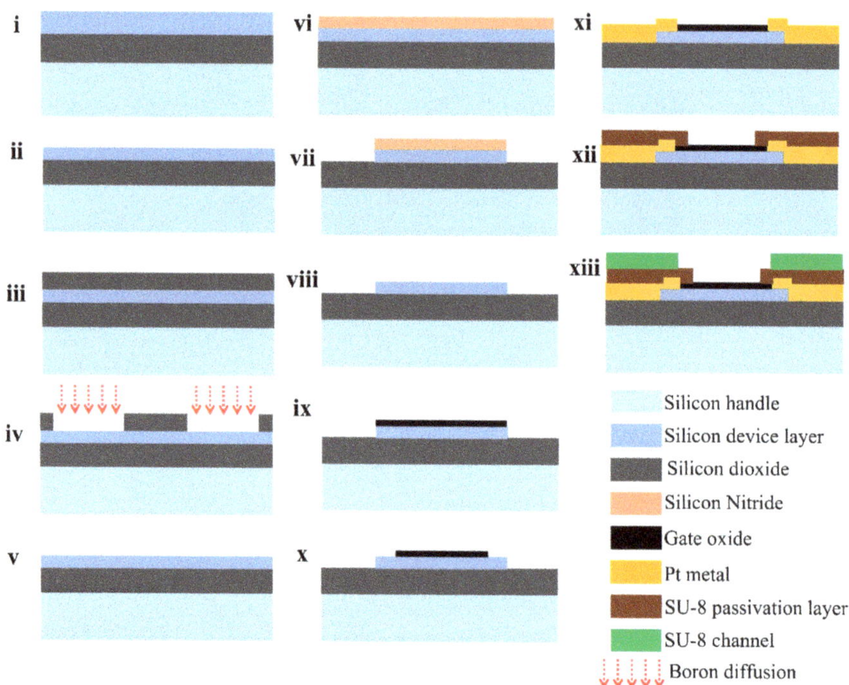

Figure 2. Fabrication process flow of the planar junctionless FETs. (**i**) cleaning SOI wafer, (**ii**) thinning of device layer using oxidation and etching, (**iii**) silicon dioxide growth as a masking layer, (**iv**) patterning silicon dioxide and boron diffusion, (**v**) etching silicon dioxide, (**vi**) silicon nitride deposition, (**vii**) patterning silicon nitride to define silicon islands, (**viii**) etching silicon nitride, (**ix**) gate oxide growth, (**x**) patterning gate oxide to define source and drain regions, (**xi**) Ta/Pt Metal lift-off, (**xii**) patterning SU-8 passivation layer, and (**xiii**) patterning SU-8 channels.

2.3.1. Cleaning SOI Wafer

The process started with the pre-furnace cleaning of SOI wafers in 99% HNO_3 for 10 min to remove organic traces, followed by rinsing in DI water for the removal of traces of chemical agents. The rinsed SOI wafers were further cleaned in 69% HNO_3 at 95 °C for 10 min to remove metallic traces. The wafer was further rinsed in DI water and dried with nitrogen. To remove the native oxide, the wafer was transferred to a 1% HF etching chamber at room temperature. Within several seconds, the surface became hydrophobic, which is an indication of the removal of native oxide from the surface. The wafer was further rinsed in DI water and dried with nitrogen, which was then loaded into the wet oxidation furnace (step (i) in the process flow).

2.3.2. Thinning Silicon Device Layer

The first wet oxidation was done at 1150 °C for 15 h to obtain a thickness of 2.6 μm. The oxide was etched in a 50% HF solution for approximately 3 min until the surface became completely hydrophobic. Similar to this, the second step of oxidation to get an oxide thickness of 1.1 μm was done for 3 h at 1150 °C. This 1.1 μm thick oxide layer was thinned down to a 300 nm oxide layer by etching in buffered HF acid solution to be used as a mask for doping the source and drain regions (steps (ii) and (iii) in the process flow).

2.3.3. Doping Source and Drain Regions

The source and drain regions were first opened by photolithography. This process started with HMDS priming on a spin coater at 4000 rpm for 30 s. The wafer was further spin coated with photoresist Olin OiR 907-17 at 4000 rpm for 30 s, followed by pre-baking at 95 °C for 90 s to remove the residual solvent from the resist film after spin coating. The spin-coated wafer was exposed to UV-LED light with an exposure dose of 100 mJ/cm^2. The exposed wafer was then post-exposure baked at 120 °C for 60 s on a hot plate. The wafer was developed for 60 s, followed by rinsing in DI water and drying with nitrogen. The patterns were inspected using an optical microscope. The developed wafer was then baked at 120 °C for 10 min, followed by UV-ozone cleaning for 5 min to remove any residue of the photoresist. The patterned oxide layer was etched in BHF solution, and the resist was stripped in HNO$_3$. Boron doping was done using Plasma Enhanced Chemical Vapor Deposition of a 100 nm boron doped oxide, covered with a 250 nm undoped capping oxide layer, followed by drive-in at 1100 °C for 30 min. The oxide was then removed in a 50% HF solution (steps (iv) and (v) in the process flow).

2.3.4. Defining Silicon Islands

Next, silicon nitride was deposited to act as a mask to define the silicon islands. The silicon islands were defined using photolithography. The nitride was then removed by dry etching and, subsequently, the silicon was etched in TMAH at 70 °C [32]. The color change was observed as proof of the complete etching of silicon. Next, the nitride layer on top of the silicon islands was removed by etching it in a phosphoric acid solution at 180 °C for 10 min (steps (vi), (vii), and (viii) in the process flow).

2.3.5. Source and Drain Patterning

Next, 10 nm of gate oxide was grown on the islands by dry oxidation for 25 min at 900 °C before defining the source/drain area. Source and drain regions were defined using another photolithography step followed by BHF etching of oxide (steps (ix) and (x) in the process flow).

2.3.6. Metal Contacts Lift-Off

Metal lift-off patterns were defined using photolithography on double layer photoresist: LOR5A and Olin OiR 907-17. Ta/Pt of 2 nm/100 nm was sputtered and lift off in acetone solution with an ultrasonic bath [33]. The LOR5A photoresist was then removed in a 99% HNO$_3$ solution, followed by rinsing in DI water and drying with nitrogen. The Ta/Pt patterned wafer was then annealed at 350 °C for 10 min to improve the electrical contact as it reduces the interface trap density at the metal-semiconductor interface. Although the forming of a thin layer of PtSi alloy after annealing is supposed to ensure ohmic contacts between metal leads and source/drain regions, it is, in practice, rather challenging due to the sensitivity of the PtSi formation to the annealing temperature [23,34]. Moreover, the thin device layer makes the process of annealing prone to irreproducibility because annealing time and temperature can considerably affect the thickness of the device layer due to its being consumed during PtSi formation. To address these challenges and to ensure robustness of the fabrication process, the source and drain regions were doped (see Supplementary Figures S1–S5 for more details) [step (xi) in the process flow].

2.3.7. SU-8 Passivation Layer and Channels

SU-8 patterns were defined in SU-8-2005 (thin SU-8 layer opening at gate and contact area) and SU-8-100 (thick layer, SU-8 channels) using photolithography. First, the SU-8 layer was spin coated at 500 rpm for 10 s (step I), and then at 5000 rpm for 30 s (step II). The spin coated wafer was soft baked at 95 °C for 2 min. The wafer is then exposed at

90 mJ/cm^2 using UV-LED. The exposed wafer was then post exposer baked at 95 °C for 2 min. The wafer was then developed in RER600 developer for 1 min, followed by rinsing in isopropanol and drying with nitrogen. The SU-8 channels were defined using another lithography step. First, the SU-8 layer was spin coated at 500 rpm for 10 s (step I), and then at 3000 rpm for 30 s (step II). The spin coated wafer was soft baked at 95 °C for 15 min. The wafer was then exposed at 400 mJ/cm^2 using UV-LED. The exposed wafer was then post exposure baked at 95 °C for 10 min. The wafer was developed in RER600 developer for 10 min, followed by rinsing in isopropanol and drying with nitrogen. The wafer was then hard baked at 135 °C for 30 min before being diced into chips. One wafer consists of several chips with test patterns and junctionless FETs. Therefore, the wafer was diced into chips using a dicing saw (Disco DAD3220) before use [steps (xii) and (xiii) in the process flow].

2.4. Chip Design and Encapsulation

The planar junctionless FET chip was designed using CleWin software for a 100 mm wafer mask. The dimension of a single chip is 1×1 cm^2, which consists of 15 metal contact pads and three microfluidic channels (Figure 3A). Figure 3B shows SU-8 microfluidic channels. There are 12 FET devices in total (four devices inside each microfluidic channel) with a common source along with a pseudo reference electrode available in this chip. The channel length and width of the device are 4 μm and 12 μm. Figure 3C shows a single junctionless FET device with an open gate area. The pseudo reference electrode has three terminals for each microfluidic channel, which are supposed to be electroplated with silver/silver chloride in future applications. In this work we have used an external silver/silver chloride reference electrode for the simplicity of the measurement set-up. This three-channel based design is adopted by considering a long-term goal of capturing biomolecular interactions where these FETs will be functionalized with different sequences of peptides and their interactions with proteins will be tested. The diced chips were wire bonded for electrical connections to a PCB and insulated using epoxy glue. For proper insulation and hardening of the epoxy glue on the chip, the PCB with epoxy glue was heated on a hot plate for 2 h (Figure 3D). After that, the PCB connected chip was cleaned using plasma for 5 min. Prior to the pH characterization of these devices, the leakage test of the PCB-connected device was performed by putting the device in water and buffer solution and connecting it to the power supply. No leakage current between different electrodes was observed over several hours, which is indicative of the proper insulation of the device with epoxy glue. A microscopic inspection was done to make sure that there was no water that leaked through the SU-8 layer. After having a detailed test of the devices, the pH characterization was done.

2.5. pH Measurement Setup

The encapsulated device was submerged in the buffer solution along with a reference electrode (REF201, red-rod reference saturated in 3M KCl solution, Radiometer Analytical,) with a connection to the source meter. The measurement started in a mixture solution of 200 μL of 0.5 M UBM and 50 mL of 0.1 M TBACl (pH 2.7). The pH of the solution was changed by adding 400 μL of 0.1M TBAOH in steps followed by stirring the solution to make it homogeneous mixture. After stabilization, the pH was measured before recording the pH response. The pH meter (Mettler-Toledo B.V., S-400 basic) was used to measure the pH of the solution, and it was calibrated before use.

2.6. V_{gs} vs. pH and Ids vs. pH Characterization

First, the drain current, I_{ds}, was measured as a function of drain-source voltage, V_{ds}, and gate voltage, V_{gs}. These characterizations were done at a fixed pH to find out the set-point for the device operation. After establishing the optimal setpoint, the V_{gs} was measured as a function of pH as well as I_{ds} to show the sensitivity of the device towards pH change. The drift characterizations were done using the same set-point. For the voltage sensitivity analysis, the V_{gs} was adjusted to maintain the constant I_{ds} for every pH of the electrolyte solution. V_{ds} was kept at 0.05 V. For current sensitivity analysis, the I_{ds} was recorded for a varying pH of the electrolyte solution at a V_{gs} and V_{ds} of −0.5 V and 0.05 V. The voltage and current sensitivities were calculated from the linear fit of V_{gs} vs. pH and Ids vs. pH characteristics.

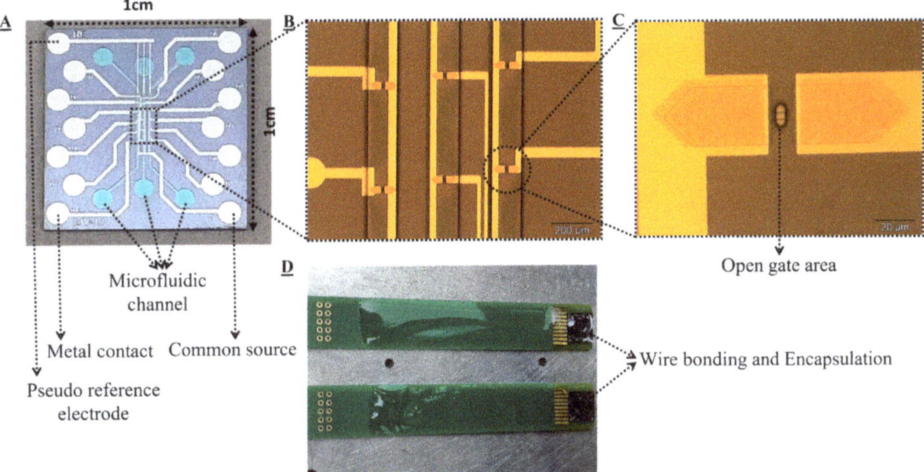

Figure 3. Fabricated chip design and encapsulation. (**A**) image of the single chip, (**B**) SU-8 channels, (**C**) gate opening of a single FET in a thin SU-8 layer. Color change is observed at doped source and drain regions due to the formation of a PtSi alloy, and (**D**) encapsulation with epoxy glue.

2.7. Drift Analysis

I_{ds} was recorded for 100 min with a varying pH every 10 min and then for 2 h at a constant pH to check the drift over time. Current drift over a time period of 2 h was calculated by subtracting the current value at the start and after 2 h at a constant pH. The pH drift over time was calculated using the current drift over time at a fixed pH and the current sensitivity of the pH response.

3. Results and Discussion

3.1. pH Characterization of 2D Planar JUNCTIONLESS FETs

Before we started the pH characterization of the FETs, the set-point (or working point) of the device was decided such that the device operates in a linear region of operation to include the ohmic contribution in the current variation. To decide the set-point of the junctionless FETs for pH characterization and sensitivity analysis, we measured the I_{ds} vs. V_{ds} and I_{ds} vs. V_{gs} characteristics in a constant pH solution of 4.91 (Figure 4A,B). These characterizations provide the working voltage range (set-point) for these devices, which is V_{gs} = −0.5 V and V_{ds} = 0.05 V.

Next, the device was characterized for the voltage and current sensitivities as a function of the pH. The surface potential is changed by V_{gs} and pH, and V_{gs} is related to the threshold voltage V_{th} [35]. Therefore, the shift in the V_{th} is observed as a change in V_{gs}. As the pH changes from acidic to basic, the shift in the V_{th} moves towards a less negative value. The shift of the V_{th} with increasing pH must be compensated for by increasing V_{gs} to keep the concentration of carriers in our p-type channel the same. This can further be detailed by the relationship between the surface potential and the pH, which is derived by combining the electrostatic interactions at the dielectric surface and the distribution of ions inside the electrolyte (Equation (1)) [35].

$$\frac{\partial \psi_0}{\partial pH_B} = 2.303 \frac{k_B T}{q} \alpha \tag{1}$$

where ψ_0, pH_B, k_B, T and q represent the surface potential, bulk pH of the electrolyte, the Boltzmann constant, the absolute temperature, and the elementary charge, respectively. α is a sensitivity parameter with a value varying between 0 and 1, depending on the intrinsic properties of the oxide. For $\alpha = 1$, the sensor shows maximum sensitivity called Nernstian sensitivity which is 59.2 mV/pH at 298 K.

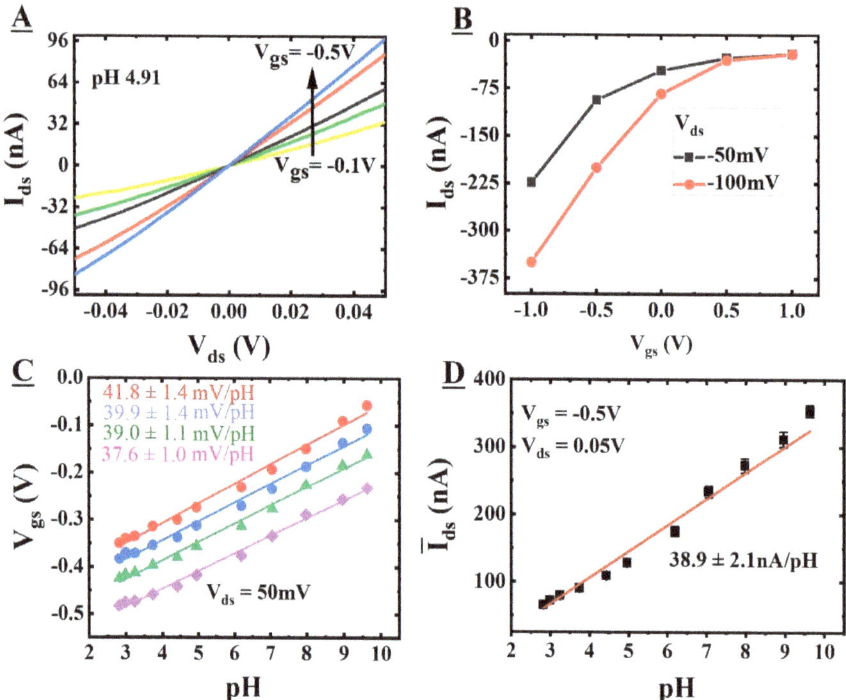

Figure 4. I-V characteristics at fixed pH of 4.91 (**A**) I_{ds} vs. V_{ds} for a varying gate voltage, applied via the reference electrode (-0.1 to -0.5 V in steps of -0.1 V) with V_{ds} ranging from -0.05 V to 0.05 V, and (**B**) I_{ds} vs. V_{gs} for input gate voltage range of -1 to 1 V, applied via the reference electrode for a V_{ds} of 0.05 and 0.1 V. Voltage and current sensitivity analysis (**C**) V_{gs} vs. pH, and (**D**) average I_{ds} vs. pH.

For voltage sensitivity analysis, the change in the V_{gs} (as a result of the shift in V_{th}) was recorded for a constant I_{ds} as a change in pH at a V_{ds} of 0.05 V. Figure 4C shows the variation of V_{gs} for different pH values at a fixed V_{ds} of 0.05 V for constant currents of 5 (red circles), 10 (blue circles), 20 (green triangles), and 50 nA (purple squares). From the V_{gs} vs. pH characteristics, the sensitivities were calculated from the linear fitting. It is found that for constant currents of 5, 10, 20, and 50 nA, voltage sensitivities are 41.8 ± 1.4, 39.9 ± 1.4, 39.0 ± 1.1, and 37.6 ± 1.0 mV/pH at a fixed drain to source voltage of 0.05 V. This shows that our junctionless FET devices with 10 nm of SiO_2 are sensitive to pH change, as expected and reported in the literature [36]. The slight change in voltage sensitivity values for different constant current values is due to the dominant effect of noise current levels at lower constant current values. The sensitivity (α) is calculated using equation 1. The average value of the sensitivity factor (α) calculated for all the constant currents was found to be 0.70 ± 0.03. The calculated sensitivity value is in good agreement with the literature values for pH response at the SiO_2 surface [37,38]. The calculated current sensitivity from the I_{ds} vs pH characteristics was found to be 38.9 ± 2.1 nA/pH, with a wide range of current response (50 to 400 nA; Figure 4D). This wide current range provides us with an insight of almost complete depleted channel to a fully conducting channel. The current sensitivity was further used to calculate the stability of the sensor for pH change in drift analysis. We have plotted the I_{ds} vs. charge density using an analytical model which shows a sensitivity for change in charge density of 0.20 ± 0.01 nA/C/cm^2 (see Supplementary Figure S6 for more details). For a positively charged surface on the oxide electrolyte interface, the channel is almost closed, and a minimum current is observed. As the charge density changes to a negatively charged interface, we observe the small change of that charge effect in terms of drain to source current. These results show promising proof-of-concept device characteristics to be used for sensing interactions of biological molecules, which is one of the long-term project goals with chemistries that can generate different charge densities due to different surface sites on the surface [39,40].

3.2. Calculating Surface States

To evaluate the nature of our surface oxide, we have used the site-binding model with the Gouy–Chapman–Stern (GCS) model. Matlab (R2022a) has been used for analytical modeling and simulation of the experimental data and to calculate the surface states present on the oxide layer. Using the linear regime of our sensors, we can obtain the number of silanol groups that exchange protons with the electrolyte, and thus contribute to the sensitivity, providing a good value of the quality of our oxide. We have considered a stern capacitance of 0.8 F/m^2. Equating the site-binding model with the Gouy–Chapman–Stern theory for the double layer capacitance provides an equation of the 5th order, which results in possible saturation at a higher pH range due to an approximation of $4kb/ka \ll 1$ with unwanted ripples even with the iterative method solution of the 5th order equation [39–41]. Thus, we solved both equations independently with an assumption of the same surface potential values and then equated them later on with a tolerance of 10–50 as explained in detail in [40]. Such an approach is more accurate and flexible enough to be used for all types of oxides or even at the surface with more than two affinity sites with different dissociation constants. In this approach, zeta potential was indirectly considered as an experimental index of the surface states by correlating the zeta potential to the surface potential with a potential drop across the stern layer. As per the Gouy–Chapman–Stern model, the stern layer (uncharged dielectric) between the diffuse layer and the oxide–electrolyte interface decreases the effective potential at the shear plane (zeta potential). Figure 5A shows the scheme of surface cites present in SiO$_2$. The electrolyte concentration and device parameters were kept the same as in the experimental setup. Figure 5B shows the calibration of the simulated model with the experimental data in terms of reference gate bias with respect to the electrolyte pH. The graphs for different current values of 5 nA, 10 nA, 20 nA, and 50 nA have been plotted while representing the possible root mean square error (RMSE) using surface states as the fitting parameter. The following equations were used to calculate different parameters in the model.

$$\sigma_{DL1} = qN_S \left(\frac{cH_s^2 - K_aK_b}{K_aK_b + K_bcH_s + cH_s^2} \right) \tag{2}$$

$$cH_s = cH_Bexp\frac{-\Psi_0}{2V_T}, \ cH_B = 10^{-pH_B} \tag{3}$$

$$\Psi_0 = \Psi_{stern} + \Psi_\zeta = \frac{Q_0sinh\left(\Psi_\zeta/V_T\right)}{C_{stern}} + \Psi_\zeta \tag{4}$$

where σ_{DL1} is the surface charge density, N_S is the number of surface states, $K_a = 10^{-pK_a}$ and $K_b = 10^{-pK_b}$ are the dissociation constants, cH_s is the surface proton concetration, cH_B is the bulk proton concentration, Ψ_0 is the surface potential, V_T is the thermal voltage, Ψ_{stern} is the potential drop across the stern layer and Ψ_ζ is the potential drop across the diffuse layer. We have used the dissociation constants reported in the literature for silanol groups: pK$_a$ = 6 and pK$_b$ = -2 [42]. Keeping the constant affinity of the silanol sites, the density of surface states is varied as a fitting parameter. A surface potential-pH curve is extracted by self-consistently solving the site-binding and GCS models. As an assumption, the curves (four samples for different current values) are supposed to have the same potential near the isoelectric point (pH = 2) and it was considered a starting point to decide the slope of the curve. Every sample was compared with each simulated surface potential-pH curve for different surface states, and the RMSE was calculated. The closest curve to the corresponding sample was extracted with the minimum RMSE value. Assuming constant affinity values, the possible induced doping for different current values may be the reason behind the variation of surface states that can be counted as an error. The obtained mean value of the number of surface states is $3.8 \times 10^{15}/\mathrm{cm}^2$ with a standard deviation of $3.6 \times 10^{15}/\mathrm{cm}^2$. The obtained value of the surface potential used to get these surface states is in good agreement with the simulated and experimental work [40]. Such a high value of surface states signifies the quality of deposited oxide, resulting in high sensitivity for SiO_2 FETs. These values for the number of surface sites were further used to compare the surface sites due to different surface functionalization in the next sections.

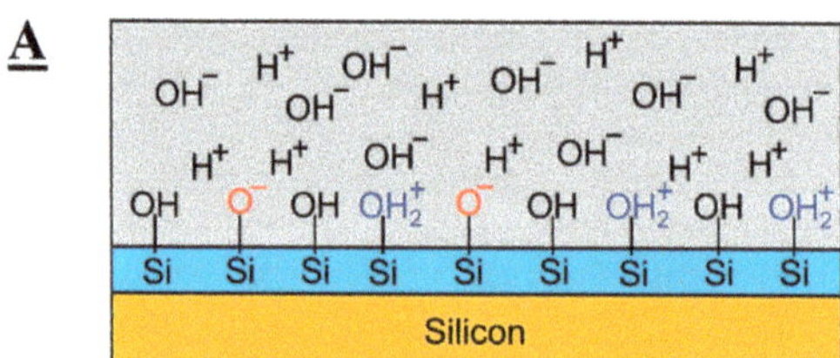

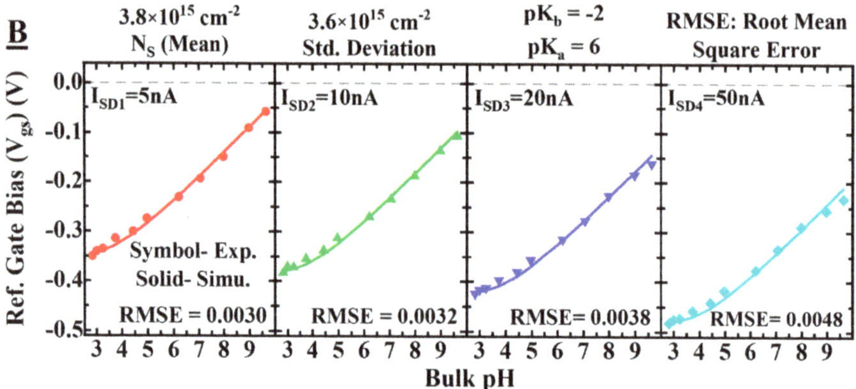

Figure 5. Simulation and modeling of the junctionless FETs. (**A**) Scheme for the surface sites available in SiO_2. (**B**) Calibration of the simulated model with the experimental data in terms of reference gate bias with respect to the electrolyte pH. Separated graphs for different current values of 5 nA, 10 nA, 20 nA, and 50 nA while representing the possible RMSE error using surface states as the fitting parameter. The used color codes are the same for corresponding current values.

3.3. Measuring APTES Functionalization and Monitoring Peptide-Protein Interactions Using Surface Plasmon Resonance (SPR)

Gold SPR chips with a silicon oxide coating (Au/SiO_2) were used to study peptide-protein interactions [43,44]. The use of the SiO_2 surface allows mimicking the conditions to be encountered on the silicon oxide surface of the FET sensors (for more information about the APTES functionalization protocol and steps, see supplementary information Section S2. Surface functionalization of SiO_2 with APTES, peptides, and proteins). The verification of the APTES layer has been done by XPS analysis on bare SiO_2 and silanized (APTES coated SiO_2 chip) (see supplementary information Section S2.2: Surface characterization using XPS). The molecular densities at different steps of functionalization are presented in Table 1. The APTES density on the sensor surface was found to be $1.3 \times 10^{15}/\text{cm}^2$, which corresponds to the presence of a monolayer of APTES. The peptide layers with non-specific antibodies showed no antibody retention on the surface, confirming the specificity of the sensor to the specific peptide antibody interactions. The experimental value of APTES density is comparable with the surface sites present in SiO_2 calculated using an analytical model showing complete coverage of oxide surface sites after functionalization. The density of the peptides and proteins is much lower than the number of silanol groups/APTES groups. Based on the SPR experiments to measure the number of peptide protein interactions, we expect a significant sensitivity that at least will be equivalent to the number of neutralized peptides interacting with proteins. (Table 1).

Table 1. Surface molecular densities with respect to the surface mass absorption.

Surface Groups	Surface Concentration (ng/cm^2)	Calculated Molecular Density (/cm^2)
APTES	470	1.3×10^{15}
Peptide	70	2.3×10^{13}
*p*Ab (solution concentrations in the range of 0.1–20 µg/mL)	7–934	2.8×10^{10}–1.8×10^{12}

3.4. Stability of the Sensor

Figure 6A shows the drain-to-source current, I_{dS}, vs. time at different pH values. The gate-to-source voltage, V_{gS}, and the drain-to-source voltages, V_{dS}, were fixed at -0.5 V and 0.05 V. From Figure 6A, it can be clearly seen that the device is responsive to the pH change happening at the dielectric–electrolyte interface. As the pH increases, the OH-concentration in the solution increases and that is why the charge carriers in the channel regions for a p-type channel also increase and that is why there is an increase in the current. The measurement for each pH was recorded for 10 min and after changing the pH, the solution was stirred to mix the ions and make a homogenous solution for the pH measurement. The short time for stabilization restricts the charging of the electrical double layer due to higher screening with increased ion concentration, resulting in a drift in the response.

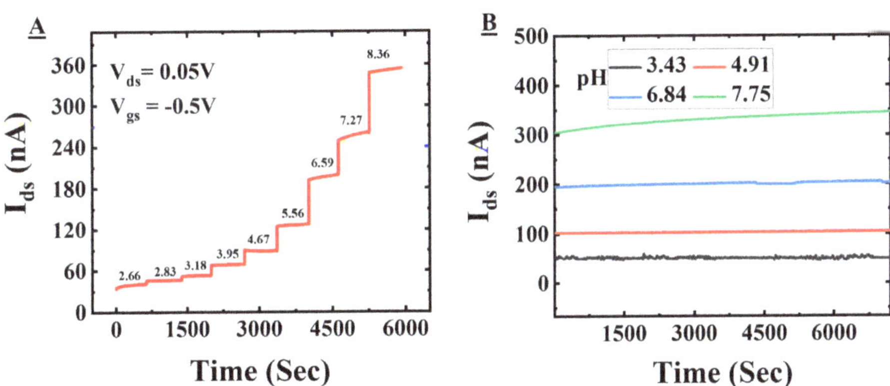

Figure 6. Drift characterization. (**A**) Current vs time step response for different pH and (**B**) current vs. time for a longer time for several constant pH values.

Figure 6B shows the drain-to-source current vs. time for several pH values. The response of the device was recorded for a time period of 2 h for each pH value (3.43,

4.91, 6.84, and 7.75). We have calculated the drift over time, which shows $-18\,\text{nA/hour}$ (pH 7.75), $-3.5\,\text{nA/hour}$ (pH 6.84), $-0.5\,\text{nA/h}$ (pH 4.91), and $0.5\,\text{nA/h}$ (pH 3.43) for the corresponding drift in pH value over time calculated using the current sensitivity obtained from Figure 4D of -0.45, -0.09, -0.01, and 0.01 per hour. It can be seen that the device has a stable response at lower pH values with a small drift, which is expected due to the fluctuation in the electric field because of the larger surface area of the device. The response at a higher pH (7.75) value takes time to stabilize as some of the mobile charges at the dielectric surface take time to charge/discharge at the surface. That is why the drift is observable for a longer time as compared to lower pH values. From this analysis, it is clear that working at a low pH value has a gain in stability at the cost of sensitivity. We have used these sensors for several days and they have shown stable sensitivity. We have tested the response of this device in phosphate buffered saline (PBS) as well. It is found that the drift is $-6\,\text{nA/h}$ for pH 7.74, which is less than the buffer used in Figure 6B (Please see Supplementary Figure S12 for more details). This signifies that the drift is buffer solution dependent as the ion concentration is different and it can be minimized by using an appropriate buffer solution. The calculated values of drift (% $\Delta I_{ds}/\text{h} = 6\%$ for TBACl and 0.2% for phosphate buffer solution) at pH 7.8 are below the change in the current observations for detecting the interaction of biological molecules (e.g., protein–protein [15,38,45]), which shows that our proposed device is suitable for such measurements.

4. Conclusions

Here, we have presented the fabrication and characterization of junctionless FETs. We have optimized different process parameters, e.g., annealing temperature and time, to achieve ohmic contacts in these devices. To this end, we have doped the source and drain regions. The fabricated device has shown the expected voltage and current sensitivity for pH measurements as per the literature. Theoretical simulation and modeling have shown that the calculated number of surface sites of an oxide surface is comparable with the experimentally obtained results of the APTES surface functionalization. Further, the peptide and protein surface density were calculated using SPR experiments, which shows that the numbers of peptides and proteins are very close; therefore, we expect a minimum significant sensitivity. Further, the stability of the device was tested using drift analysis that shows stability of the device in the range required for detecting peptide–protein interactions. This rational design of junctionless FET chips will later be used as a multiplexed set-up of immunosensors to detect the interaction between proteins, which is selective for different peptide sequences functionalized on the sensor surface [46]. Later on, these devices will be integrated into a microfluidic setup with a more automated setup for detecting peptide and protein interactions.

Supplementary Materials: The following supporting information can be downloaded at: https://www.mdpi.com/article/10.3390/s22155783/s1, Figure S1: Annealing optimization of contacts. (A) Electrical characterization of the devices on SOI wafer annealed at 4500 °C, (B) Electrical characterization of the metal patterns on Si wafers at different annealing temperatures, (C,D) resistance vs. annealing temperature; Figure S2: Height profile measurement of the Pt-etched samples annealed at different temperatures showing the height profile of the silicide formations; Figure S3: Scanning electron microscopy of the silicide surface. annealed at 200 °C. (A) low and (B) high resolution image, and (C) interface of Pt-Si. For annealing temperature of 350 °C, (D) low and (E) high resolution image and (F) interface of Pt-Si. For annealing temperature of 500 °C, (G) low and (H) high resolution image and (I) interface of Pt-Si; Figure S4: (A) Electrical I–V characteristics of the FETs device, and (B) Electrical I–V characteristics of the FET device with source and drain doped with boron impurity using PECVD process followed by drive in; Figure S5: I–V characteristics of the FETs devices with doping source and drain regions; Figure S6: Drain current vs. surface charge density; Figure S7: Schematic representation of the different steps of the Au/SiO$_2$ surface of the SPR sensor to study peptide-protein interactions; Figure S8: XPS spectra showing the presence of elements on bare SiO$_2$ sensor; Figure S9: XPS spectra showing the presence of elements on APTES grafted SiO$_2$ sensor;

Figure S10: XPS spectra for Silicon and primary amine presence arrangement after APTES grafting on bare SiO_2 sensor; Figure S11: Representative SPR response showing real time binding events of peptide and Ab (1 µg/mL); Figure S12: Ids vs. time for a pH of 7.74 PBS buffer; Table S1: Elemental and chemical composition recorded with XPS on bare SiO_2, and APTES grafted SiO_2 sensor.

Author Contributions: Conceptualization, W.O., V.P.G., S.K. and C.P.G.; methodology, R.P.S., J.G.B., D.W., C.P.G., W.O., S.P., N.K., V.P.G., A.C.S. and S.K.; software, R.P.S. and N.K.; validation, R.P.S., N.K., A.C.S. and S.K.; formal analysis, R.P.S., S.P., W.O., C.P.G., N.K., V.P.G., A.C.S. and S.K.; data curation, R.P.S., N.K. and A.C.S.; writing—original draft preparation, R.P.S.; writing—review and editing, R.P.S., C.P.G., W.O., S.P., N.K., V.P.G., A.C.S. and S.K.; supervision, S.P., W.O., C.P.G., V.P.G. and S.K.; project administration, C.P.G.; funding acquisition, C.P.G.; visualization, R.P.S., S.P., N.K. and A.C.S. All authors have read and agreed to the published version of the manuscript.

Funding: This project has received funding from the European Union's Horizon 2020 research and innovation program under grant agreement No. 862539-Electromed-FET OPEN.

Institutional Review Board Statement: Not applicable.

Informed Consent Statement: Not applicable.

Data Availability Statement: The data presented in this study are available on request from the corresponding author.

Acknowledgments: The authors gratefully acknowledge the financial support of the European Union's Horizon 2020 research and innovation program under grant agreement No. 862539-Electromed-FET OPEN for funding the project. The authors also thank the staff members and process engineers at MESA+ Institute of Nanotechnology for their support during the device fabrication and seamless access to the clean room facilities. The authors also thank IDS group led by Jurriaan Schmitz for their support during the electrical characterization in the EEMC department at the University of Twente. The authors extend their thanks to Ray Hueting (IDS group) for the fruitful discussions and guidance during the fabrication of the device.

Conflicts of Interest: The authors declare no conflict of interest.

References

1. Timp, W.; Timp, G. Beyond mass spectrometry, the next step in proteomics. *Sci. Adv.* **2020**, *6*, eaax8978. [CrossRef]
2. Dallas, D.C.; Guerrero, A.; Parker, E.A.; Robinson, R.C.; Gan, J.N.; German, J.B.; Barile, D.; Lebrilla, C.B. Current peptidomics: Applications, purification, identification, quantification, and functional analysis. *Proteomics* **2015**, *15*, 1026–1038. [CrossRef] [PubMed]
3. Szymczak, L.C.; Kuo, H Y.; Mrksich, M. Peptide Arrays: Development and Application. *Anal. Chem.* **2018**, *90*, 266–282. [CrossRef] [PubMed]
4. Singh, R.P.; Oh, B.K.; Choi, J.W. Application of peptide nucleic acid towards development of nanobiosensor arrays *Bioelectrochemistry* **2010**, *79*, 153–161. [CrossRef] [PubMed]
5. Lu, Y.C.; Robbins, P.F. Cancer immunotherapy targeting neoantigens. *Semin. Immunol.* **2016**, *28*, 22–27. [CrossRef] [PubMed]
6. Harndahl, M.; Rasmussen, M.; Roder, G.; Pedersen, I.D.; Sorensen, M.; Nielsen, M.; Buus, S. Peptide-MHC class I stability is a better predictor than peptide affinity of CTL immunogenicity. *Eur. J. Immunol.* **2012**, *42*, 1405–1416. [CrossRef] [PubMed]
7. Ray, S.; Mehta, G.; Srivastava, S. Label-free detection techniques for protein microarrays: Prospects, merits and challenges. *Proteomics* **2010**, *10*, 731–748. [CrossRef] [PubMed]
8. Kim, A.; Ah, C.S.; Yu, H.Y.; Yang, J.H.; Baek, I.B.; Ahn, C.G.; Park, C.W.; Jun, M.S.; Lee, S. Ultrasensitive, label-free, and real-time immunodetection using silicon field-effect transistors. *Appl. Phys. Lett.* **2007**, *91*, 103901. [CrossRef]
9. Bange, A.; Halsall, H.B.; Heineman, W.R. Microfluidic immunosensor systems. *Biosens. Bioelectron.* **2005**, *20*, 2488–2503. [CrossRef] [PubMed]
10. Avolio, R.; Grozdanov, A.; Avella, M.; Barton, J.; Cocca, M.; De Falco, F.; Dimitrov, A.T.; Errico, M.E.; Fanjul-Bolado, P.; Gentile, G.; et al. Review of pH sensing materials from macro- to nano-scale: Recent developments and examples of seawater applications. *Crit. Rev. Environ. Sci. Technol.* **2022**, *52*, 979–1021. [CrossRef]
11. Pfattner, R.; Foudeh, A.M.; Chen, S.C.; Niu, W.J.; Matthews, J.R.; He, M.Q.; Bao, Z.N. Dual-Gate Organic Field-Effect Transistor for pH Sensors with Tunable Sensitivity. *Adv. Electron. Mater.* **2019**, *5*, 1800381. [CrossRef]
12. Rigante, S.; Scarbolo, P.; Wipf, M.; Stoop, R.L.; Bedner, K.; Buitrago, E.; Bazigos, A.; Bouvet, D.; Calame, M.; Schonenberger, C.; et al. Sensing with Advanced Computing Technology: Fin Field-Effect Transistors with High-k Gate Stack on Bulk Silicon. *Acs Nano* **2015**, *9*, 4872–4881. [CrossRef] [PubMed]
13. Toumazou, C.; Georgiou, P.; Bergveld, P. Piet Bergveld-40 years of ISFET technology: From neuronal sensing to DNA sequencing. *Electron. Lett.* **2011**, *47*, S7–S12. [CrossRef]

14. Stern, E.; Wagner, R.; Sigworth, F.J.; Breaker, R.; Fahmy, T.M.; Reed, M.A. Importance of the Debye screening length on nanowire field effect transistor sensors. *Nano Lett.* **2007**, *7*, 3405–3409. [CrossRef] [PubMed]

15. Espinosa, F.M.; Uhlig, M.R.; Garcia, R. Molecular Recognition by Silicon Nanowire Field-Effect Transistor and Single-Molecule Force Spectroscopy. *Micromachines* **2022**, *13*, 97. [CrossRef] [PubMed]

16. Arjmand, T.; Legallais, M.; Nguyen, T.T.T.; Serre, P.; Vallejo-Perez, M.; Morisot, F.; Salem, B.; Ternon, C. Functional Devices from Bottom-Up Silicon Nanowires: A Review. *Nanomaterials* **2022**, *12*, 1043. [CrossRef] [PubMed]

17. Rollo, S.; Rani, D.; Olthuis, W.; Pascual Garcia, C. The influence of geometry and other fundamental challenges for bio-sensing with field effect transistors. *Biophys. Rev.* **2019**, *11*, 757–763. [CrossRef]

18. Rollo, S.; Rani, D.; Leturcq, R.; Olthuis, W.; Pascual Garcia, C. High Aspect Ratio Fin-Ion Sensitive Field Effect Transistor: Compromises toward Better Electrochemical Biosensing. *Nano Lett.* **2019**, *19*, 2879–2887. [CrossRef]

19. Rajan, N.K.; Duan, X.X.; Reed, M.A. Performance limitations for nanowire/nanoribbon biosensors. *Wires Nanomed. Nanobi.* **2013**, *5*, 629–645. [CrossRef] [PubMed]

20. Narang, R.; Saxena, M.; Gupta, M. Analytical Model of pH sensing Characteristics of Junctionless Silicon on Insulator ISFET. *IEEE Trans. Electron Devices* **2017**, *64*, 1742–1750. [CrossRef]

21. Kutovyi, Y.; Madrid, I.; Zadorozhnyi, I.; Boichuk, N.; Kim, S.H.; Fujii, T.; Jalabert, L.; Offenhaeusser, A.; Vitusevich, S.; Clement, N. Noise suppression beyond the thermal limit with nanotransistor biosensors. *Sci. Rep.* **2020**, *10*, 12678. [CrossRef] [PubMed]

22. Grove, A.S. *Physics and Technology of Semiconductor Devices*; Wiley: New York, NY, USA, 1967; 366p.

23. Poate, J.M.; Tisone, T.C. Kinetics and Mechanism of Platinum Silicide Formation on Silicon. *Appl. Phys. Lett.* **1974**, *24*, 391–393. [CrossRef]

24. Cohen, S.S.; Piacente, P.A.; Gildenblat, G.; Brown, D.M. Platinum Silicide Ohmic Contacts to Shallow Junctions in Silicon. *J. Appl. Phys.* **1982**, *53*, 8856–8862. [CrossRef]

25. Naem, A.A. Platinum Silicide Formation Using Rapid Thermal-Processing. *J. Appl. Phys.* **1988**, *64*, 4161–4167. [CrossRef]

26. Faber, E.J.; Wolters, R.A.M.; Schmitz, J. On the kinetics of platinum silicide formation. *Appl. Phys. Lett.* **2011**, *98*, 082102. [CrossRef]

27. Chizh, K.V.; Dubkov, V.P.; Senkov, V.M.; Pirshin, I.V.; Arapkina, L.V.; Mironov, S.A.; Orekhov, A.S.; Yuryev, V.A. Low-temperature formation of platinum silicides on polycrystalline silicon. *J. Alloys Compd.* **2020**, *843*, 155908. [CrossRef]

28. Idczak, K.; Owczarek, S.; Markowski, L. Platinum silicide formation on selected semiconductors surfaces via thermal annealing and intercalation. *Appl. Surf. Sci.* **2022**, *572*, 151345. [CrossRef]

29. Gueye, R.; Akiyama, T.; Briand, D.; de Rooij, N.F. Fabrication and formation of Ta/Pt-Si ohmic contacts applied to high-temperature Through Silicon Vias (TSVs). *Sens. Actuators A Phys.* **2013**, *191*, 45–50. [CrossRef]

30. Schasfoort, R.B.M.; Bergveld, P.; Kooyman, R.P.H.; Greve, J. Possibilities and Limitations of Direct Detection of Protein Charges by Means of an Immunological Field-Effect Transistor. *Anal. Chim. Acta* **1990**, *238*, 323–329. [CrossRef]

31. Kanai, Y.; Ohmuro-Matsuyama, Y.; Tanioku, M.; Ushiba, S.; Ono, T.; Inoue, K.; Kitaguchi, T.; Kimura, M.; Ueda, H.; Matsumoto, K. Graphene Field Effect Transistor-Based Immunosensor for Ultrasensitive Noncompetitive Detection of Small Antigens. *ACS Sens.* **2020**, *5*, 24–28. [CrossRef] [PubMed]

32. Thong, J.T.L.; Choi, W.K.; Chong, C.W. TMAH etching of silicon and the interaction of etching parameters. *Sens. Actuat. A-Phys.* **1997**, *63*, 243–249. [CrossRef]

33. Guarnieri, V.; Biazi, L.; Marchiori, R.; Lago, A. Platinum metallization for MEMS application. Focus on coating adhesion for biomedical applications. *Biomatter* **2014**, *4*, e28822. [CrossRef] [PubMed]

34. Sinha, A.K. Electrical Characteristics and Thermal-Stability of Platinum Silicide-to-Silicon Ohmic Contacts Metalized with Tungsten. *J. Electrochem. Soc.* **1973**, *120*, 1767–1771. [CrossRef]

35. Medina-Bailon, C.; Kumar, N.; Dhar, R.P.S.; Todorova, I.; Lenoble, D.; Georgiev, V.P.; Garcia, C.P. Comprehensive Analytical Modelling of an Absolute pH Sensor. *Sensors* **2021**, *21*, 5190. [CrossRef] [PubMed]

36. Tarasov, A.; Wipf, M.; Stoop, R.L.; Bedner, K.; Fu, W.; Guzenko, V.A.; Knopfmacher, O.; Calame, M.; Schonenberger, C. Understanding the electrolyte background for biochemical sensing with ion-sensitive field-effect transistors. *ACS Nano* **2012**, *6*, 9291–9298. [CrossRef]

37. vanHal, R.E.G.; Eijkel, J.C.T.; Bergveld, P. A general model to describe the electrostatic potential at electrolyte oxide interfaces. *Adv. Colloid. Interfac.* **1996**, *69*, 31–62. [CrossRef]

38. Rani, D.; Pachauri, V.; Mueller, A.; Vu, X.T.; Nguyen, T.C.; Ingebrandt, S. On the Use of Scalable NanoISFET Arrays of Silicon with Highly Reproducible Sensor Performance for Biosensor Applications. *Acs Omega* **2016**, *1*, 84–92. [CrossRef]

39. Kumar, N.; Dhar, R.P.S.; García, C.P.; Georgiev, V. Discovery of Amino Acid fingerprints transducing their amphoteric signatures by field-effect transistors. *Nanoscience* **2022**. [CrossRef]

40. Dhar, R.; Kumar, N.; Pascual Garcia, C.; Georgiev, V. Assessing the effect of Scaling High-Aspect-Ratio ISFET with Physical Model Interface for Nano-Biosensing Application. *Solid-State Electron.* **2022**, *195*, 108374. [CrossRef]

41. Parizi, K.B.; Xu, X.Q.; Pal, A.; Hu, X.L.; Wong, H.S.P. ISFET pH Sensitivity: Counter-Ions Play a Key Role. *Sci. Rep.* **2017**, *7*, 41305. [CrossRef]

42. Bandiziol, A.; Palestri, P.; Pittino, F.; Esseni, D.; Selmi, L. A TCAD-Based Methodology to Model the Site-Binding Charge at ISFET/Electrolyte Interfaces. *IEEE Trans. Electron Devices* **2015**, *62*, 3379–3386. [CrossRef]

43. Rani, D.; Rollo, S.; Olthuis, W.; Krishnamoorthy, S.; Garcia, C.P. Combining Chemical Functionalization and FinFET Geometry for Field Effect Sensors as Accessible Technology to Optimize pH Sensing. *Chemosensors* **2021**, *9*, 20. [CrossRef]

44. Ghorbanpour, M.; Falamaki, C. A novel method for the fabrication of ATPES silanized SPR sensor chips: Exclusion of Cr or Ti intermediate layers and optimization of optical/adherence properties. *Appl. Surf. Sci.* **2014**, *301*, 544–550. [CrossRef]
45. Wipf, M.; Stoop, R.L.; Navarra, G.; Rabbani, S.; Ernst, B.; Bedner, K.; Schonenberger, C.; Calame, M. Label-Free FimH Protein Interaction Analysis Using Silicon Nanoribbon BioFETs. *ACS Sens.* **2016**, *1*, 781–788. [CrossRef]
46. Ozkumur, E.; Needham, J.W.; Bergstein, D.A.; Gonzalez, R.; Cabodi, M.; Gershoni, J.M.; Goldberg, B.B.; Unlu, M.S. Label-free and dynamic detection of biomolecular interactions for high-throughput microarray applications. *Prac. Natl. Acad. Sci. USA* **2008**, *105*, 7988–7992. [CrossRef] [PubMed]

Article

Rational Design of Field-Effect Sensors Using Partial Differential Equations, Bayesian Inversion, and Artificial Neural Networks

Amirreza Khodadadian [1,*], Maryam Parvizi [1,2], Mohammad Teshnehlab [3], Clemens Heitzinger [4,5]

1   Institute of Applied Mathematics, Leibniz University Hannover, Welfengarten 1, 30167 Hannover, Germany; parvizi@ifam.uni-hannover.de
2   Cluster of Excellence PhoenixD (Photonics, Optics, and Engineering-Innovation Across Disciplines), Leibniz University Hannover, 30167 Hannover, Germany
3   Faculty of Electrical Engineering, K. N. Toosi University of Technology, Tehran 19697, Iran; teshnehlab@eetd.kntu.ac.ir
4   Institute of Analysis and Scientific Computing, TU Wien, Wiedner Hauptstrasse 8–10, 1040 Vienna, Austria; clemens.heitzinger@tuwien.ac.at
5   Center for Artificial Intelligence and Machine Learning (CAIML), TU Wien, 1040 Vienna, Austria
*   Correspondence: khodadadian@ifam.uni-hannover.de

Abstract: Silicon nanowire field-effect transistors are promising devices used to detect minute amounts of different biological species. We introduce the theoretical and computational aspects of forward and backward modeling of biosensitive sensors. Firstly, we introduce a forward system of partial differential equations to model the electrical behavior, and secondly, a backward Bayesian Markov-chain Monte-Carlo method is used to identify the unknown parameters such as the concentration of target molecules. Furthermore, we introduce a machine learning algorithm according to multilayer feed-forward neural networks. The trained model makes it possible to predict the sensor behavior based on the given parameters.

Keywords: field-effect sensors; biosensors; charge transport; neural networks; Bayesian inversion; inverse modeling

Citation: Khodadadian, A.; Parvizi, M.; Teshehlab, M.; Heitzinger, C. Rational Design of Field-Effect Sensors Using Partial Differential Equations, Bayesian Inversion, and Artificial Neural Networks. *Sensors* **2022**, *22*, 4785. https://doi.org/10.3390/s22134785

Academic Editors: Michael J. Schöning and Sven Ingebrandt

Received: 31 May 2022
Accepted: 21 June 2022
Published: 24 June 2022

Publisher's Note: MDPI stays neutral with regard to jurisdictional claims in published maps and institutional affiliations.

1. Introduction

Silicon nanowire (SiNW) field-effect transistors (FETs) are typically used to detect proteins [1], cancer cells [2], DNA and miRNA strands [3,4], enzymes [5], and toxic gases such as carbon monoxide [6,7]. The sensors have several advantages including fast response, very high sensitivity, and low power consumption; they do not need labeling and can be used to detect subpicomolar concentrations of biological species [8–13]. The functioning of the sensors is based on the field effect due to the (partial) charges of the target molecules. When they are selectively bound to probe molecules and close enough to the semiconducting transducer, they affect the charge concentration inside the nanowire, which changes the current through the nanowire.

Using mathematical models based on partial differential equations (PDEs) enables us to model physically relevant quantities such as electrostatic potential, electron and hole current density, device sensitivity to the target molecule and signal-to-noise ratio [14–18]. The three-dimensional simulations give rise to more reliable models compared to two-dimensional cross-sections, since all target molecules bound to bio-receptors will be included [19,20]. We couple a charge transport model (the drift-diffusion equations) and the nonlinear Poisson–Boltzmann equation (PBE) for fully self-consistent simulations. The system of equations is a comprehensive model to compute the electrical current and study the nonlinear effects of different semiconductor parameters (e.g., doping concentration) and device parameters such as nanowire type (radial, trapezoidal, radial, or rectangular), its dimensions, contact voltages, and insulator thickness on device performance (output and sensitivity).

Having an accurate model enables the rational design of field-effect sensors. However, in the model equations, there are several material parameters that cannot be (easily) measured. The surface charge density of the insulator has an essential effect on the device and also affects the probe and target molecules. The doping concentration has a crucial effect on the device and the model. Due to the nonlinear effect of these parameters, an efficient parameter estimation framework will enhance the accuracy and reliability of the model.

Markov-chain Monte-Carlo (MCMC) techniques are among the most efficient probabilistic methods to extract information by comparison between measurements and simulations by updating available prior knowledge and estimating the posterior densities of unknown quantities of interest. Here, we use a forward model, and a backward, inverse setting is used to determined the unknown parameters using the experiments. The classical algorithm was introduced in 1970 and is called the *Metropolis–Hastings (MH) algorithm* [21]. There are several improvements in the algorithm, e.g., adaptive-proposal Metropolis [22], delayed-rejection Metropolis [23], and delayed rejection adaptive Metropolis (DRAM) [24], as well as using ensemble Kalman filters [25]. In all techniques, different candidates are proposed based on a proposal distribution, and the algorithm decides whether they are rejected or accepted. A review of the MCMC methods is given in [26]. For SiNW-FETs, the DRAM algorithm has been used to identify the doping concentration and the amount of target molecules [14]. Considering the selective functionalization of SiNW, the authors of [1] used the MH algorithm to estimate the probe-target density at the surface.

Neural networks (also known as artificial neural networks (ANNs)) as the subset of machine learning are frameworks to analyze the available data and discover patterns that can not be observed independently. The ANNs have been inspired by the human brain and are suitable for complicated and nonlinear cases. Here, we split the prior data into two categories, namely training and testing data. The training set (between 60% and 80%) is used to extract useful information from the data, and the test set (between 20% and 40%) is employed to monitor the algorithm performance. In SiNW-FETs, there are a large amount of simulation and experimental data concerning different input (physical, chemical, and device) parameters that should be analyzed to ensure their accuracy and reliability. Of course, this process is time consuming and reduces the efficiency. Furthermore, the sensors are developed to detect specific biological species with the highest sensitivity. In the design process, using neural networks enables us to optimize the design parameters to enhance the sensor performance [27–32].

This article is structured as follows. In Section 2, we present the model equations and explain how the electrical current is computed. In Section 3, we discuss the parameter estimation methods and explain how MCMC can be used to determine the unknown parameters. In Section 4, we introduce the developed neural networks algorithm for SiNW-FETs. In Section 5, we first verify the model response with the experimental data; then, Bayesian inversion is used to identify the material parameters. Afterward, the developed machine-learning algorithm is employed in training and testing. Finally, the conclusions are summarized in Section 6.

2. The Model Equations

The drift–diffusion–Poisson system is used to describe the electrochemical interactions (Poisson–Boltzmann equation) and the charge transport (drift–diffusion equations) in field-effect sensors. The convex and bounded domain $\Omega \subset \mathbb{R}^3$ consists of four subdomains, namely the insulator (SiO_2, Ω_{Si}), the silicon substrate and transducer (Ω_{Si}), the aqueous solution (Ω_{liq}), and the charged molecules (Ω_{mol}). To model the potential interactions, we use the Poisson–Boltzmann equation

$$-\nabla \cdot (A(x)\nabla V(x)) = \begin{cases} q(C_{\text{dop}}(x) + p(x) - n(x)) & \text{in } \Omega_{\text{Si}}, \\ 0 & \text{in } \Omega_{\text{ox}}, \\ \rho(x) & \text{in } \Omega_{\text{M}}, \\ -2\varphi(x)\sinh(\beta(V(x) - \Phi_F)) & \text{in } \Omega_{\text{liq}}, \end{cases} \tag{1}$$

where A indicates the dielectric constant, which is a function of the material, V is the electrostatic potential, C_{dop} is the doping concentration, ρ is the surface charge of the molecules, ϕ_F denotes the Fermi level, and φ is the ionic concentration. Regarding the electrical constants, we use the relative values $A_{\text{Si}} = 11.7$, $A_{\text{ox}} = 3.9$, $A_{\text{M}} = 3.7$, and $A_{\text{liq}} = 78.4$. Considering the Boltzmann constant k_B, the temperature T and the elementary charge q, we define $\beta = q/(k_B T)$. In the simulations, a thermal voltage of 0.021 V will be used.

A two-dimensional cross-section of the device is given in Figure 1.

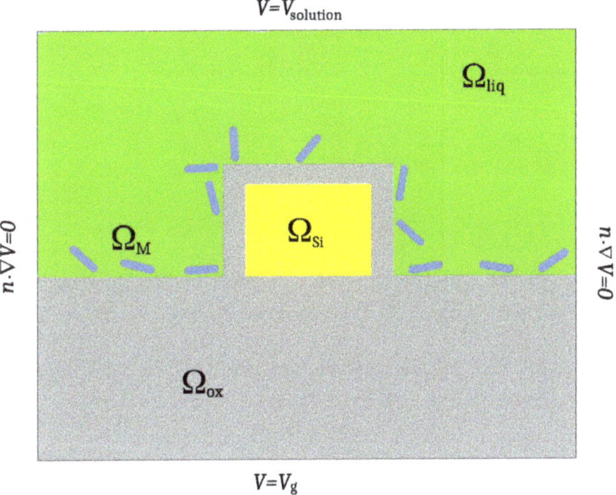

Figure 1. A schematic cross-section of a SiNW-FET depicting the subdomains, i.e., the transducer Ω_{Si}, SiO$_2$ insulator (Ω_{ox}), the aqueous solution Ω_{liq}, the binding of the target molecules to the immobilized receptor molecules (Ω_{mol}), and the boundary conditions.

At the interface between the insulator and the liquid (i.e., $\Gamma := \Omega_{\text{ox}} \cap \Omega_{\text{liq}}$), we impose the interface conditions

$$A(0+)(V(0+, y, z) - V(0-, y, z)) = \alpha(y, z) \qquad \text{on } \Gamma, \tag{2a}$$
$$A(0+)\partial_x V(0+, y, z) - A(0-)\partial_x V(0-, y, z) = \gamma(y, z) \qquad \text{on } \Gamma \tag{2b}$$

for V_I. Here, $0+$ and $0-$ denote the limit at the interface on the side of liquid and insulator. Furthermore, α is macroscopic dipole moment density, and γ is the macroscopic surface-charge density.

In Ω_{Si}, we solve the drift–diffusion system

$$-\nabla \cdot (A\nabla V) = q(p(x) - n(x) + C_{\text{dop}}(x)), \tag{3a}$$
$$\nabla \cdot J_n = qR(n, p), \tag{3b}$$
$$\nabla \cdot J_p = -qR(n, p), \tag{3c}$$
$$J_n = q(D_n \nabla n - \mu_n n \nabla V), \tag{3d}$$
$$J_p = q(-D_p \nabla p - \mu_p p \nabla V) \tag{3e}$$

to model the charges in the transistor, where D_n and D_p are the electron and hole diffusion coefficients. The concentrations of electrons and holes are given by

$$p =: n_i \exp\left(\frac{q}{K_B T}(\Phi_1 - V)\right), \qquad n =: n_i \exp\left(\frac{-q}{K_B T}(\Phi_2 - V)\right), \tag{4}$$

where n_i is the intrinsic carrier density and Φ_1 and Φ_1 are the Fermi levels. In order to compute the electron and hole current densities, we use the Shockley–Read–Hall recombination rate, i.e.,

$$R(n, p) := \frac{np - n_i^2}{\tau_n(p + n_i) + \tau_p(n + n_i)},$$

where τ_n and τ_p denote the lifetimes of the electrons and holes.

For solving the nonlinear system of equations, we use the Scharfetter–Gummel iteration. For this, we write the concentrations n and p in terms of the two Slotboom variables u and v as

$$n(x, \omega) =: n_i e^{V(x,\omega)/U_T} u(x, \omega), \tag{5a}$$

$$p(x, \omega) =: n_i e^{-V(x,\omega)/U_T} v(x, \omega). \tag{5b}$$

Therefore, the model problem (3) can be rewritten as

$$-\nabla \cdot (A(x)\nabla V(x)) = q\left(C_{\text{dop}}(x) - n_i\left(e^{V(x)/U_T}u(x) - e^{-V(x)/U_T}v(x)\right)\right), \tag{6a}$$

$$U_T n_i \nabla \cdot (\mu_n e^{V/U_T}\nabla u(x)) = R(x), \tag{6b}$$

$$U_T n_i \nabla \cdot (\mu_p e^{-V/U_T}\nabla v(x)) = R(x), \tag{6c}$$

where U_T is the thermal voltage and the Shockley–Read–Hall recombination rate takes the form

$$R_{\text{SRH}}(x) = n_i \frac{u(x)v(x) - 1}{\tau_p(e^{V/U_T}u(x) + 1) + \tau_n(e^{-V/U_T}v(x) + 1)}.$$

At the ohmic contacts (backgate, source, and drain) and the solution gate, we have a Dirichlet boundary condition $V_{\partial\Omega} = V_D$ consisting of

$$V|_{\partial\Omega_G} = V_g \qquad V|_{\partial\Omega_S} = V_S \qquad V|_{\partial\Omega_D} = V_D \qquad V|_{\partial\Omega_{\text{sol}}} = V_{\text{solution}}. \tag{7}$$

At the source and drain contacts (on $\partial\Omega_{\text{Si}}$), we apply

$$u(x) = u_D(x), \quad v(x) = v_D(x). \tag{8}$$

For the remaining part of the domain, we impose a zero Neumann boundary condition to guarantee the self-isolation. We refer the interested reader to [15,19,33,34] for theoretical discussions about the model including the Slotboom variables. The existence and uniqueness of the solutions for deterministic and stochastic model problems are given in [15,35]. Finally, the computation of J_n and J_p enables us to calculate the electrical current as

$$\mathcal{I} := \int (J_n + J_p)\,\mathrm{d}x, \tag{9}$$

where we take the integral on a cross-section of the transducing part.

In this work, we use the finite element method (FEM) to solve the coupled system of equations. We define the spaces

$$X^1 = \left\{ V \in H^1(\Omega) \quad | \quad V|_{\partial\Omega} = V_D \, , \quad V|_\Gamma = V_I \right\}, \tag{10a}$$

$$X^2 = \left\{ u \in H^1(\Omega_{\mathrm{Si}}) \quad | \quad u|_{\partial\Omega_{\mathrm{Si}}} = u_D \right\}, \tag{10b}$$

$$X^3 = \left\{ v \in H^1(\Omega_{\mathrm{Si}}) \quad | \quad v|_{\partial\Omega_{\mathrm{Si}}} = v_D \right\}. \tag{10c}$$

Therefore, we define the continuous solution space $X := X^1 \times X^2 \times X^3$ for the DDP system. Regarding the space discretization, we assume $\mathcal{T}_h = \{T_1, T_2, \ldots, T_n\}$ denotes a quasi-uniform mesh defined on $\Omega_h \approx \Omega$ with mesh width $h := \max_{T_j \in \mathcal{T}_h} \mathrm{diam}(T_j)$. We define

$$\mathcal{S}_V^1(\mathcal{T}_h) := \{ V \in H^1(\Omega) \quad | \quad V|_T \in \mathcal{P}_1(T) \ \forall T \in \mathcal{T}_h \},$$

$$\mathcal{S}_u^1(\mathcal{T}_h) := \{ u \in H^1(\Omega) \quad | \quad u|_T \in \mathcal{P}_1(T) \ \forall T \in \mathcal{T}_h \},$$

$$\mathcal{S}_v^1(\mathcal{T}_h) := \{ v \in H^1(\Omega) \quad | \quad v|_T \in \mathcal{P}_1(T) \ \forall T \in \mathcal{T}_h \},$$

where $\mathcal{P}_1$ is the space of first-order polynomials. Then, we have

$$X_h^1 := \left\{ V_h \in \mathcal{S}_V^1(\mathcal{T}_h) \quad | \quad V_h|_{\partial\Omega} = V_D \, , \quad V_h|_\Gamma = V_I \right\}, \tag{11a}$$

$$X_h^2 := \left\{ u_h \in \mathcal{S}_u^1(\mathcal{T}_h) \quad | \quad u_h|_{\partial\Omega_{\mathrm{Si}}} = u_D \right\}, \tag{11b}$$

$$X_h^3 := \left\{ v_h \in \mathcal{S}_v^1(\mathcal{T}_h) \quad | \quad v_h|_{\partial\Omega_{\mathrm{Si}}} = v_D \right\}. \tag{11c}$$

The discrete solution is defined as $X_h := X_h^1 \times X_h^2 \times X_h^3$, which is a subset of X. The weak form of the model equations can be found in [15,33]. The a prior and a posterior estimations are proved in [33]. More theoretical works regarding the finite elements analysis are given in [36–38].

3. Parameter Estimation Based on Bayesian Inference

In different experimental situations, an accurate estimation of the effective parameters and constants cannot be easily estimated. Bayesian inversion techniques based on Markov chain Monte Carlo methods are efficient and straightforward probabilistic techniques to estimate these unknowns. We initiate the algorithm using the available information, named *prior* knowledge (which may not be sufficiently accurate), and during several iterations, we can update the information and provide more reliable data (i.e., the posterior density). Then, we can extract valuable information from the posterior density, and its mean/median can be used as the solution of the interference. A very strong agreement with the experimental values and the model response can be achieved. We start a statistical model

$$\mathcal{M} = \mathcal{P}(x, \chi) + \varepsilon, \tag{12}$$

where $\mathcal{M}$ is the experimental observation (normally $n-$ dimensional), while $\mathcal{P}$ is the solution of the model problem which depends on the set of parameters χ (i.e., $\chi = (\chi_1, \chi_2, \ldots, \chi_k)$) and the Cartesian coordinates x. Here, ε is the measurement error, and we assume that it is normally distributed, i.e., $\varepsilon \sim \mathcal{N}(0, \sigma^2 I)$, including the parameter σ^2. Having an experimental observation, for instance electrical current (i.e., $\mathcal{M} = \mathrm{obs}$), we define the probability function

$$\pi(\mathrm{obs}) = \int_{\mathbb{R}^n} \pi(\mathrm{obs}|\chi)\pi_0(\chi)\,\mathrm{d}\chi. \tag{13}$$

Our aim is to estimate the posterior density $\pi(\chi|m)$, considering the measured observation m and the available prior information. For this, we compute the likelihood function

$$\pi(\mathcal{M}|\chi) = L(\chi, \sigma^2|\mathcal{M}) = \frac{1}{(2\pi\sigma^2)^{n/2}} \exp\left(-\mathcal{M}_\mathcal{P}/2\sigma^2\right) \tag{14}$$

where

$$\mathcal{M}_\mathcal{P} = \sum_{j=1}^{n} [\mathcal{M}_j - \mathcal{P}_j(x,\chi)]^2 \tag{15}$$

is the sum of square errors. Obviously, if the model response with respect to the (set of) parameters χ will be closer to the measured value, the square error (15) will converge to zero, and its relative probability (computed by the likelihood function) will converge to 1. Inaccurate estimation of χ will increase the error term, and the probability will converge to zero.

In the Metropolis algorithm, we initiate the process using an initial guess χ^0 based on the prior density. According to the proposal distribution, a new candidate χ^* is proposed. We compute the acceptance rate by

$$\lambda(\chi^{j-1}, \chi^\star) = \min\left(1, \frac{\pi(\chi^\star)}{\pi(\chi^{j-1})}\right). \tag{16}$$

If the new candidate χ^* is accepted, we continue the MCMC chain with that; otherwise, (χ^{j-1} has a higher probability concerning $\chi^\star$), we follow the chain with the previous candidate. Using a non-symmetric proposal density is a generalization of the Metropolis algorithm, introduced by Hastings [21], where the probability of the forward jump is not equal to the backward one. A summary of the algorithm is given in Algorithm 1.

Algorithm 1 The Metropolis–Hastings algorithm.

Initialization: Start the process with the initial guess χ^0 and number of samples N.
while $j < N$

   1. Propose a new sample according to the proposal density $\chi^* \sim \mathcal{T}(\chi^*|\chi^{j-1})$.

   2. Compute the acceptance/rejection ratio

$$\zeta(\chi^*|\theta^{j-1}) = \min\left(1, \frac{\pi(\chi^*|m)}{\pi(\chi^{j-1}|m)} \frac{\mathcal{T}(\chi^{j-1}|\chi^*)}{\mathcal{T}(\chi^*|\chi^{j-1}))}\right).$$

   3. Sample $\mathcal{R} \sim \text{Uniform}\,(0,1)$.

   4. **if** $\mathcal{R} < \zeta$ **then**

      accept χ^* and set $\chi^j := \chi^*$

    else

      reject χ^* and set $\chi^j := \chi^{j-1}$

    end if

   5. Set $j = j + 1$.

The Metropolis–Hastings algorithm is a simple and versatile technique and has been widely used for several problems in applied science. However, for the high-dimensional cases (different parameters should be inferred simultaneously), the algorithm does not work appropriately, since the rejection rate increases significantly. To improve its computational drawbacks, different improvements, such as the adaptive Metropolis algorithm [22], delayed rejection Metropolis [23], and their combination, namely delayed rejection adaptive

Metropolis (DRAM) [24]. We refer the interested readers to [26] as a review paper about the methods.

Mcmc with Ensemble-Kalman Filter (EnKF-MCMC)

In EnKF-MCM [25], we use a Kalman gain employing the mean and the covariance of the prior distribution and the cross-covariance between parameters and observations. It will be used to compute the proposal distribution and make the convergence to the target density faster. Here, the new candidate is computed as the jump of the Kalman-inspired proposal $\Delta\chi$ as

$$\chi^\star = \theta^{j-1} + \Delta\chi. \tag{17}$$

In order to update the candidates, we compute $\Delta\chi$ by

$$\Delta\chi = \mathcal{K}\left(y^{j-1} + s^{j-1}\right), \tag{18}$$

where $\mathcal{K}$ denotes the so-called Kalman gain,

$$\mathcal{K} = \mathcal{C}_{\chi M}(\mathcal{C}_{MM} + \mathcal{R})^{-1}. \tag{19}$$

Here, $\mathcal{C}_{\theta M}$ indicates the covariance matrix between the identified unknowns and model response, $\mathcal{C}_{MM}$ points out the covariance matrix of the model response, and $\mathcal{R}$ denotes the measurement noise covariance matrix [39]. In addition, y^{j-1} is the residual of the proposed values concerning the model and $s^{j-1} \sim \mathcal{N}(0, \mathcal{R})$ relates to the density of measurement. A summary of the relative algorithm is given in Algorithm 2. Finally, Figure 2 shows the implementation of EnKF-MCMC and Schafetter–Gummel iteration for parameter estimation and solving the model equations.

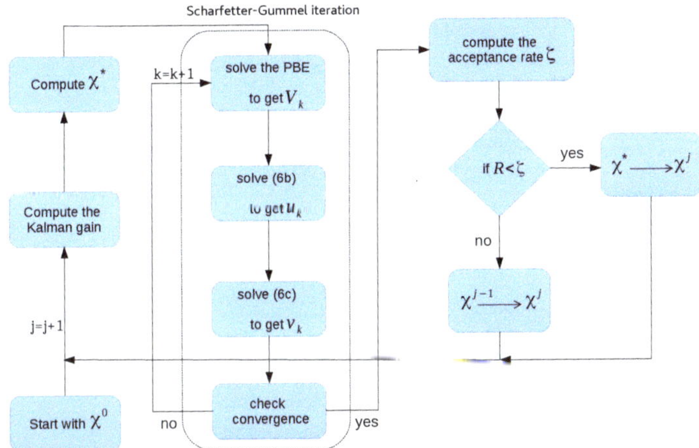

Figure 2. Bayesian inversion using EnKF-MCMC to identify the unknown material parameters, where the Scharfetter–Gummel iteration is used to solved the coupled system of equations.

Algorithm 2 Bayesian inference using EnKF-MCMC

Initialization ($j = 0$): Start the process with the initial guess χ^0 and number of samples N.
while $j < N$

    1. Estimate the model response with respect to χ^{j-1}

    2. Compute the Kalman gain $\mathcal{K} = \mathcal{C}_{\chi M}(\mathcal{C}_{MM} + \mathcal{R})^{-1}$

    3. Produce the new proposal using the shift $\chi^\star = \chi^{j-1} + \mathcal{K}(y^{j-1} + s^{j-1})$

    4. Accepted/rejected $\chi^\star$

    5. Set $j = j + 1$.

---

4. Multilayer Feed-Forward Neural Networks

Neural networks are efficient, flexible, and robust simulation tools specifically for nonlinear and complicated problems. They consist of three effective components, including neurons, structures, and weights, which all affect the response and behavior of the network. Artificial neural networks (ANNs) are supervised machine learning algorithms consisting of neurons and hidden layers. The input data are processed into the hidden layers, the output is compared with the target trajectory, and the relative error is computed. The neural networks strive to minimize this error.

Typically, there are two common classes of neural networks, namely feed-forward neural networks (single or multilayers) and recurrent dynamics neural networks. Single-layer neural networks [40] have less complexity; however, they are more suitable for linear problems. In multilayer feed-forward neural networks (MFNNs) [41,42], more than one layer of the artificial neurons will be used to enhance the capability to learn nonlinear patterns, which is more appropriate for BIO-FETs. In MFNNs, the neurons are organized in different non-recurrent layers, where in the first layer, we have the input vector (here are the parameters of the sensor), and the output is given to the first hidden layer. After the data processing, the data are transferred to the next layers using the weights; the procedure is followed until the latest MFNNs layer. These networks are also named multilayer perceptrons, and their structure is shown in Figure 3.

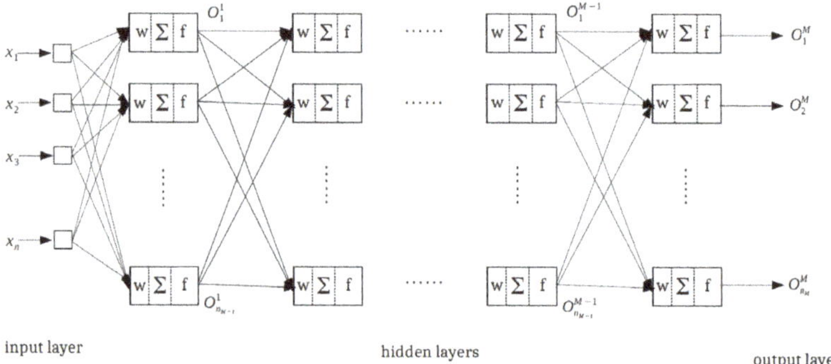

Figure 3. The structure of multilayer feed-forward neural networks (MFNNs).

Let us assume d denotes the desired trajectory (i.e., the device output); for M-layer neural networks, we have

$$\nabla w_j^s(k) = \eta^s \frac{\partial E}{\partial w_j^s}(k) = -\eta^s \delta_i^s(k) \nabla w_j^s(k) \atop (n_{s-1}) \times 1 \quad\quad (n_{s-1}) \times 1$$

$$= -\eta^s e_j^s(k) f_j^{s'}\left(net_j^s(k)\right) \underset{(n_{s-1}) \times 1}{x^{s-1}} \qquad s = 1, 2, \ldots, M, \quad j = 1, 2, \ldots, n_s, \tag{20}$$

$$\delta_j^s(k) := -\frac{\partial E}{\partial net_j^s}(k) = e_j^s(k) f_j^{s'}\left(net_j^s(k)\right), \tag{21}$$

$$e_j^s(k) = \sum_{l=1}^{n_{s+1}} \delta_j^{s+1}(k) w_{lj}^{s+1}(k), \tag{22}$$

where w is the weights, η is the training rate, E is the network mean square error (MSE), δ is the sensitivity function (here, δ^s indicates the network error in the jth layer), net^s is the weighted input, n_s is the number of neurons in the sth layer, x^0 is the network input, x^{s-1} is the output of the $s-1$th layer, and it is also the input of the sth layer. We also have the following initial conditions for the recurrent process

$$\delta_j^M(k) = e_j^M(k) f_j^{M'}\left(net_j^M(k)\right), \tag{23}$$

$$e_j^M(k)) \triangleq d_j(k) - O_j^M(k). \tag{24}$$

Figure 4 shows the jth neuron in the ith layer in the learning algorithm. In the recurrent process, in order to adjust the weights from the first layer, we follow as

$$\delta_{l_i}^s(k) = -\frac{\partial E(k)}{\partial net_{l_i}^s} \sum_{l_M=1}^{n_M} \sum_{l_{M-1}=1}^{n_{M-1}} \cdots \sum_{l_{i+2}=1}^{n_{i+2}} \sum_{l_{i+1}=1}^{n_{i+1}} \frac{\partial E}{\partial net_{l_m}^s} \frac{\partial net_{l_m}^s}{\partial net_{l_m-1}^{s-1}} \cdots \frac{\partial net_{l_{i+2}}^{s+2}}{\partial net_{l_{i+1}}^{s+1}} \frac{\partial net_{l_{i+1}}^{s+1}}{\partial net_{l_i}^s}(k) \tag{25}$$

For $i = 1, 2, \ldots, M-1$ and $s = 1, 2, \ldots, M$, the relation $net_{l_i}^s$ and $net_{l_{i+1}}^{s+1}$ takes

$$net_{l_{i+1}}^{s+1}(k) = \sum_{p=1}^{n_i} w_{l_{i+1}p}^{s+1}(k) f_p^s\left(net_p^s(k)\right), \tag{26}$$

therefore

$$\frac{\partial net_l^{s+1}}{\partial net_{l_i}^s}(k) = w_{l_{i+1}l_i}^{s+1}(k) f_{l_i}^{s'}\left(net_{l_i}^{s'}(k)\right). \tag{27}$$

So, we can write $\delta_{l_i}^s$ as

$$\delta_{l_i}^s(k) = \left(\sum_{l=1}^{n_{i+1}} \delta_l^{s+1}(k) w_{l\,l_i}^{s+1}(k)\right) f_{l_i}^{s'}\left(net_{l_i}^s(k)\right) = e_{l_i}^s(k) f_{l_i}^{s'}\left(net_{l_i}^s(k)\right), \tag{28}$$

where

$$e_{l_i}^s(k) = \sum_{l=1}^{n_{i+1}} \delta_l^{s+1}(k) w_{l\,l_i}^{s+1}(k). \tag{29}$$

The gradient of E (the difference between desired trajectory and the neural networks's output) with respect to the weight vector is given by

$$\frac{\partial E}{\partial w_{l_i}^s}(k) = \sum_{l=1}^{n} \frac{\partial E}{\partial net_{l_i}^s}(k) \frac{\partial net_{l_i}^s}{\partial w_{l_i}^s}(k), \tag{30}$$

where the second term depends only on the neurons features and takes

$$\frac{\partial net_l^s}{\partial w_{l_i}^s}(k) = \begin{cases} x^{s-1}(k) & l = l_i, \\ 0 & \text{otherwise,} \end{cases} \tag{31}$$

$$\frac{\partial E}{\partial w_{l_i}^s}(k) = -\delta_{l_i}^s(k)x^{s-1}(k). \tag{32}$$

Using the back-propagation error algorithm enables us to adjust the weight functions in order to minimize the network error. This training process is also named the supervised learning algorithm.

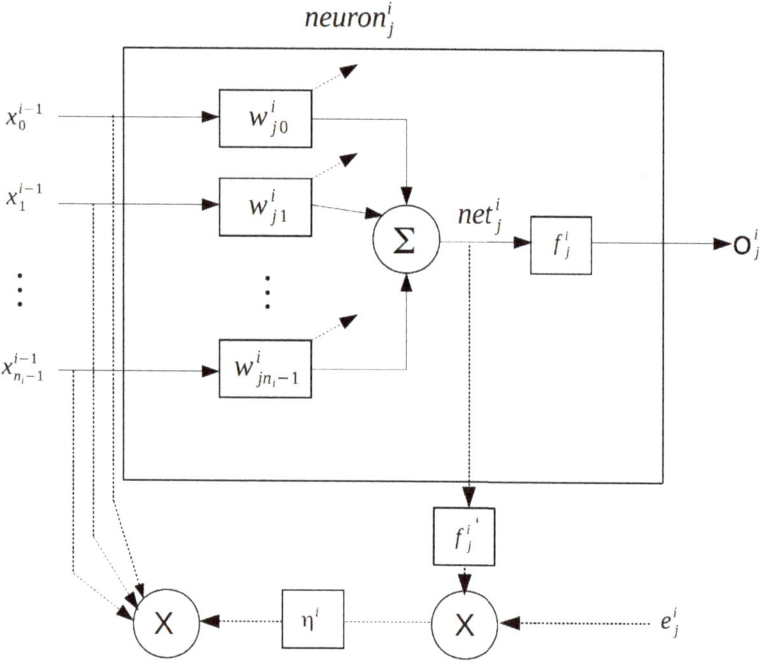

Figure 4. The back-propagation algorithm for the adjustment of neuron weights.

5. Numerical Experiments

As we already mentioned, the DDP system is a roust and reliable system of equations to model the electrical behavior of the FET devices. We use a prostate-specific antigen (PSA) sensitive sensor which is used to diagnose prostate cancer. For the simulations, we use a sensor device with the nanowire length of 1000 nm, width of 100 nm and height of 50 nm, which is coated with SiO_2 with 8 nm thickness. We use the P_1 finite element to solve the model problem, and tetrahedral meshes are employed to discretize the domain. A schematic of the bio-FET including dimensions using 6622 nodes and 45,735 tetrahedra is shown in Figure 5. The sensor is developed for the detection of 2ZCH (https://www.rcsb.org/structure/2ZCH). The PROPKA algorithm predicts the pK_a values of ionizable groups in proteins and protein–ligand complexes based on the 3D structure. The values are the basis for understanding the pH-dependent characteristics of proteins and catalytic mechanisms of many enzymes [43]. To compute the net charge, we performed a PROPKA algorithm [44–46] to detect the net charge for different pH values. The simulations are completed using a pH value of 9, giving rise to the net charge of $-15\,$q [14]. In field-effect sensors, surface reactions at the oxide surface depending on the pH value and the binding of charged target molecules result in changes in the charge concentration at and near the surface, and subsequently in changes in the electrostatic potential, which then modulates the current through the transducer. Since the molecules are negatively charged, the binding of the target molecules to the bio-receptors will enhance the charge conductance and increase the response of the sensor (i.e., the electrical current).

The system of equations is capable of modeling the surface charges at the surface. In a previous work, we developed a Monte-Carlo approach to simulate the charges around a charged biomolecule at a charged surface [47]. Furthermore, in [48], a nonlinear Poisson model was used to calculate the free energies of various molecule orientations in dependence of the surface charge. Based on the free energies, the probabilities of the orientations were calculated, and hence, the biological noise was simulated.

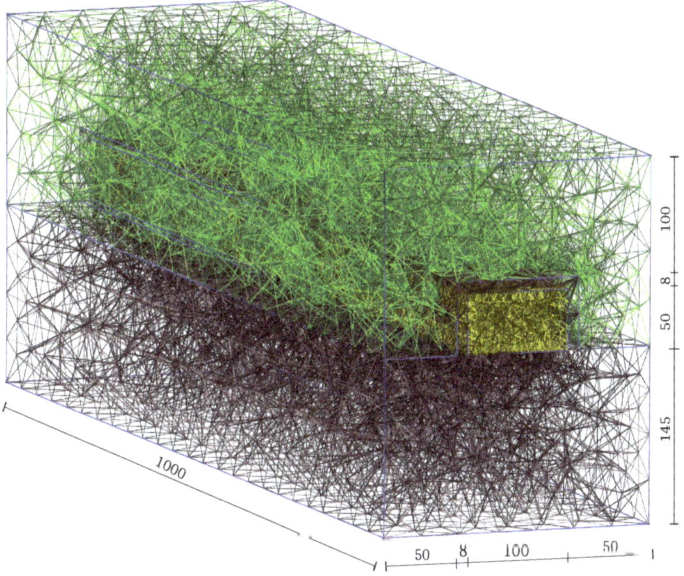

Figure 5. A 3D schematic of the sensor device including the dimensions and tetrahedral meshes for the discretization. All values are in nanometers.

5.1. Model Verification

As the first step, we verify the model accuracy with the experiments. We compute the electrical current I with respect to different gate voltages V_G where the source-to-drain voltage $V_{SD} = 0.2$ V, doping concentration $C_{dop} = 1 \times 10^{16}\,\text{cm}^{-3}$, and the thermal voltage $U_T = 0.021$ V. The experimental data are taken form [20]. In order to solve the nonlinear coupled system of equations, a Scharfetter–Gummel-type iteration is used. Figure 6 shows the current as a function gate voltage varying between $V_G = -1$ V and $V_G = -3.5$ V for experimental and simulation values. These results indicate that the DDP system is reliable and will be used for the next simulations.

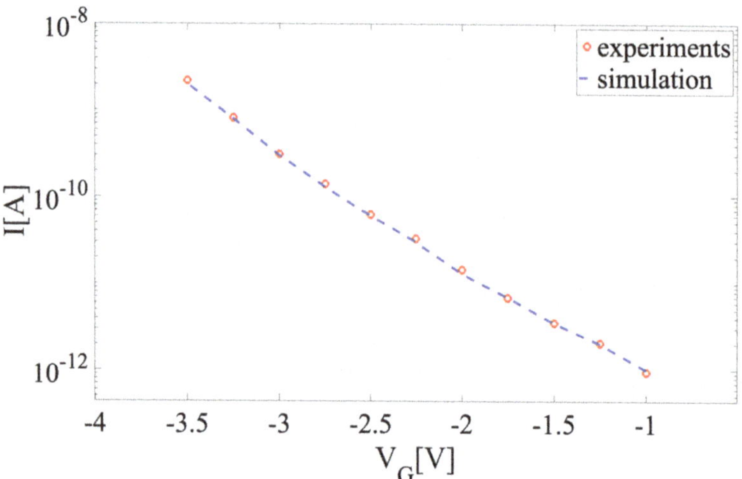

Figure 6. A comparison between the experimental [20] and simulation current.

5.2. Bayesian Inversion

The molecules are negatively charged (here, $-15\,q$ is used); however, an accurate estimation of the molecule charge density will be necessary. In semiconductor devices, in order to enhance the conductivity, impurity atoms are added to the silicon lattice, namely the doping process. Higher doping concentration will improve the transistor conductivity; however, the device will be less sensitive to the charged molecules. Physically, doping concentration (as a macroscopic quantity) denotes the average amount of the dopants. We implemented a delayed rejection adaptive Metropolis (DRAM) [14] and the Metropolis–Hastings algorithm [1] to infer doping concentration, molecule charge density, and probe–target density. The efficiency of the EnKF-MCMC compared to these algorithms is studied in [26]. Therefore, we employ the Kalman filter for the proposal adaptation. We performed the MCMC algorithm with $N = 10{,}000$ iterations, and a uniform prior density is used. The computational aspects are summarized in Table 1.

The back-propagation error is an efficient algorithm for the training of neural networks where we compute the gradient of the loss function with respect to the weights of the network.

Table 1. The computational features and the results of the Bayesian inversion.

Parameter	Min	Max	EnKF (Median)	True Values	Acceptance Rate
C_{dop} (cm^3)	1×10^{15}	5×10^{16}	9.4×10^{15}	1×10^{16}	91%
ρ (q/nm^2)	-5	1	-1.55	-1.5	86%

Employing a footprint of 10 nm for the molecules [20,49] gives rise to a surface charge of -1.5 q/nm^2. In the experiments, a doping concentration of 1×10^{16} is used in the transducer (both values are selected as the true values). The posterior densities are shown in Figure 7. As expected, the posterior densities are around the true values. Regarding the surface charge, we have a normal distribution, and the charge cannot be positive (which is reasonable due to using P-type FET). For the doping concentration, the distribution points out that for C_{dop} more than 2×10^{16}, the sensitivity will reduce significantly, and almost all of the candidates are rejected.

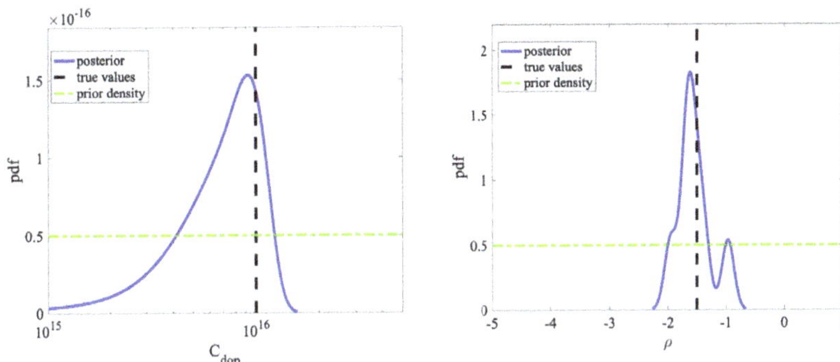

Figure 7. The posterior density of doping concentration (**left**) and surface charge density (**right**) using EnKF-MCMC. The units are C_{dop} (cm^3) and ρ (q/nm^2).

5.3. Machine Learning Based on MFNNs

In this section, we employ MFNNs to train the machine according to available information from the sensors. The effective physical/geometrical parameters will have a nonlinear effect on the device output. For instance, for a doping concentration of more than $C_{dop} = 2 \times 10^{16}$, the current will increase sharply, which is compatible with the results in Bayesian inversion (Figure 7). Due to this nonlinear behavior, the MFNNs algorithm is chosen to monitor the data accuracy and reliability and predict the sensor behavior.

More hidden layers will facilitate the convergence to the desired trajectory; however, it will increase dramatically the computational costs (e.g., computational time). In this work, we use two hidden layers for the MFNNs algorithm to strike a balance between complexity and efficiency. The procedure is shown in Figure 8. We define five specific scenarios according to the number of inputs. In Case 1, we only have one input (V_g) varying between -1 V and -5 V, where other parameters including insulator thickness, nanowire width (N_W), doping concentration, and nanowire height (N_H) are constant. In Case 5, we have five inputs, and the output is the calculated electrical current. Table 2 shows the range of the parameters used for different cases.

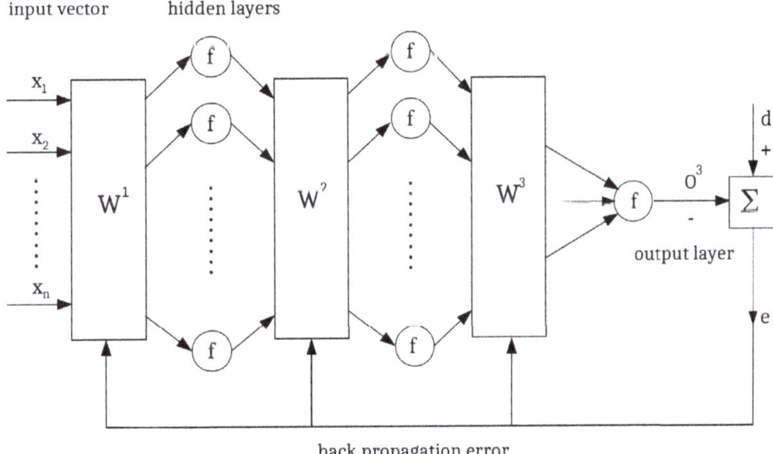

Figure 8. The structure of the MFNNs algorithm.

Table 2. The range of parameters used to compute the electrical current in different cases.

Cases	Inputs	V_g [V]	SiO_2 [nm]	N_W [nm]	C_{dop} [cm^3]	N_H [nm]
Case 1	1	$\mathcal{U}(-1,-5)$	8	100	1×10^{16}	50
Case 2	2	$\mathcal{U}(-1,-5)$	$\mathcal{U}(5,15)$	100	1×10^{16}	50
Case 3	3	$\mathcal{U}(-1,-5)$	$\mathcal{U}(5,15)$	$\mathcal{U}(80,120)$	1×10^{16}	50
Case 4	4	$\mathcal{U}(-1,-5)$	$\mathcal{U}(5,15)$	$\mathcal{U}(80,120)$	$\mathcal{U}(1\times10^{15},5\times10^{16})$	50
Case 5	5	$\mathcal{U}(-1,-5)$	$\mathcal{U}(5,15)$	$\mathcal{U}(80,120)$	$\mathcal{U}(1\times10^{15},5\times10^{16})$	$\mathcal{U}(40,60)$

The MFNNs algorithm is trained with two learning rates (i.e., $\eta = 0.1$ and $\eta = 0.2$) and different numbers of epochs. Here, we use 75% of the samples for data training and 25% of the samples for data testing. The numbers of epochs and neurons in the 1st and 2nd hidden layers are given in Table 3. The sigmoid function is used as an activation function in hidden and output layers. In order to verify the efficiency/accuracy of the MFNNs structure algorithm, for different cases, we compare the output of the machine learning algorithm with the desired trajectories (computed currents). We have the relative MSE for the test and training process and performed a linear regression test to explain the relation between the targets and MFNNs output. Figures 9 and 10 show the results for Cases 1–5, where in all cases, there is a good agreement between the machine learning output and the sensor data.

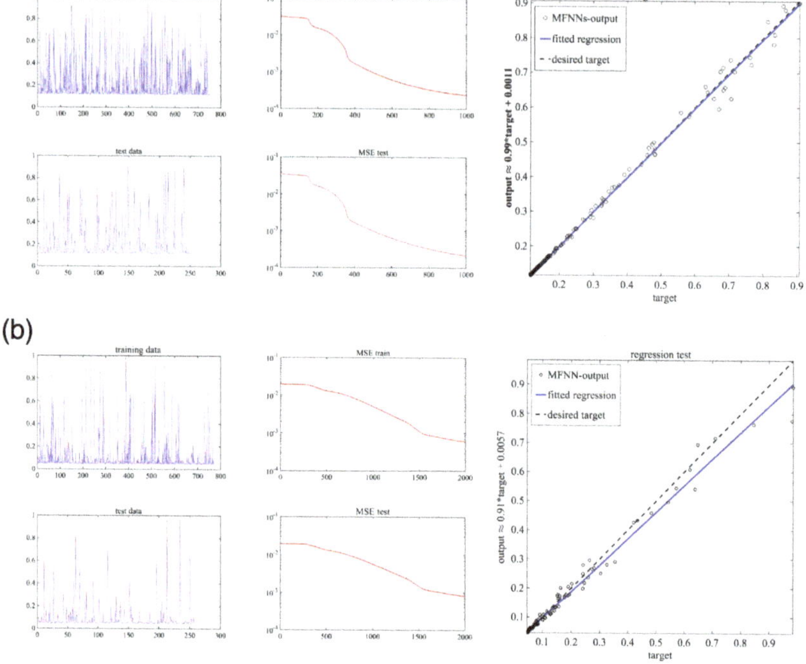

Figure 9. *Cont.*

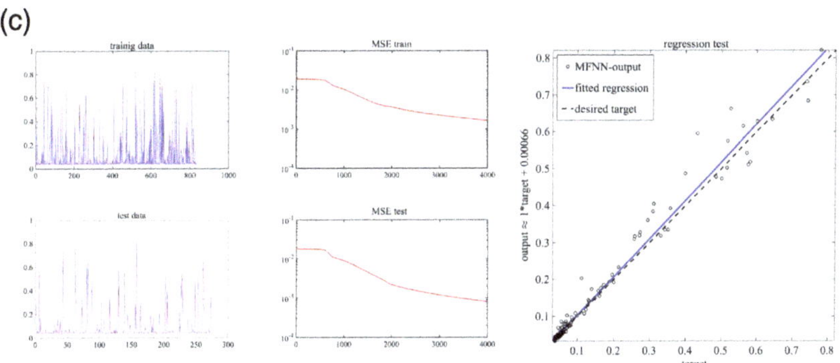

Figure 9. The performance of MFNNs algorithm for Case 1 (**a**), Case 2 (**b**), and Case 3 (**c**). In the first column, the desired trajectories (shown in blue) are compared with the MFNN output (shown in red). In the second column, we have the relative MSE, and the regression test is given in the third column.

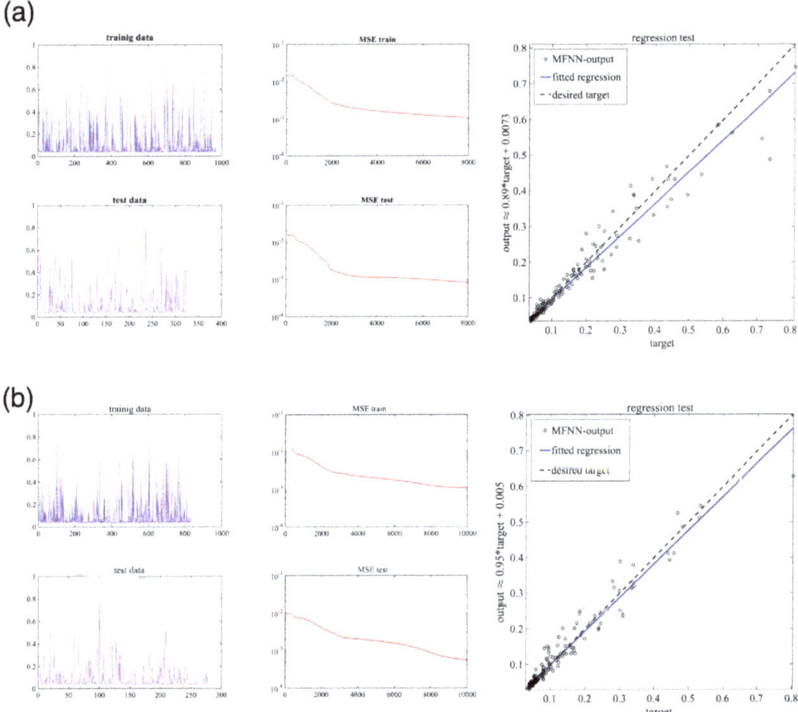

Figure 10. The performance of the MFNNs algorithm for Case 4 (**a**) and Case 5 (**b**). In the first column, the desired trajectories (shown in blue) are compared with the MFNN output (shown in red). In the second column, we have the relative MSE, and the regression test is given in the third column.

Table 3. The features of the MFNNs algorithm including the MSE of training and test processes.

Case	No. Neurons in 1st Hidden Layer	No. Neurons in 2nd Hidden Layer	MSE-Train	MSE-Test	No. Epochs	η
1	10	4	0.00057	0.00061	1000	0.1
2	20	7	0.00147	0.00184	2000	0.2
3	20	7	0.00181	0.000836	4000	0.2
4	20	7	0.000842	0.000517	8000	0.2
5	20	7	0.0011	0.000058	10000	0.2

6. Conclusions

In this work, we introduced a computational framework for modeling charge transport and electrostatic potential distribution in SiNW-FETs in order to enable the rational design of this sensor technology. The PDE-based model has been verified with the experimental data and showed its accuracy. Bayesian inversion can be used to determine quantities of interest such as molecule concentrations, surface charges, and doping concentrations.

Our approach and results can be extended to different types of sensors including plasma resonance-based biosensors, fluorescence-based sensors, and electrochemiluminescence-based biosensors that are used to detect biomarkers.

Finally, machine learning algorithms based on MFNNs have been developed for SiNW-FETs. Here, we use two hidden layers to deal with the nonlinear behavior of the current (with respect to the input parameters), where the method shows its computational efficiency. We used 75% of the data to train the machine and the remaining 25% for testing. In both cases, the obtained MSE shows the convergence to the desired trajectory. The results indicate that MFNNs are a suitable machine learning algorithm for SiNW-FETs and can be used to predict the sensor output behavior as a compact model.

Author Contributions: Writing—review & editing, A.K., M.P., M.T. and C.H. All authors have read and agreed to the published version of the manuscript.

Funding: C. Heitzinger and A. Khodadadian acknowledge support by FWF START project no. Y660 *PDE Models for Nanotechnology*. M. Parvizi acknowledges the financial support of the *Alexander von Humbold Foundation* project named $\mathcal{H}-$matrix approximability of the inverses for FEM, BEM and FEM–BEM coupling of the electromagnetic problems. She is affiliated to the Cluster of Excellence PhoenixD (EXC 2122, Project ID 390833453).

Institutional Review Board Statement: Not applicable.

Informed Consent Statement: Not applicable.

Data Availability Statement: Not applicable

Conflicts of Interest: The authors declare no conflict of interest.

References

1. Mirsian, S.; Khodadadian, A.; Hedayati, M.; Manzour-ol Ajdad, A.; Kalantarinejad, R.; Heitzinger, C. A new method for selective functionalization of silicon nanowire sensors and Bayesian inversion for its parameters. *Biosens. Bioelectron.* **2019**, *142*, 111527. [CrossRef]
2. Kuang, T.; Chang, L.; Peng, X.; Hu, X.; Gallego-Perez, D. Molecular beacon nano-sensors for probing living cancer cells. *Trends Biotechnol.* **2017**, *35*, 347–359. [CrossRef]
3. Hahm, J.i.; Lieber, C.M. Direct ultrasensitive electrical detection of DNA and DNA sequence variations using nanowire nanosensors. *Nano Lett.* **2004**, *4*, 51–54. [CrossRef]
4. Zhang, G.J.; Chua, J.H.; Chee, R.E.; Agarwal, A.; Wong, S.M. Label-free direct detection of MiRNAs with silicon nanowire biosensors. *Biosens. Bioelectron.* **2009**, *24*, 2504–2508. [CrossRef] [PubMed]
5. Choi, J.H.; Kim, H.; Kim, H.S.; Um, S.H.; Choi, J.W.; Oh, B.K. MMP-2 detective silicon nanowire biosensor using enzymatic cleavage reaction. *J. Biomed. Nanotechnol.* **2013**, *9*, 732–735. [CrossRef] [PubMed]
6. De Santiago, F.; Trejo, A.; Miranda, A.; Salazar, F.; Carvajal, E.; Pérez, L.; Cruz-Irisson, M. Carbon monoxide sensing properties of B-, Al-and Ga-doped Si nanowires. *Nanotechnology* **2018**, *29*, 204001. [CrossRef]
7. Song, X.; Hu, R.; Xu, S.; Liu, Z.; Wang, J.; Shi, Y.; Xu, J.; Chen, K.; Yu, L. Highly sensitive ammonia gas detection at room temperature by integratable silicon nanowire field-effect sensors. *ACS Appl. Mater. Interfaces* **2021**, *13*, 14377–14384. [CrossRef]

8. Duan, X.; Li, Y.; Rajan, N.K.; Routenberg, D.A.; Modis, Y.; Reed, M.A. Quantification of the affinities and kinetics of protein interactions using silicon nanowire biosensors. *Nat. Nanotechnol.* **2012**, *7*, 401–407. [CrossRef]
9. Patolsky, F.; Zheng, G.; Lieber, C.M. Fabrication of silicon nanowire devices for ultrasensitive, label-free, real-time detection of biological and chemical species. *Nat. Protoc.* **2006**, *1*, 1711–1724. [CrossRef]
10. Stern, E.; Vacic, A.; Rajan, N.K.; Criscione, J.M.; Park, J.; Ilic, B.R.; Mooney, D.J.; Reed, M.A.; Fahmy, T.M. Label-free biomarker detection from whole blood. *Nat. Nanotechnol.* **2010**, *5*, 138–142. [CrossRef]
11. Chua, J.H.; Chee, R.E.; Agarwal, A.; Wong, S.M.; Zhang, G.J. Label-free electrical detection of cardiac biomarker with complementary metal-oxide semiconductor-compatible silicon nanowire sensor arrays. *Anal. Chem.* **2009**, *81*, 6266–6271. [CrossRef]
12. Chen, K.I.; Li, B.R.; Chen, Y.T. Silicon nanowire field-effect transistor-based biosensors for biomedical diagnosis and cellular recording investigation. *Nano Today* **2011**, *6*, 131–154. [CrossRef]
13. Gao, A.; Lu, N.; Wang, Y.; Dai, P.; Li, T.; Gao, X.; Wang, Y.; Fan, C. Enhanced sensing of nucleic acids with silicon nanowire field effect transistor biosensors. *Nano Lett.* **2012**, *12*, 5262–5268. [CrossRef]
14. Khodadadian, A.; Stadlbauer, B.; Heitzinger, C. Bayesian inversion for nanowire field-effect sensors. *J. Comput. Electron.* **2020**, *19*, 147–159. [CrossRef]
15. Taghizadeh, L.; Khodadadian, A.; Heitzinger, C. The optimal multilevel Monte-Carlo approximation of the stochastic drift–diffusion-Poisson system. *Comput. Methods Appl. Mech. Eng.* **2017**, *318*, 739–761. [CrossRef]
16. Khodadadian, A.; Hosseini, K.; Manzour-ol Ajdad, A.; Hedayati, M.; Kalantarinejad, R.; Heitzinger, C. Optimal design of nanowire field-effect troponin sensors. *Comput. Biol. Med.* **2017**, *87*, 46–56. [CrossRef]
17. Pittino, F.; Selmi, L. Use and comparative assessment of the CVFEM method for Poisson–Boltzmann and Poisson–Nernst–Planck three dimensional simulations of impedimetric nano-biosensors operated in the DC and AC small signal regimes. *Comput. Methods Appl. Mech. Eng.* **2014**, *278*, 902–923. [CrossRef]
18. Khodadadian, A.; Heitzinger, C. Basis adaptation for the stochastic nonlinear Poisson–Boltzmann equation. *J. Comput. Electron.* **2016**, *15*, 1393–1406. [CrossRef]
19. Khodadadian, A.; Taghizadeh, L.; Heitzinger, C. Three-dimensional optimal multi-level Monte–Carlo approximation of the stochastic drift–diffusion–Poisson system in nanoscale devices. *J. Comput. Electron.* **2018**, *17*, 76–89. [CrossRef]
20. Baumgartner, S.; Heitzinger, C.; Vacic, A.; Reed, M.A. Predictive simulations and optimization of nanowire field-effect PSA sensors including screening. *Nanotechnology* **2013**, *24*, 225503. [CrossRef]
21. Hastings, W.K. Monte Carlo sampling methods using Markov chains and their applications *Biometrika* **1970**, *57*, 97–109. [CrossRef]
22. Haario, H.; Saksman, E.; Tamminen, J. Adaptive proposal distribution for random walk Metropolis algorithm. *Comput. Stat.* **1999**, *14*, 375–395. [CrossRef]
23. Green, P.J.; Mira, A. Delayed rejection in reversible jump Metropolis–Hastings. *Biometrika* **2001**, *88*, 1035–1053. [CrossRef]
24. Haario, H.; Laine, M.; Mira, A.; Saksman, E. DRAM: Efficient adaptive MCMC. *Stat. Comput.* **2006**, *16*, 339–354. [CrossRef]
25. Evensen, G. The ensemble Kalman filter for combined state and parameter estimation. *IEEE Control Syst. Mag.* **2009**, *29*, 83–104. [CrossRef]
26. Noii, N.; Khodadadian, A.; Ulloa, J.; Aldakheel, F.; Wick, T.; François, S.; Wriggers, P. Bayesian Inversion with Open-Source Codes for Various One-Dimensional Model Problems in Computational Mechanics. *Arch. Comput. Methods Eng.* **2022**, 1–34. [CrossRef]
27. Schackart, K.E.; Yoon, J.Y. Machine learning enhances the performance of bioreceptor-free biosensors. *Sensors* **2021**, *21*, 5519. [CrossRef] [PubMed]
28. Cui, F.; Yue, Y.; Zhang, Y.; Zhang, Z.; Zhou, H.S. Advancing biosensors with machine learning. *ACS Sensors* **2020**, *5*, 3346–3364. [CrossRef] [PubMed]
29. Albrecht, T.; Slabaugh, G.; Alonso, E.; Al-Arif, S.M.R. Deep learning for single-molecule science. *Nanotechnology* **2017**, *28*, 423001. [CrossRef] [PubMed]
30. Jin, X., Liu, C.; Xu, T.; Su, L.; Zhang, X. Artificial intelligence biosensors: Challenges and prospects. *Biosens. Bioelectron.* **2020**, *165*, 112412. [CrossRef] [PubMed]
31. Raji, H.; Tayyab, M.; Sui, J.; Mahmoodi, S.R.; Javanmard, M. Biosensors and machine learning for enhanced detection, stratification, and classification of cells: A review. *arXiv* **2021**, arXiv:2101.01866.
32. Rivera, E.C.; Swerdlow, J.J.; Summerscales, R.L.; Uppala, P.P.T.; Maciel Filho, R.; Neto, M.R.; Kwon, H.J. Data-driven modeling of smartphone-based electrochemiluminescence sensor data using artificial intelligence. *Sensors* **2020**, *20*, 625. [CrossRef]
33. Khodadadian, A.; Parvizi, M.; Heitzinger, C. An adaptive multilevel Monte Carlo algorithm for the stochastic drift–diffusion–Poisson system. *Comput. Methods Appl. Mech. Eng.* **2020**, *368*, 113163. [CrossRef]
34. Khodadadian, A.; Taghizadeh, L.; Heitzinger, C. Optimal multilevel randomized quasi-Monte-Carlo method for the stochastic drift–diffusion-Poisson system. *Comput. Methods Appl. Mech. Eng.* **2018**, *329*, 480–497. [CrossRef]
35. Baumgartner, S.; Heitzinger, C. Existence and local uniqueness for 3d self-consistent multiscale models of field-effect sensors. *Commun. Math. Sci.* **2012**, *10*, 693–716. [CrossRef]
36. Cockburn, B.; Triandaf, I. Convergence of a finite element method for the drift-diffusion semiconductor device equations: The zero diffusion case. *Math. Comput.* **1992**, *59*, 383–401. [CrossRef]
37. Chen, Z.; Cockburn, B. Analysis of a finite element method for the drift-diffusion semiconductor device equations: The multidimensional case. *Numer. Math.* **1995**, *71*, 1–28. [CrossRef]

38. Zlámal, M. Finite element solution of the fundamental equations of semiconductor devices. I. *Math. Comput.* **1986**, *46*, 27–43. [CrossRef]
39. Zhang, J.; Vrugt, J.A.; Shi, X.; Lin, G.; Wu, L.; Zeng, L. Improving Simulation Efficiency of MCMC for Inverse Modeling of Hydrologic Systems with a Kalman-Inspired Proposal Distribution. *Water Resour. Res.* **2020**, *56*, e2019WR025474. [CrossRef]
40. Bebis, G.; Georgiopoulos, M. Feed-forward neural networks. *IEEE Potentials* **1994**, *13*, 27–31. [CrossRef]
41. Svozil, D.; Kvasnicka, V.; Pospichal, J. Introduction to multi-layer feed-forward neural networks. *Chemom. Intell. Lab. Syst.* **1997**, *39*, 43–62. [CrossRef]
42. Sazli, M.H. A brief review of feed-forward neural networks. *Commun. Fac. Sci. Univ. Ank. Ser. A2-A3 Phys. Sci. Eng.* **2006**, *50*. [CrossRef]
43. Dolinsky, T.J.; Czodrowski, P.; Li, H.; Nielsen, J.E.; Jensen, J.H.; Klebe, G.; Baker, N.A. PDB2PQR: Expanding and upgrading automated preparation of biomolecular structures for molecular simulations. *Nucleic Acids Res.* **2007**, *35*, W522–W525. [CrossRef] [PubMed]
44. Li, H.; Robertson, A.D.; Jensen, J.H. Very fast empirical prediction and rationalization of protein pKa values. *Proteins: Struct. Funct. Bioinform.* **2005**, *61*, 704–721. [CrossRef] [PubMed]
45. Søndergaard, C.R.; Olsson, M.H.; Rostkowski, M.; Jensen, J.H. Improved treatment of ligands and coupling effects in empirical calculation and rationalization of pKa values. *J. Chem. Theory Comput.* **2011**, *7*, 2284–2295. [CrossRef] [PubMed]
46. Olsson, M.H.; Søndergaard, C.R.; Rostkowski, M.; Jensen, J.H. PROPKA3: Consistent treatment of internal and surface residues in empirical pKa predictions. *J. Chem. Theory Comput.* **2011**, *7*, 525–537. [CrossRef] [PubMed]
47. Bulyha, A.; Heitzinger, C. An algorithm for three-dimensional Monte-Carlo simulation of charge distribution at biofunctionalized surfaces. *Nanoscale* **2011**, *3*, 1608–1617. [CrossRef]
48. Heitzinger, C.; Liu, Y.; Mauser, N.J.; Ringhofer, C.; Dutton, R.W. Calculation of fluctuations in boundary layers of nanowire field-effect biosensors. *J. Comput. Theor. Nanosci.* **2010**, *7*, 2574–2580. [CrossRef]
49. Punzet, M.; Baurecht, D.; Varga, F.; Karlic, H.; Heitzinger, C. Determination of surface concentrations of individual molecule-layers used in nanoscale biosensors by in situ ATR-FTIR spectroscopy. *Nanoscale* **2012**, *4*, 2431–2438. [CrossRef]

Article

An Array of On-Chip Integrated, Individually Addressable Capacitive Field-Effect Sensors with Control Gate: Design and Modelling

Arshak Poghossian [1,*], Rene Welden [2,3], Vahe V. Buniatyan [4] and Michael J. Schöning [2,5,*]

1   MicroNanoBio, Liebigstr. 4, 40479 Düsseldorf, Germany
2   Institute of Nano- and Biotechnologies (INB), FH Aachen, Campus Jülich, Heinrich-Mußmannstr. 1, 52428 Jülich, Germany; welden@fh-aachen.de
3   Laboratory for Soft Matter and Biophysics, KU Leuven, Celestijnenlaan 200D, 3001 Leuven, Belgium
4   Department of Microelectronics and Biomedical Devices, National Polytechnic University of Armenia (NPUA), 105 Teryan St., NPUA, Yerevan 0009, Armenia; vbuniat@seua.am
5   Institute of Biological Information Processing (IBI-3), Forschungszentrum Jülich, 52425 Jülich, Germany
*   Correspondence: a.poghossian@gmx.de (A.P.); schoening@fh-aachen.de (M.J.S.)

Abstract: The on-chip integration of multiple biochemical sensors based on field-effect electrolyte-insulator-semiconductor capacitors (EISCAP) is challenging due to technological difficulties in realization of electrically isolated EISCAPs on the same Si chip. In this work, we present a new simple design for an array of on-chip integrated, individually electrically addressable EISCAPs with an additional control gate (CG-EISCAP). The existence of the CG enables an addressable activation or deactivation of on-chip integrated individual CG-EISCAPs by simple electrical switching the CG of each sensor in various setups, and makes the new design capable for multianalyte detection without cross-talk effects between the sensors in the array. The new designed CG-EISCAP chip was modelled in so-called floating/short-circuited and floating/capacitively-coupled setups, and the corresponding electrical equivalent circuits were developed. In addition, the capacitance-voltage curves of the CG-EISCAP chip in different setups were simulated and compared with that of a single EISCAP sensor. Moreover, the sensitivity of the CG-EISCAP chip to surface potential changes induced by biochemical reactions was simulated and an impact of different parameters, such as gate voltage, insulator thickness and doping concentration in Si, on the sensitivity has been discussed.

Keywords: capacitive field-effect sensor; on-chip integrated addressable EISCAP sensors; control gate; multianalyte detection; modelling; equivalent circuit

Citation: Poghossian, A.; Welden, R.; Buniatyan, V.V.; Schöning, M.J. An Array of On-Chip Integrated, Individually Addressable Capacitive Field-Effect Sensors with Control Gate: Design and Modelling. *Sensors* **2021**, *21*, 6161. https://doi.org/10.3390/s21186161

Academic Editor: Ligia Maria Moretto

Received: 10 August 2021
Accepted: 11 September 2021
Published: 14 September 2021

Publisher's Note: MDPI stays neutral with regard to jurisdictional claims in published maps and institutional affiliations.

1. Introduction

Biosensors for multianalyte detection attracted much attention in many fields of application, including point-of-care and clinical diagnostics, food and drug screening, environmental monitoring, etc. Electrolyte-gated field-effect devices (EG-FED) have been recognized as a promising transducer in designing chemical and biological sensors because of their small size and weight, fast response time, real-time monitoring, label-free and multiplexed biomolecular detection, possibility of on-chip integration of EG-FEDs and signal-processing circuit and compatibility to micro- and nanofabrication technologies with the future prospect of large-scale production at relatively low costs [1–8]. In addition, miniaturized analysis systems (e.g., lab-on-a-chip devices or electronic tongues) based on on-chip integrated EG-FEDs in an array format have received tremendous attention due to their ability for multiplexed and (quasi)simultaneous assaying of multiple chemical or biological species [9–14]. Such multiplexed biochemical sensing systems may offer several advantages over devices for single-analyte detection, such as reduced assay time and sample volume, reduced costs and high throughput.

The electrolyte-insulator-semiconductor capacitor (EISCAP) belongs to the family of EG-FEDs and represents a biochemically sensitive capacitor [15]. In contrast to ISFETs (ion-sensitive field-effect transistor) or Si nanowire transistors, EISCAPs have a simple structure (see Figure 1a) and are easy and low-cost in fabrication; typical preparation steps do not require photolithographic patterning, and the implementation of a simple O-ring provides sufficient protection of the conductive regions of the EISCAP from the electrolyte solution. At the same time, the results achieved with EISCAPs are fully transferable to other EG-FEDs, thereby circumventing the need for fabrication of complicated transistor structures. At present, a lot of single EISCAP sensors modified with particular recognition elements have been developed and successfully proved for the detection of pH [16], concentration of ions [17], enzyme-substrate reactions [18–21], charged biomolecules (nucleic acids, proteins, biomarkers, nanoparticle/molecule hybrids) [22–29], plant virus particles [30], as well as for realizing biomolecular logic gates [31–33]. For recent progress in research and development of chemical sensors and biosensors based on EISCAPs, see [15].

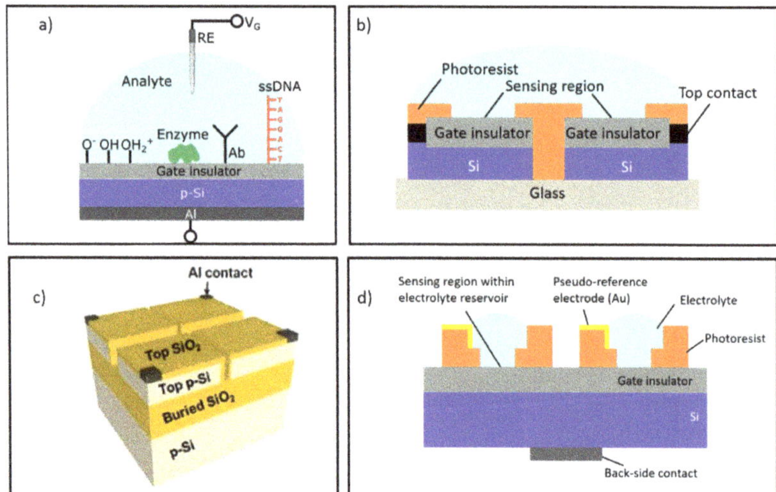

Figure 1. (**a**) Schematic structure of a conventional single EISCAP biochemical sensor with different receptor functionalities (reproduced from [15], open access publication under CC BY license); (**b**) layout of an EISCAP sensor array fabricated on a Si wafer anodically bonded to the glass substrate; (**c**) schematic of a chip combining a 2 × 2 array of nanoplate EISCAPs prepared on a SOI substrate (reproduced from [23] with permission from John Wiley and Sons); (**d**) design of an array of EISCAPs separated via the electrolyte reservoirs fabricated on the gate surface (schematically). RE: reference electrode, V_G: gate voltage, Ab: antibody, ssDNA: single-strand deoxyribonucleic acid.

In spite of successful experiments with single EISCAP sensors, however, the on-chip integration of multiple EISCAPs for multiplexed detection of multiple target analytes seems to be problematic, challenging the fabrication of electrically isolated, individually addressable capacitive structures: EISCAPs prepared on the same Si chip will stay interconnected via the common Si substrate. This may result in an unwanted cross-talk between the different EISCAPs in the array, thereby limiting the possibility to realize on-chip integrated multisensor systems. Only a few studies addressed this task in the literature. For example, Taing realized an EISCAP sensor array fabricated on a Si wafer that is anodically bonded to a glass substrate [34]. To obtain separate electrically decoupled EISCAPs, the Si wafer was diced by means of a saw cutter and, subsequently, the edges of the separated EISCAP chips were protected from contact with solution using a photoresist layer as schematically shown in Figure 1b. Another approach was proposed in [23], where an array of individually addressable nanoplate EISCAPs for chemical/biological sensing was developed using a

SOI (silicon-on-insulator) wafer (Figure 1c). The nanoplate EISCAPs were prepared on a thin top Si layer (two photolithographic steps were needed). For isolation of the individual nanoplate capacitors, the top Si layer was anisotropically etched using the patterned top SiO_2 layer as a mask. However, due to the large series lateral resistance of the top nanoplate Si, the frequency-dependent $C–V$ curves of the nanoplate EISCAPs were deformed; they significantly differed from typical $C–V$ plots of conventional EISCAPs [35]. Finally, a 2×2 array of on-chip integrated EISCAPs was demonstrated in [36], where the gate area of each sensor was separated by means of fabrication of individual electrolyte reservoirs, schematically illustrated in Figure 1d. Each EISCAP was addressed through an individual Au pseudo-reference electrode integrated onto the chip, which induced a large drift and instable sensor signal.

The above discussed examples demonstrate the possibility of realization of on-chip integrated EISCAPs. However, the price to be paid was the loss of the substantial advantages of EISCAP devices—their simple layout, as well as easy and cost-efficient preparation. In this work, we present a new and simple design, as well as the operational setup for an array of on-chip integrated, individually electrically addressable EISCAPs with a so-called control gate (CG) (further referred to as CG-EISCAP) as an alternative transducer structure for the multiplexed (quasi)simultaneous detection of multiple analytes without cross-talk effect between the individual sensors.

2. Design of On-Chip Integrated, Individually Addressable CG-EISCAPs

Figure 2 shows the schematic structure of the new designed sensor chip for detecting of multiple analytes, exemplarily combining three individual electrically addressable CG-EISCAPs (CG-EISCAP-1, CG-EISCAP-2 and CG-EISCAP-3). In comparison to conventional EISCAPs, which are based on an electrolyte-insulator-semiconductor system, the new designed CG-EISCAPs are composed of an electrolyte-insulator-metal-insulator-semiconductor structure. Here, the patterned metal layer (e.g., Au, Al) between the two insulators (insulator-1 and insulator-2) plays the role of the particular CG, in addition to the sensing gate (SG) using the common reference electrode (RE), similar to CMOS (complementary metal-oxide-semiconductor) floating- and programmable-gate ISFETs [37,38]. In order to protect the CG from contact with solution, it is covered with the top insulator-2 (e.g., Al_2O_3, Ta_2O_5) or stacked insulators (e.g., SiO_2-Si_3N_4), which may also serve as a biochemical sensing layer (e.g., being pH-sensitive). In addition, the surface areas (spots) of insulator-2 above the CGs can be modified with various recognition elements (ionophores, enzymes, antibodies, nucleic acids, etc.), thereby making the CG-EISCAP chip sensitive to multiple analytes. Both insulator layers are assumed to be ideal, that is, no current passes through the insulator. For the measurement, the RE (e.g., a conventional Ag/AgCl RE) should provide a stable potential independent of pH or concentration of the analyte solution. The distance between the metal CGs should be sufficiently small to decrease parasitic capacitances associated with the surface areas between the sensors uncovered with the metal CG layer. Conversely, this distance should be sufficiently large to prevent overlapping of the depletion regions in the semiconductor due to the fringing effect and, thereby, practically eliminate possible cross-talk effects between the on-chip integrated CG-EISCAPs.

The CG-EISCAP represents a dual-gate device combining SG and CG, which are coupled with a common floating gate (FG). Thus, the FG potential (V_{FG}) can be modulated by either SG or CG. CG has a multi-purpose function and is connected with the multiplexer by three positions (floating "F", short-circuited "SC" and capacitively-coupled "CC"), which enables an activation or deactivation of the particular CG-EISCAP. In contrast to conventional EISCAPs, the proposed design allows independent biasing and tuning of the operating point of each CG-EISCAP sensor in the desired region of the capacitance-voltage ($C–V$) curve (accumulation, depletion or inversion) by means of applying an additional voltage on the respective CG. This way, possible device-to-device differences in the flat-band voltage of various CG-EISCAPs caused from technological factors (e.g., inhomogeneously distributed trapped charges on the floating gate or non-uniform thickness of the gate insulator) can be compensated, too. In addition, both typical characterization modes of EISCAPs, namely the *C–V* curve and the ConCap (constant-capacitance) mode response can be recorded for each sensor separately in two ways: by means of applying an AC (alternating current) voltage (a) between the RE and the rear-side contact (as for conventional EISCAP sensors) or (b) between the CG and the rear-side contact. Finally, beside the field-effect measurement setup, the proposed structure can also be used as an impedimetric sensor or as capacitively-coupled contactless electrolyte-conductivity detection (so-called C^4D [39]) sensor.

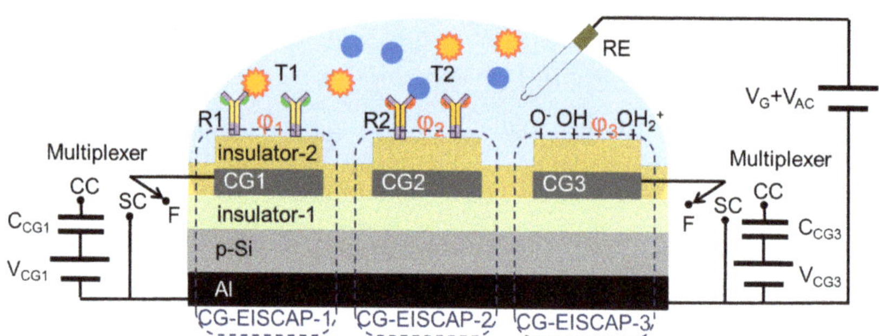

Figure 2. Schematic structure of the designed sensor chip for detecting multiple analytes, combining an array of three individually addressable CG-EISCAPs and measurement setup. RE: reference electrode (sensing gate, SG); V_G: gate voltage; V_{AC}: alternating current voltage; Al: rear-side contact; CG1, CG2 and CG3: control gates; position "F": floating; position "SC": short-circuited; position "CC": capacitively-coupled; C_{CG1} and C_{CG3}: control gate capacitances; V_{CG1} and V_{CG3}: voltage applied to the control gate; R1 and R2: receptors; T1 and T2: target species to be detected; φ_1, φ_2 and φ_3: potential at the insulator-2/electrolyte interface related to CG-EISCAP-1, CG-EISCAP-2 and CG-EISCAP-3, respectively. For better visibility, the multiplexer for CG-EISCAP-2 is not shown.

It is worth to mention, that in contrast to on-chip integrated EISCAP arrays reported in [36], our design uses one common RE for all sensors in the array. On the other hand, a conventional Si wafer is utilized instead of a costly SOI [23,35] or an anodically bonded Si wafer [34]. In addition, the fabrication of CG-EISCAPs is easy; it requires only one photolithographic step when depositing the CG layer via a shadow mask or two photolithographic steps in case of structuring of the CG layer by lift-off process or etching. In some embodiments, the technological process steps could also include front-side contacting to the Si instead of rear-side contacting or the preparation of an on-chip integrated common pseudo-RE.

3. Modelling of On-Chip Integrated, Individually Addressable CG-EISCAPs

For the development of the electrical equivalent circuit and modelling of the CG-EISCAP chip in different setups, let us assume that CG-EISCAP-1 in Figure 2 is modified with receptor R1 for the detection of target analyte T1, while CG-EISCAP-2 is modified with receptor R2 for the detection of target analyte T2. CG-EISCAP-3 is unmodified and serves for pH control of the analyte solution or as reference sensor. Since generally EG-FEDs (particularly EISCAPs) are charge-sensitive devices, any specific electrochemical interaction between the immobilized receptor and target analyte (e.g., affinity reaction, DNA hybridization, local pH changes due to enzymatic reactions, etc.) that occurs at or immediately near the gate surface (within the so-called Debye length from the surface) will induce changes in the surface charge/potential of gate insulator-2 that will consequently modulate the overall capacitance of the EISCAP sensor (see e.g., recent review [15]).

The complete electrical equivalent circuit of the CG-EISCAP chip is complex and involves components associated with the resistance of the RE (R_{RE}), resistance of the bulk solution, double-layer capacitance at the electrolyte/insulator-2 interface, capacitances of insulator-1 and insulator-2, the space-charge capacitance in the semiconductor, and resistances of the bulk semiconductor and the metal-semiconductor rear-side contact. However, as discussed in [15,40,41], for typical gate insulator films used for EISCAPs (e.g., SiO_2, Si_3N_4, Al_2O_3, Ta_2O_5) and their usual thickness range (10–100 nm) as well as appropriate experimental conditions (ionic strength of the solution >0.1 mM; measurement frequencies of <1 kHz), the interferences from several components, such as double-layer capacitance and electrolyte resistance, are negligible. In addition, the resistances of bulk Si and Al-Si rear-side contact are much smaller than R_{RE} and therefore, can be also neglected. Hence, the equivalent circuit of the individual CG-EISCAP sensor with a floating CG can be simplified as a series connection of capacitances of insulator-2, insulator-1 and the variable space-charge capacitance of the semiconductor.

Figure 3 represents the simplified equivalent circuit of the chip composed of three CG-EISCAPs. Here, C_{i1}, C_{i2} and C_{nsc} (n = 1, 2, 3) are the capacitances of insulator-1, insulator-2 and space-charge region in the semiconductor associated with CG-EISCAP-1, CG-EISCAP-2 and CG-EISCAP-3, respectively. In this work, the surface-sensing areas (spots) of all three CG-EISCAPs are assumed to be equal and all capacitances are defined per unit surface area: $C_{i1} = \varepsilon_1/d_1$, $C_{i2} = \varepsilon_2/d_2$, where ε_1, ε_2 and d_1, d_2 are permittivities and thicknesses of insulator-1 and insulator-2, respectively. The space-charge capacitances of CG-EISCAP-1, CG-EISCAP-2 and CG-EISCAP-3 are given as: $C_{1sc} = \varepsilon_s/w_1$, $C_{2sc} = \varepsilon_s/w_2$, $C_{3sc} = \varepsilon_s/w_3$, respectively, where ε_s is the permittivity of the semiconductor, and w_1, w_2 and w_3 are the widths of the corresponding depletion regions. The width of the depletion region and consequently, the space-charge capacitance of each sensor will be determined— among others—by the applied voltage on the gate (in this case, SG and/or CG) and by the respective electrolyte/insulator-2 interfacial potentials. We assume that the surface areas of insulator-1 covered with metal CGs are much larger than that of metal-free areas. Therefore, the parasitic capacitance associated with the surface areas between the individual CG-EISCAPs (uncovered with metal CG layer) has not been included in the equivalent circuit.

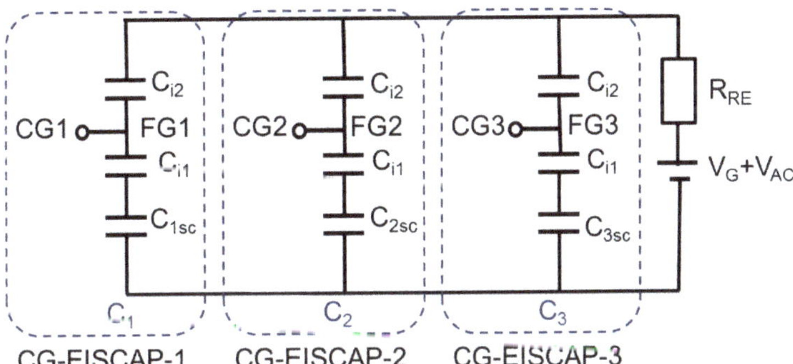

Figure 3. Electrical equivalent circuit of the chip consisting of three CG-EISCAPs. C_{i1} and C_{i2}: capacitances of insulator-1 and insulator-2, respectively; C_{1sc}, C_{2sc} and C_{3sc}: capacitances of the space-charge region in the semiconductor associated with CG-EISCAP-1, CG-EISCAP-2 and CG-EISCAP-3, respectively; CG1, CG2 and CG3: control gates; V_G: gate voltage; V_{AC}: alternating current voltage; R_{RE}: resistance of RE; FG1, FG2 and FG3: floating gates; C_1, C_2 and C_3: overall capacitances of CG-EISCAP-1, CG-EISCAP-2 and CG-EISCAP-3, respectively.

The equivalent capacitance of the chip (C_{eq}) is determined as:

$$C_{eq} = C_1 + C_2 + C_3 \tag{1}$$

where C_1, C_2, and C_3 are the overall capacitances of CG-EISCAP-1, CG-EISCAP-2 and CG-EISCAP-3, respectively, which are determined by the combination of the capacitances C_{i1}, C_{i2} and C_{nsc} in series:

$$\frac{1}{C_1} = \frac{1}{C_{i1}} + \frac{1}{C_{i2}} + \frac{1}{C_{1sc}} \tag{2}$$

$$\frac{1}{C_2} = \frac{1}{C_{i1}} + \frac{1}{C_{i2}} + \frac{1}{C_{2sc}} \tag{3}$$

$$\frac{1}{C_3} = \frac{1}{C_{i1}} + \frac{1}{C_{i2}} + \frac{1}{C_{3sc}} \tag{4}$$

As can be seen in Figure 3, the individual CG-EISCAPs in the array are still interconnected via the common Si substrate, which may result in an unwanted cross-talk between the on-chip integrated sensors (see Introduction). For example, if the chip is exposed to the solution containing both T1 and T2 target analytes, the interaction of target T1 with the immobilized receptor R1 will modulate the interfacial potential (φ_1) as well as the space-charge (C_{1sc}) and overall (C_1) capacitance of CG-EISCAP-1. Analogously, the interaction of target T2 with the immobilized receptor R2 and/or possible pH changes will modulate the overall capacitances of CG-EISCAP-2 (C_2) and CG-EISCAP-3 (C_3), respectively, all resulting in a change of the equivalent capacitance, C_{eq}, of the chip. As a consequence, such a chip is unable to selectively distinguish between particular target analytes in a multicomponent solution. However, the existence of CGs enables addressable activation/deactivation of individual CG-EISCAPs by switching the CG of each sensor in various setups, such as floating/short-circuited CG or floating/capacitively-coupled CG, which are discussed below. This feature of an addressable activation or deactivation of on-chip integrated individual CG-EISCAPs by simple electrical switching the respective CG (instead of fabricating an array of separate EISCAPs) makes the new design capable for multiplexed operation. Detection of multiple analytes is possible, eliminating cross-talk effects between the sensors in the array.

3.1. Setup with Floating/Short-Circuited CG

Figure 4 shows the electrical equivalent circuit and measurement setup of the chip with floating/short-circuited CG. The chip is exposed to the solution containing multiple target analytes (exemplarily, T1 and T2) and the gate voltage V_G is applied to the structure via the RE to set the working points of all three sensors in the depletion region.

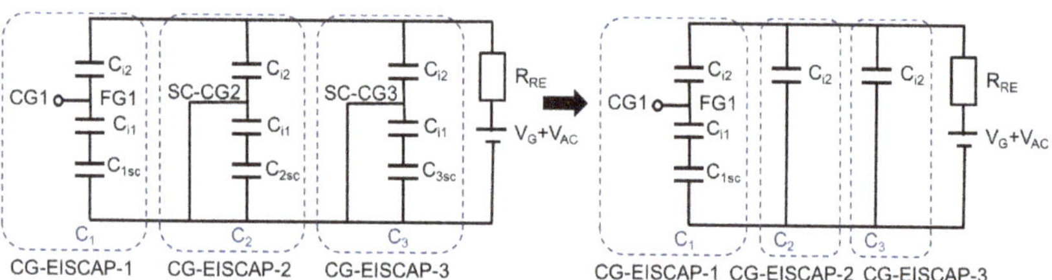

Figure 4. Equivalent circuit of the chip in floating/short-circuited CG setup. SC-CG2 and SC-CG3: short-circuited CG2 and CG3, respectively. All other terms are described in Figure 3.

To detect target analyte T1 with the CG-EISCAP-1, CG1 is kept floating (CG1 is switched to position "F" of the multiplexer), while CG2 and CG3 should be short-circuited (switched to position "SC", Figure 2) to exclude an impact of possible gate-surface potential changes of CG-EISCAP-2 and CG-EISCAP-3 on the total capacitance of the chip and, therefore, on the output signal. FG1 transfers the signal from the electrolyte/insulator-2 interface to the semiconductor in an electrostatic way: the floating gate potential of CG-EISCAP-1, V_{FG1}, will follow the changes in both the gate voltage (V_G) and the interfacial potential (φ_1). The term φ_1 can be represented as $\varphi_1 = \varphi_{01} \pm \Delta\varphi_1$, where φ_{01} is the potential at the insulator-2/electrolyte interface before the biochemical interaction of target T1 with the immobilized receptor R1 and $\Delta\varphi_1$ is the potential change induced via the biochemical interaction. The expression for V_{FG1} of CG-EISCAP-1 can be obtained using the capacitive voltage divider model:

$$V_{FG1} = \frac{C_{i2}}{C_{t1}} V_{G1-eff} \tag{5}$$

where $V_{G1\text{-}eff}$ is the effective gate voltage and is given by [42,43]:

$$V_{G1-eff} = V_G - V_{op} = V_G - E_{ref} + \varphi_1 - \chi_{sol} + W_m/q \tag{6}$$

Here, V_{op} is the overall potential drop through the RE/electrolyte/insulator system, E_{ref} is the potential of the RE relative to vacuum, χ_{sol} is the surface-dipole potential of the solvent, W_m is the metal electron work function, q is the elementary charge (1.6×10^{-19} C) and C_{t1} is the sum of all capacitances coupled to the floating node with:

$$C_{t1} = C_{i2} + \frac{C_{i1}C_{1sc}}{C_{i1} + C_{1sc}} \tag{7}$$

The simplified equivalent circuit corresponding to the floating/short-circuited CG setup is shown in Figure 4 (right), where the equivalent capacitance (C_{eq}) of the chip is determined as:

$$C_{eq} = C_1 + C_2 + C_3 = C_{i1}C_{i2}C_{1sc}/[C_{1sc}(C_{i1} + C_{i2}) + C_{i1}C_{i2}] + 2C_{i2} \tag{8}$$

In general, in the presence of a series resistance (e.g., resistance of the RE, R_{RE}), the measured capacitance (C_m) will be given by [35,44,45]:

$$C_m = C_{eq}/\left[1 + \left(2\pi f R_{RE} C_{eq}\right)^2\right] \tag{9}$$

where f is the measurement frequency. C_m will be equal to C_{eq}, if $(2\pi f R_{RE} C_{eq})^2 \ll 1$. Otherwise, C_m will be affected by the series resistance, resulting in frequency-dependent C–V curves and a much smaller C_m than the real capacitance of the system.

In Equation (8), all terms are constant except C_{1sc}, which at a constant V_G presumably will depend on the T1 concentration in solution. Thus, the chip will detect explicitly potential changes on the gate surface of CG-EISCAP-1 resulting from the interaction of T1 with R1. Although, the gate surface potential of CG-EISCAP-2 will be also altered due to the interaction of T2 with R2, this has no impact on the results of the detection of T1 with CG-EISCAP-1 (because it is short-circuited). Consequently, there are no cross-talk effects between the individual CG-EISCAP sensors in the array. Similarly, for the detection of the target analyte T2 with CG-EISCAP-2, CG2 should be held as floating, while CG1 and CG3 should be short-circuited. Finally, for the pH control with the CG-EISCAP-3, CG1 and CG2 should be short-circuited, while CG3 should be switched to position "F".

To compare the shape of the expected C–V curve and the potential sensitivity of the CG-EISCAP chip and the single EISCAP (without control gate), let us determine C_{eq} in the accumulation, depletion and inversion region, respectively. In the accumulation region ($V_G < 0$), $C_{1sc} \gg C_{i1}$ and $C_{1sc} \gg C_{i2}$. Then, the equivalent capacitance of the chip in the accumulation region ($C_{eq\text{-}acc}$) can be derived from Equation (8) as:

$$C_{eq-acc} = C_{i1}C_{i2}/(C_{i1} + C_{i2}) + 2C_{i2} \tag{10}$$

By strong inversion, the depletion-layer width reaches a maximum, w_m [46]:

$$w_m = \sqrt{\frac{4\varepsilon_s kT ln(N_A/n_i)}{q^2 N_A}} \tag{11}$$

where k is the Boltzmann's constant, T is the temperature, N_A is the density of ionized acceptors (p-Si) and n_i is the electron density in the intrinsic semiconductor. The corresponding high-frequency capacitance of EISCAP-1 in the inversion range (C_{1inv}) reaches its minimum. The equivalent capacitance of the chip in the inversion range ($C_{eq\text{-}inv}$) can be obtained from Equation (8) by replacing C_{1sc} with $C_{1inv} = \varepsilon_s / w_m$:

$$C_{eq-inv} = C_{i1}C_{i2}C_{1inv}/[C_{1inv}(C_{i1} + C_{i2}) + C_{i1}C_{i2}] + 2C_{i2} \tag{12}$$

Typically, $C_{1inv} << C_{i1}$ and $C_{1inv} << C_{i2}$, hence, Equation (12) can be simplified as

$$C_{eq-inv} = C_{1inv} + 2C_{i2} \tag{13}$$

For biochemical sensor applications, more interestingly in the depletion region, the space-charge capacitance in the semiconductor and, therefore, the overall capacitance of the chip depends on both the gate voltage and the interfacial potential. In Equation (8) for C_{eq}, the only variable term is the space-charge capacitance (C_{1sc}), which can be deduced from the expression for the depletion capacitance of a MOS (metal-oxide-semiconductor) capacitor (C_{scMOS}) [46]:

$$C_{scMOS} = \frac{1}{\sqrt{\frac{1}{C_i^2} + \frac{2(V_G - V_{FB})}{qN_A\varepsilon_s}} - \frac{1}{C_i}} \tag{14}$$

For this, the flat-band voltage V_{FB} (the externally applied voltage needed to make energy bands in the semiconductor flat from bulk to the surface and the net charge density in the semiconductor to zero) and the gate-insulator capacitance (C_i) of the MOS structure are replaced by the flat-band voltage (V_{fb1}) and series capacitances C_{i1} and C_{i2} ($C_i = C_{i1}C_{i2}/(C_{i1}+C_{i2})$) of the CG-EISCAP-1, respectively:

$$C_{1sc} = \frac{1}{\sqrt{\left(\frac{C_{i1}+C_{i2}}{C_{i1}C_{i2}}\right)^2 + \frac{2(V_G - V_{fb1})}{qN_A\varepsilon_s}} - \frac{C_{i1}+C_{i2}}{C_{i1}C_{i2}}} \tag{15}$$

Generally, the flat-band voltage of the EISCAP is given by [47]:

$$V_{fb1} = E_{ref} - \varphi_1 + \chi_{sol} - \frac{W_s}{q} - \frac{Q_i + Q_{ss}}{C_i} \tag{16}$$

where W_s is the silicon electron work function, and Q_i and Q_{ss} are the charges located in the oxide and the surface and interface states, respectively. By assuming that Q_i and Q_{ss}, and the charge at the floating gate are zero and grouping analyte-concentration independent potentials in $V_{ip} = E_{ref} + \chi_{sol} - W_s/q$, the expression (16) for the flat-band voltage can be simplified as:

$$V_{fb1} = V_{ip} - \varphi_1 \tag{17}$$

By substituting expressions (15) and (17) into Equation (8), we obtain the following equation for the equivalent capacitance of the chip in the depletion region, $C_{eq\text{-}dep}$:

$$C_{eq-dep} = \frac{1}{\sqrt{\left(\frac{C_{i1} + C_{i2}}{C_{i1}C_{i2}}\right)^2 + \frac{2(V_G - V_{ip} + \varphi_1)}{qN_A\varepsilon_s}}} + 2C_{i2} \tag{18}$$

At a constant V_G, all terms in Equation (18) can be considered as constant except for φ_1, which is analyte-concentration dependent. The combination of Equations (10), (12) and (18) gives the complete description of the C–V curve. The sensitivity (S_φ) of the chip to surface potential changes induced by the receptor-target analyte interaction onto the CG-EISCAP-1 surface can be obtained by differentiation of $C_{eq\text{-}dep}$ with respect to φ_1:

$$S_\varphi = \left| \frac{dC_{eq-dep}}{d\varphi_1} \right| = \cfrac{1}{q\varepsilon_s N_a \left[\left(\dfrac{C_{i1} + C_{i2}}{C_{i1} C_{i2}} \right)^2 + \dfrac{2(V_G - V_{ip} + \varphi_1)}{q\varepsilon_s N_A} \right]^{\frac{3}{2}}} \tag{19}$$

The analysis of Equations (10), (13), (18) and (19) reveals that the *C–V* curve of the chip will have the same shape as for a single conventional EISCAP sensor (without control gate) with the same stacked double-gate insulators and gate surface area as the CG-EISCAP-1, but will be shifted parallel along the capacitance axis with the amount of $2C_{i2}$, as shown in Figure 5. The overall capacitance and output signals of other CG-EISCAPs will remain unchanged. The chip combining an array of CG-EISCAPs will respond in exactly the same manner as the single conventional EISCAP sensor. Therefore, no loss in sensitivity of the CG-EISCAP chip to surface potential changes in comparison with a single EISCAP sensor will be observed. Similar expressions can be obtained in the case of measurements with the CG-EISCAP-2 and the CG-EISCAP-3 sensors. Equations (18) and (19) describe the equivalent capacitance in the depletion region and potential sensitivity of the CG-EISCAP chip without defining the origin of the potential generation at the analyte/insulator-2 interface. If the applied gate voltage, V_G, is fixed, the only variable component is the interfacial potential φ, which is analogous to the effect of applying an additional voltage to the gate. The sensitivity of the chip to analyte concentration variations will be determined by the particular mechanism of the interfacial potential generation (e.g., pH or ion-concentration change, antibody-antigen affinity reaction, DNA hybridization, enzymatic reactions, etc.) and many other experimental factors (e.g., effective charge of the target analyte, distance of bound analyte charge from the gate surface, density of receptors, buffer capacity and ionic strength of the sample, and so on). Therefore, corresponding expressions for the analyte sensitivity, derived from other kinds of EG-FEDs, are fully transferable to CG-EISCAPs. For example, the pH sensitivity of such CG-EISCAPs can be determined as changes of the interfacial potential (φ) in response to a change in the bulk pH [48]:

$$\frac{\delta\varphi}{\delta pH} = -2.3 \frac{kT}{q} \alpha \tag{20}$$

$$\text{with } \alpha = \frac{1}{(2.3 \, kT C_{Dl} / q^2 \beta_{int}) + 1} \tag{21}$$

Here, α is a dimensionless sensitivity parameter, varying between 0 and 1, β_{int} is the surface intrinsic buffer capacity that characterizes the ability of the oxide surface to release or bind protons, and C_{DL} is the double-layer capacitance.

3.2. Setup with Floating/Capacitively-Coupled CGs

Figure 6 shows the equivalent circuit and measurement setup of the chip with floating/capacitively-coupled CGs. First, the CGs of all three sensors are floating (switched to position "F", Figure 2) and the working points of all three sensors are fixed in the depletion region by applying the gate voltage, V_G, via the RE. To detect target analyte T1 with the CG-EISCAP-1, CG1 is kept floating, while CG2 and CG3 are capacitively coupled (switched to position "CC", Figure 2) to the floating gates of FG2 and FG3 via external (or technologically on-chip integrated) capacitances C_{CG2} and C_{CG3}, respectively. In this setup, in addition to SG, the capacitively coupled CG2 and CG3 can also be used to modulate the space-charge capacitances of the CG-EISCAP-2 and the CG-EISCAP-3. The floating gate voltage of the CG-EISCAP-2 (V_{FG2}) or the CG-EISCAP-3 (V_{FG3}) is established by a weighted sum of the two input voltages, namely, the effective gate voltage ($V_{G2\text{-}eff}$ or $V_{G3\text{-}eff}$) and the control gate voltage (V_{CG2} or V_{CG3}). The expressions for V_{FG2} and V_{FG3}

can be obtained by taking into account that each weight is determined by the capacitance of its input normalized by the total capacitance (C_{t2} or C_{t3}) coupled to the floating node:

$$V_{FG2} = \frac{C_{i2}V_{G2-eff} + C_{CG2}V_{CG2}}{C_{2t}} \tag{22}$$

$$V_{FG3} = \frac{C_{i2}V_{G3-eff} + C_{CG3}V_{CG3}}{C_{3t}} \tag{23}$$

$$C_{t2} = C_{i2} + \frac{C_{i1}C_{2sc}}{C_{i1} + C_{2sc}} + C_{CG2} \tag{24}$$

$$C_{t3} = C_{i2} + \frac{C_{i1}C_{3sc}}{C_{i1} + C_{3sc}} + C_{CG3} \tag{25}$$

where $V_{G2\text{-}eff}$ and $V_{G3\text{-}eff}$ are determined by Equation (6) by replacing φ_1 with φ_2 or φ_3, respectively.

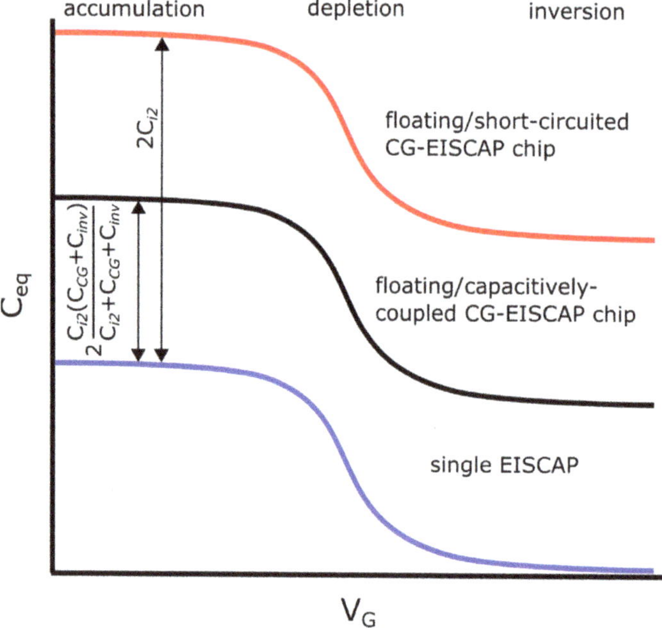

Figure 5. *C–V* curves of a single EISCAP sensor (blue), a floating/short-circuited (red) and a floating/capacitively-coupled (black) CG-EISCAP chip (schematically).

To exclude an impact of possible gate-surface potential changes of the CG-EISCAP-2 and the CG-EISCAP-3 on the total capacitance of the chip, the CG-EISCAP-2 and the CG-EISCAP-3 have to be deactivated by switching to position "CC" (Figure 6) and applying a voltage on CG2 and CG3. The CG-EISCAP-2 and the CG-EISCAP-3 can be deactivated by shifting the operation point either to the accumulation or strong inversion state, where the overall capacitance of these sensors is independent of the gate voltage (V_G) or respective interfacial potentials (φ_2, φ_3). In the following, exemplarily, the expressions for the equivalent capacitance of the CG-EISCAP chip are obtained by assuming that the CG-EISCAP-2 and the CG-EISCAP-3 are deactivated by shifting their operation point in the strong inversion region.

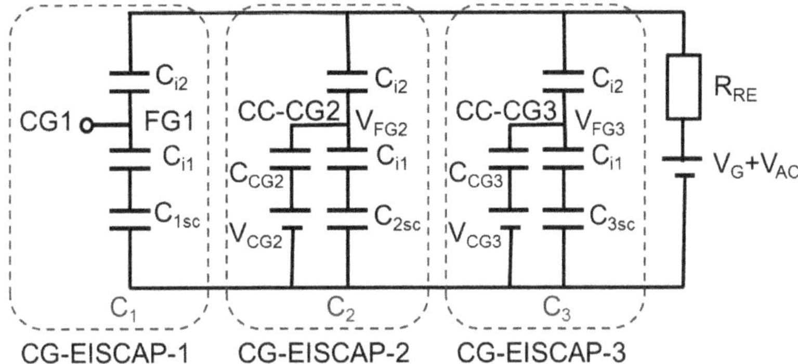

Figure 6. Equivalent circuit and measurement setup of the chip with a floating/capacitively-coupled CG setup. V_{CG2} and V_{CG3}: voltage applied to CG2 and CG3, respectively; V_{FG2} and V_{FG3}: voltage on the floating gate FG2 and FG3, respectively; C_{CG2} and C_{CG3}: control-gate capacitances; CC-CG2 and CC-CG3: capacitively-coupled control gate of CG-EISCAP-2 and CG-EISCAP-3, respectively.

By assuming that $C_{CG2} = C_{CG3} = C_{CG}$ and $C_{2sc} = C_{3sc} = C_{inv}$, the overall capacitance of the CG-EISCAP-2 (C_2) or the CG-EISCAP-3 (C_3) in the inversion region is given by:

$$C_2 = C_3 = \frac{C_{i2}[C_{CG}(C_{i1} + C_{inv}) + C_{i1}C_{inv}]}{(C_{i2} + C_{CG})(C_{i1} + C_{inv}) + C_{i1}C_{inv}} \tag{26}$$

Typically, $C_{inv} \ll C_{i1}$, hence, Equation (26) can be simplified as:

$$C_2 = C_3 = \frac{C_{i2}(C_{CG} + C_{inv})}{C_{i2} + C_{CG} + C_{inv}} \tag{27}$$

The equivalent capacitance of the chip in the accumulation, inversion and depletion regions can be derived from Equations (10), (13) and (18) by replacing C_{i2} by C_2, Equation (27):

$$C_{eq-acc} = \frac{C_{i1}C_{i2}}{C_{i1} + C_{i2}} + 2\frac{C_{i2}(C_{CG} + C_{inv})}{C_{i2} + C_{CG} + C_{inv}} \tag{28}$$

$$C_{eq-inv} = C_{1inv} + 2\frac{C_{i2}(C_{CG} + C_{inv})}{C_{i2} + C_{CG} + C_{inv}} \tag{29}$$

$$C_{eq-dep} = \frac{1}{\sqrt{\left(\frac{C_{i1} + C_{i2}}{C_{i1}C_{i2}}\right)^2 + \frac{2(V_G - V_{ip} + \varphi_1)}{qN_A\varepsilon_s}}} + 2\frac{C_{i2}(C_{CG} + C_{inv})}{C_{i2} + C_{CG} + C_{inv}} \tag{30}$$

Note, since in Equation (30) the inversion capacitance, C_{inv}, of the high-frequency C–V curve is independent of the gate voltage or interfacial potentials, the sensitivity (S_φ) of the chip using the setup with floating/capacitively-coupled CG will be defined by the same Equation (19) as for the setup with floating/short-circuited CG. Thus, the chip combining an array of CG-EISCAPs will respond only to surface potential changes induced by the receptor-target analyte interactions occurring onto the surface of the CG-EISCAP-1 and in exactly the same manner as the single conventional EISCAP sensor. However, the C–V curve will be shifted parallel along the capacitance axis with the amount of $2C_2$ (Equation (27)), as shown in Figure 5 (black curve). Similar expressions can be obtained in the case of detection with the CG-EISCAP-2 (CG2 is floating, while CG1 and CG3 are capacitively coupled) or the CG-EISCAP-3 (CG3 is floating and CG1 and CG2 are capacitively-coupled).

4. Simulation Results

We simulated C–V curves of the CG-EISCAP chip in floating/short-circuited and floating/capacitively-coupled arrangements and compared them with C–V curves of a single EISCAP using Python 3.8 simulation software (Python Software Foundation). In addition, the potential-sensitivity of the CG-EISCAP chip as a function of V_G, d_1 and N_A has been calculated. The simulation parameters are: $\varepsilon_1 = 3.9$ (SiO_2); $d_1 = 10$–60 nm, $\varepsilon_2 = 25$ (Ta_2O_5), $d_2 = 30$ and 60 nm, $\varepsilon_s = 11.7$ (Si), $N_A = 10^{14}$–10^{16} cm^{-3}, $n_i = 1.5 \times 10^{10}$ cm^{-3}, CG-EISCAP sensor area of $A = 3$ mm $\times$ 3 mm $= 9$ mm^2, $q = 1.6 \times 10^{-19}$ C, $T = 300$ K, $k = 1.38 \times 10^{-23}$ J/K, $V_{ip} = 0$ V, $V_G = -0.45$–1 V, $C_{CG} = 50$ nF, $\varphi_{01} = 0$ V, $\varphi_1 = -0.05, 0$ and 0.05 V.

Figure 7 illustrates C–V curves of the single EISCAP and the CG-EISCAP chip in floating/short-circuited and floating/capacitively-coupled setups simulated at different values of φ_1.

The C–V curves of the CG-EISCAP chip in floating/short-circuited and floating/capacitively-coupled setups were calculated for the accumulation, inversion and depletion regions using Equations (10), (12), (18), (28), (29) and (30), respectively. The C–V curves of the single EISCAP were simulated using Equations (10), (12) and (18) without the second term ($2C_{i2}$). To depict the course of the equivalent capacitance of the CG-EISCAP chip and the single EISCAP, also in the transition range from depletion to accumulation region, all C–V curves in Figure 7 were extrapolated (dotted curves).

As predicted in Section 3.1, these C–V curves have the same shape, independent of the CG-EISCAP chip or the single EISCAP setup. However, in comparison with the C–V curves of the single EISCAP, the C–V curves of the CG-EISCAP chip are shifted along the capacitance axis in the direction of larger capacitance values due to the additional parallel constant capacitances of the other sensors in the array. The amount of these shifts is defined by the second term in Equations (10) and (28) (see also Figure 5). As expected, at a constant C_{eq}, the C–V curves are also shifted along the voltage axis. The direction and amount of these shifts (ΔV_G, see top C–V curves in Figure 7) depend on the sign and amplitude of additional potential changes induced by any biochemical interaction on the sensor surface: $\Delta V_G = \pm \Delta \varphi_1$. In case of a p-type EISCAP, an additional positive potential generated by the biochemical interactions on the EISCAP surface will lead to an increase in the width of the depletion layer (correspondingly, the depletion capacitance decreases). As a consequence, the overall capacitance of the CG-EISCAP chip or the single EISCAP will also decrease, resulting in a shift of the C–V curve towards more negative (less positive) gate voltages (Figure 7, blue curve).

Conversely, an additional negative potential generated by the biochemical interaction on the EISCAP surface will decrease the width of the depletion layer in the Si and consequently, increase the depletion capacitance. The overall capacitance of the CG-EISCAP chip or the single EISCAP will also increase, resulting in a shift of the C–V curve in the direction of more positive (less negative) gate voltages (Figure 7, red curve). Such shifts of the C–V curve along the voltage axis upon biochemical interaction was observed in many experiments on conventional single EISCAP-based pH sensors or biosensors (e.g., [17,19,20,24,26]). Often, these sensors work in the ConCap mode, by which gate-surface potential shifts induced upon biochemical interactions can directly be determined from the dynamic sensor response (see [15] and references therein).

Figure 8 shows the calculated curves of the sensitivity of the CG-EISCAP chip on the gate voltage (a), the thickness of the insulator-1 and insulator-2 (b) and the doping concentration (c). With increasing gate voltage, the sensitivity of the CG-EISCAP chip to surface potential changes induced by the receptor-target analyte interaction is decreased. Maximum sensitivity will be achieved at the inflection point of the C–V curve, which corresponds to the flat-band condition as it has been discussed in [49]. In the transition range from depletion to accumulation region (i.e., at gate voltages of $V_G < V_{fb1}$), the sensitivity of the CG-EISCAP chip will again decrease (not shown), similar to conventional EISCAPs. As expected, the calculations, depicted in Figure 8b, show that the sensitivity is increased with decreasing the layer thicknesses of both insulator-1 and insulator-2. This is due to the increase in the $C_{eq\text{-}acc}/C_{eq\text{-}inv}$ ratio and the steepness of the C–V curve in the depletion region.

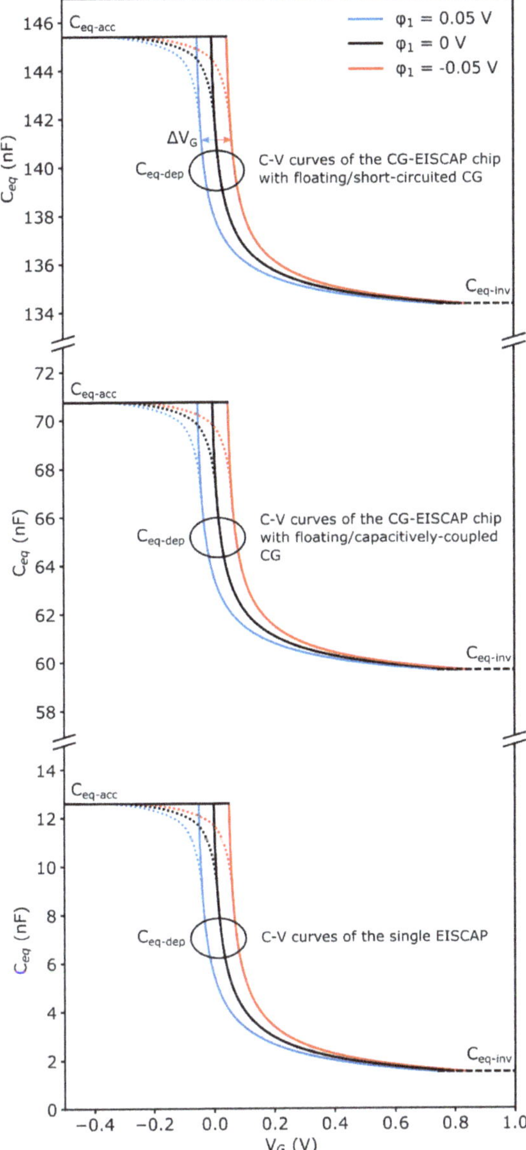

Figure 7. C–V curves of the single EISCAP and the CG-EISCAP chip in floating/short-circuited and floating/capacitively-coupled setups simulated at different values of $\varphi_1 = -0.05$, 0 and 0.05 V. The dotted curves depict the extrapolated course of the overall equivalent capacitance of the CG-EISCAP chip and the single EISCAP in the transition region from depletion to accumulation. Simulation parameters: $d_1 = 20$ nm, $d_2 = 30$ nm, $N_A = 2.76 \times 10^{15}$ cm^{-3} (that corresponds to the resistivity of Si wafer of 5 Ω·cm). ΔV_G: shift of the C–V curves induced by the biochemical interactions on the sensor surface.

Finally, Figure 8c illustrates the dependence of the sensitivity on the doping concentration (N_A) at different thicknesses of insulator-1 (d_1). The sensitivity is increased with increasing N_A, reaching its maximum value, and is decreased by further increase in N_A.

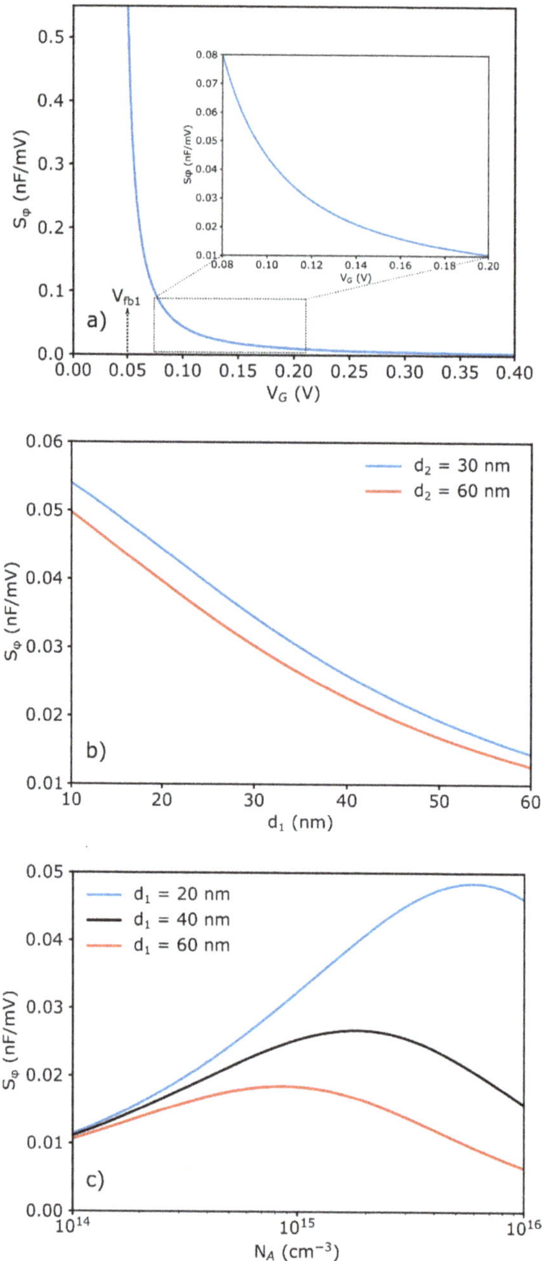

Figure 8. (**a**) Sensitivity of the CG-EISCAP chip as a function of gate voltage, (**b**) thickness of insulator-1 and insulator-2 and (**c**) doping concentration. Simulation parameters for (**a**): $d_1 = 20$ nm, $d_2 = 30$ nm, $N_A = 2.76 \times 10^{15}$ cm^{-3}, $\varphi_1 = -0.05$ V; for (**b**): $N_A = 2.76 \times 10^{15}$ cm^{-3}, $V_G = 0.1$ V, $\varphi_1 = -0.05$ V; for (**c**): $d_2 = 30$ nm, $V_G = 0.1$ V, $\varphi_1 = -0.05$ V.

Such a course of S_φ curves may be explained by taking into consideration the ratio:

$$R = \left(\frac{C_{i1} + C_{i2}}{C_{i1}C_{i2}}\right)^2 / \left[\frac{2(V_G - V_{ip} + \varphi_1)}{q\varepsilon_s V_A}\right] \tag{31}$$

in Equation (19). If $R << 1$, $S_\varphi \sim N_A^{1/2}$ and the sensitivity is increased with increasing the doping concentration. Conversely, if $R >> 1$, the $S_\varphi \sim N_A^{-1}$ and the sensitivity is decreased with increasing the doping concentration. The maximum sensitivity value and its position along the N_A-axis depends on d_1. At a constant V_G and φ_1, with decreasing d_1, the maximum sensitivity is increased and its position is shifted towards higher N_A values.

5. Conclusions

Multiplexed biochips for multianalyte detection have been increasingly recognized as powerful tools in many fields of application, including point-of-care diagnostics and personalized medicine. In this work, a new design for an array of on-chip integrated, individually electrically addressable CG-EISCAPs for a multiplexed (quasi)simultaneous detection of multiple analytes is presented. In comparison with conventional EISCAPs, CG-EISCAPs have a supplemental control gate in addition to their sensing gate, which enables the activation or deactivation of individual CG-EISCAPs inside the array, thus (practically) eliminating possible cross-talk effects between the sensors.

The new designed CG-EISCAP chip was modelled for two setups (floating/short-circuited CG and floating/capacitively-coupled CG). To validate the equivalent-circuit model of the CG-EISCAP chip, the capacitance-voltage curves were simulated for different setups and compared with that of a single EISCAP sensor (without CG). The simulation results reveal that the chip combining an array of CG-EISCAPs will respond in exactly the same manner as the single EISCAP sensor, without loss in sensitivity. Additional to the C–V curves, the sensitivity of the CG-EISCAP chip to surface potential changes induced by biochemical reactions was simulated and the impact of different parameters such as the gate voltage, the insulator thickness and the doping concentration on the sensitivity has been discussed.

In conclusion, the results achieved in this work underline a great potential of CG-EISCAPs as an alternative transducer structure for the realization of multiplexed biochips for (quasi)simultaneous detection of multiple analytes without additional process complexity and with numerous possible applications. Although in this work, an array combining three CG-EISCAPs was modelled, the proposed approach may be extended to CG-EISCAP chips consisting of N sensors, as well as to other kinds of EIS based biochemical sensors (e.g., light-addressable potentiometric sensors [50]).

Author Contributions: Conceptualization: A.P. and M.J.S. Methodology: A.P., V.V.B. and M.J.S. Validation: A.P. and M.J.S. Formal analysis: A.P., R.W., V.V.B. and M.J.S. Simulation: A.P., R.W. and V.V.B. Writing: A.P., R.W., V.V.B. and M.J.S. All authors have read and agreed to the published version of the manuscript.

Funding: This research did not receive any specific grant from funding agencies in the public, commercial, or not-for-profit sectors.

Acknowledgments: The authors thank T. Wagner for valuable discussions.

Conflicts of Interest: The authors declare no conflict of interest.

References

1. Syu, Y.-C.; Hsu, W.-E.; Lin, C.-T. Review—Field-effect transistor biosensing: Devices and clinical applications. *ECS J. Solid State Sci. Technol.* **2013**, *7*, Q3196–Q3207. [CrossRef]
2. de Moraes, A.C.M.; Kubota, L.T. Recent trends in field-effect transistors-based immunosensors. *Chemosensors* **2016**, *4*, 20. [CrossRef]
3. Poghossian, A.; Jablonski, M.; Molinnus, D.; Wege, C.; Schöning, M.J. Field-effect sensors for virus detection: From Ebola to SARS-CoV-2 and plant viral enhancers. *Front. Plant. Sci.* **2020**, *11*, 598103. [CrossRef]

4.   Poghossian, A.; Lüth, H.; Schultze, J.W.; Schöning, M.J. (Bio-)chemical and physical microsensor arrays using an identical transducer principle. *Electrochim. Acta* **2001**, *47*, 243–249. [CrossRef]

5.   Gao, A.; Chen, S.; Wang, Y.; Li, T. Silicon nanowire field-effect-transistor-based biosensor for biomedical applications. *Sens. Mater.* **2018**, *30*, 1619–1628. [CrossRef]

6.   Mu, L.; Chang, Y.; Sawtelle, S.D.; Wipf, M.; Duan, X.; Reed, M.A. Silicon nanowire field-effect transistors—A versatile class of potentiometric nanobiosensors. *IEEE Access* **2015**, *3*, 287–302. [CrossRef]

7.   Pullano, S.A.; Critello, C.D.; Mahbub, I.; Tasneem, N.T.; Shamsir, S.; Islam, S.K.; Greco, M.; Fiorillo, A.S. EGFET-based sensors for bioanalytical applications: A review. *Sensors* **2018**, *18*, 4042. [CrossRef] [PubMed]

8.   Wu, C.; Poghossian, A.; Bronder, T.S.; Schöning, M.J. Sensing of double-stranded DNA molecules by their intrinsic molecular charge using the light-addressable potentiometric sensor. *Sens. Actuators B* **2016**, *229*, 506–512. [CrossRef]

9.   Zhang, Y.; Chen, R.; Xu, L.; Ning, Y.; Xie, S.; Zhang, G.-J. Silicon nanowire biosensor for highly sensitive and multiplexed detection of oral squamous cell carcinoma biomarkers in saliva. *Anal. Sci.* **2015**, *31*, 73–78. [CrossRef] [PubMed]

10. Cheng, S.; Hideshima, S.; Kuroiwa, S.; Nakanishi, T.; Osaka, T. Label-free detection of tumor markers using field effect transistor (FET)-based biosensors for lung cancer diagnosis. *Sens. Actuators B* **2015**, *212*, 329–334. [CrossRef]

11. Si, K.; Cheng, S.; Hideshima, S.; Kuroiwa, S.; Nakanishi, T.; Osaka, T. Multianalyte detection of cancer biomarkers in human serum using label-free field effect transistor biosensor. *Sens. Mater.* **2018**, *30*, 991–999.

12. Jia, Y.-F.; Gao, C.-Y.; He, J.; Feng, D.-F.; Xing, K.-L.; Wu, M.; Liu, Y.; Cai, W.-S.; Feng, X.-Z. Unlabeled multi tumor marker detection system based on bioinitiated light addressable potentiometric sensor. *Analyst* **2012**, *137*, 3806–3813. [CrossRef]

13. Ipatov, A.; Abramova, N.; Bratov, A. Integrated multi-sensor chip with photocured polymer membranes containing copolymerised plasticizer for direct pH, potassium, sodium and chloride ions determination in blood serum. *Talanta* **2009**, *79*, 984–989.

14. Moser, N.; Leong, C.L.; Hu, Y.; Cicatiello, C.; Gowers, S.; Boutelle, M.; Georgiou, P. Complementary metal-oxide-semiconductor potentiometric field-effect transistor array platform using sensor learning for multi-ion imaging. *Anal. Chem.* **2020**, *92*, 5276–5285. [CrossRef] [PubMed]

15. Poghossian, A.; Schöning, M.J. Capacitive field-effect chemical sensors and biosensors: A status report. *Sensors* **2020**, *20*, 5639. [CrossRef] [PubMed]

16. Chen, M.; Jin, Y.; Qu, X.; Jin, W.; Zhao, J. Electrochemical impedance spectroscopy study of Ta_2O_5 based EIOS pH sensors in acid environment. *Sens. Actuators B* **2014**, *192*, 399–405. [CrossRef]

17. Cho, H.; Kim, K.; Meyyappan, M.; Baek, C.-K. LaF_3 electrolyte-insulator-semiconductor sensor for detecting fluoride ions. *Sens. Actuators B* **2019**, *279*, 183–188. [CrossRef]

18. Abouzar, M.H.; Poghossian, A.; Siqueira, J.R., Jr.; Oliveira, O.N., Jr.; Moritz, W.; Schöning, M.J. Capacitive electrolyte-insulator-semiconductor structures functionalized with a polyelectrolyte/enzyme multilayer: New strategy for enhanced field-effect biosensing. *Phys. Status Solidi A* **2010**, *207*, 884–890. [CrossRef]

19. Lin, C.F.; Kao, C.H.; Lin, C.Y.; Chen, K.L.; Lin, Y.H. NH_3 plasma-treated magnesium doped zinc oxide in biomedical sensors with electrolyte-insulator-semiconductor (EIS) structure for urea and glucose applications. *Nanomaterials* **2020**, *10*, 583. [CrossRef] [PubMed]

20. Poghossian, A.; Jablonski, M.; Koch, C.; Bronder, T.S.; Rolka, D.; Wege, C.; Schöning, M.J. Field-effect biosensor using virus particles as scaffolds for enzyme immobilization. *Biosens. Bioelectron.* **2018**, *110*, 168–174. [CrossRef]

21. Jablonski, M.; Münstermann, F.; Nork, J.; Molinnus, D.; Muschallik, L.; Bongaerts, J.; Wagner, T.; Keusgen, M.; Siegert, P.; Schöning, M.J. Capacitive field-effect biosensor applied for the detection of acetoin in alcoholic beverages and fermentation broths. *Phys. Status Solidi A* **2021**, *218*, 2000765. [CrossRef]

22. Lin, Y.-T.; Luo, J.-D.; Chiou, C.-C.; Yang, C.-M.; Wang, C.-Y.; Chou, C.; Lai, C.-S. Detection of KRAS mutation by combination of polymerase chain reaction (PCR) and EIS sensor with new amino group functionalization. *Sens. Actuators B* **2013**, *186*, 374–379. [CrossRef]

23. Abouzar, M.H.; Poghossian, A.; Cherstvy, A.G.; Pedraza, A.M.; Ingebrandt, S.; Schöning, M.J. Label-free electrical detection of DNA by means of field-effect nanoplate capacitors: Experiments and modeling. *Phys. Status Solidi A* **2012**, *209*, 925–934. [CrossRef]

24. Chand, R.; Han, D.; Neethirajan, S.; Kim, Y.-S. Detection of protein kinase using an aptamer on a microchip integrated electrolyte-insulator-semiconductor sensor. *Sens. Actuators B* **2017**, *248*, 973–979. [CrossRef]

25. Kumar, N.; Kumar, S.; Kumar, J.; Panda, S. Investigation of mechanisms involved in the enhanced label free detection of prostate cancer biomarkers using field effect devices. *J. Electrochem. Soc.* **2017**, *164*, B409–B416. [CrossRef]

26. Hlukhova, H.; Menger, M.; Offenhäusser, A.; Vitusevich, S. Highly sensitive aptamer-based method for the detection of cardiac biomolecules on silicon dioxide surfaces. *MRS Adv.* **2016**, *3*, 1535–1541. [CrossRef]

27. Bahri, M.; Baraket, A.; Zine, N.; Ali, M.B.; Bausells, J.; Errachid, A. Capacitance electrochemical biosensor based on silicon nitride transducer for TNF-α cytokine detection in artificial human saliva: Heart failure (HF). *Talanta* **2020**, *209*, 12050. [CrossRef]

28. Pan, T.-M.; Lin, T.-W.; Chen, C.-Y. Label-free detection of rheumatoid factor using YbY_xO_y electrolyte-insulator-semiconductor devices. *Anal. Chim. Acta* **2015**, *891*, 304–311. [CrossRef]

29. Poghossian, A.; Bäcker, M.; Mayer, D.; Schöning, M.J. Gating capacitive field-effect sensors by the charge of nanoparticle/molecule hybrids. *Nanoscale* **2015**, *7*, 1023–1031. [CrossRef]

30. Jablonski, M.; Poghossian, A.; Severins, R.; Keusgen, M.; Wege, C.; Schöning, M.J. Capacitive field-effect biosensor studying adsorption of *tobacco mosaic virus* particles. *Micromachines* **2021**, *12*, 57. [CrossRef]
31. Katz, E.; Poghossian, A.; Schöning, M.J. Enzyme-based logic gates and circuits-analytical applications and interfacing with electronics. *Anal. Bioanal. Chem.* **2017**, *409*, 81–94. [CrossRef]
32. Poghossian, A.; Malzahn, K.; Abouzar, M.H.; Mehndiratta, P.; Katz, E.; Schöning, M.J. Integration of biomolecular logic gates with field-effect transducers. *Electrochim. Acta* **2011**, *56*, 9661–9665. [CrossRef]
33. Poghossian, A.; Katz, E.; Schöning, M.J. Enzyme logic AND-Reset and OR-Reset gates based on a field-effect electronic transducer modified with multi-enzyme membrane. *Chem. Commun.* **2015**, *51*, 6564–6567. [CrossRef]
34. Taing, M.; Sweatman, D. Fabrication techniques for an arrayed EIS biosensor. In Proceedings of the Electronics Packaging Technology Conference (IEEE), Singapore, 9–11 December 2009; pp. 168–173.
35. Abouzar, M.H.; Moritz, W.; Schöning, M.J.; Poghossian, A. Capacitance–voltage and impedance-spectroscopy characteristics of nanoplate EISOI capacitors. *Phys. Status Solidi A* **2011**, *208*, 1327–1332. [CrossRef]
36. Dastidar, S.; Agarwal, A.; Kumar, N.; Bal, V.; Panda, S. Sensitivity enhancement of electrolyte-insulator-semiconductor sensors using mesotextured and nanotextured dielectric surfaces. *IEEE Sens. J.* **2015**, *15*, 2039–2045. [CrossRef]
37. Kaisti, M.; Zhang, Q.; Prabhu, A.; Lehmusvuori, A.; Rahman, A.; Levon, K. An ion-sensitive floating gate FET model: Operating principles and electrofluidic gating. *IEEE Trans. Electron Devices* **2015**, *62*, 2628–2635. [CrossRef]
38. Jayant, K.; Auluck, K.; Funke, M.; Anwar, S.; Phelps, J.B.; Gordon, P.H.; Rajwade, S.R.; Kan, E.C. Programmable ion-sensitive transistor interfaces. I. Electrochemical gating. *Phys. Rev. E* **2013**, *88*, 012801. [CrossRef]
39. Huck, C.; Poghossian, A.; Bäcker, M.; Chaudhuri, S.; Zander, W.; Schubert, J.; Begoyan, V.K.; Buniatyan, V.V.; Wagner, P.; Schöning, M.J. Capacitively coupled electrolyte-conductivity sensor based on high-k material of barium strontium titanate. *Sens. Actuators B* **2014**, *198*, 102–109. [CrossRef]
40. Fabry, P.; Laurent-Yvonnou, L. The C-V method for characterizing ISFET or EOS device with ion-sensitive membranes. *J. Electroanal. Chem.* **1990**, *286*, 23–40. [CrossRef]
41. Siu, W.M.; Cobbold, R.S.C. Basic properties of the electrolyte-SiO_2-Si system: Physical and theoretical aspects. *IEEE Trans. ED* **1979**, *26*, 1805–1815. [CrossRef]
42. Georgiou, P.; Toumazou, C. CMOS-based programmable gate ISFET. *Electron. Lett.* **2008**, *44*, 1289–1291. [CrossRef]
43. Shepherd, L.; Toumazou, C. Weak inversion ISFETs for ultra-low power biochemical sensing and real-time analysis. *Sens. Actuators B* **2005**, *107*, 468–473. [CrossRef]
44. del Cueto, M.E.; Altuzarra, A.C. On the analysis of C-V curves for high resistivity substrates. *Solid-State Electron.* **1996**, *39*, 1519–1521. [CrossRef]
45. Prasad, B.; Lal, R. A capacitive immunosensor measurement system with a lock-in amplifier and potentiostatic control by software. *Meas. Sci. Technol.* **1999**, *10*, 1097–1104. [CrossRef]
46. Sze, S.M.; Ng, K.K. *Physics of Semiconductor Devices*; John Wiley & Sons: Hoboken, NJ, USA, 2007.
47. Bergveld, P. A critical evaluation of direct electrical protein detection methods. *Biosens. Bioelectron.* **1991**, *6*, 55–72. [CrossRef]
48. van Hal, R.E.G.; Eijkel, J.C.T.; Bergveld, P. A novel description of ISFET sensitivity with the buffer capacity and double-layer capacitance as key parameters. *Sens. Actuators B* **1995**, *24–25*, 201–205. [CrossRef]
49. Winter, R.; Ahn, J.; McIntyre, P.C.; Eizenberg, M. New method for determining flat-band voltage in high mobility semiconductors. *J. Vac. Sci. Technol. B* **2013**, *31*, 030604. [CrossRef]
50. Poghossian, A.; Werner, C.F.; Buniatyan, V.V.; Wagner, T.; Miamoto, K.; Yoshinobu, T.; Schöning, M.J. Towards addressability of light-addressable potentiometric sensors: Shunting effect of non-illuminated region and cross-talk. *Sens. Actuators B* **2017**, *244*, 1071–1079. [CrossRef]

Article

Comprehensive Analytical Modelling of an Absolute pH Sensor

Cristina Medina-Bailon [1,*], Naveen Kumar [1], Rakshita Pritam Singh Dhar [1], Ilina Todorova [1], Damien Lenoble [2], Vihar P. Georgiev [1,*] and César Pascual García [2,*]

[1] Device Modelling Group, School of Engineering, University of Glasgow, Glasgow G12 8LT, UK; Naveen.Kumar@glasgow.ac.uk (N.K.); r.dhar.1@research.gla.ac.uk (R.P.S.D.); 2326960T@student.gla.ac.uk (I.T.)

[2] Nano-Enabled Medicine and Cosmetics Group, Materials Research and Technology Department, Luxembourg Institute of Science and Technology (LIST), L-4362 Esch-sur-Alzette, Luxembourg; damien.lenoble@list.lu

[*] Correspondence: cristina.medinabailon@glasgow.ac.uk (C.M.-B.); vihar.georgiev@glasgow.ac.uk (V.P.G.); cesar.pascual@list.lu (C.P.G.)

Abstract: In this work, we present a comprehensive analytical model and results for an absolute pH sensor. Our work aims to address critical scientific issues such as: (1) the impact of the oxide degradation (sensing interface deterioration) on the sensor's performance and (2) how to achieve a measurement of the absolute ion activity. The methods described here are based on analytical equations which we have derived and implemented in MATLAB code to execute the numerical experiments. The main results of our work show that the depletion width of the sensors is strongly influenced by the pH and the variations of the same depletion width as a function of the pH is significantly smaller for hafnium dioxide in comparison to silicon dioxide. We propose a method to determine the absolute pH using a dual capacitance system, which can be mapped to unequivocally determine the acidity. We compare the impact of degradation in two materials: SiO_2 and HfO_2, and we illustrate the acidity determination with the functioning of a dual device with SiO_2.

Keywords: nano-biosensor; analytical model; oxide degradation; depletion width; pH sensor modelling and simulations

Citation: Medina-Bailon, C.; Kumar, N.; Dhar, R.P.S.; Todorova, I.; Lenoble, D.; Georgiev, V.P.; García, C.P. Comprehensive Analytical Modelling of an Absolute pH Sensor. *Sensors* **2021**, *21*, 5190. https://doi.org/10.3390/s21155190

Academic Editors: Michael J. Schöning and Sven Ingebrandt

Received: 25 June 2021
Accepted: 27 July 2021
Published: 30 July 2021

Publisher's Note: MDPI stays neutral with regard to jurisdictional claims in published maps and institutional affiliations.

1. Introduction

Ion-Sensitive Field Effect Transistors (ISFETs) [1–3] are devices that measure acidity, which offer the best accuracy and miniaturisation. They employ semiconductor fabrication techniques that lower the cost per sensor while providing a high level of sophistication by the integration of the sensors with metal-oxide-semiconductor (MOS) circuits and microfluidics [4,5]. ISFETs transduce the ion activity in the solvent by a capacitance effect that measures the associate charge. Ions are adsorbed from the electrolyte depending on the bulk concentration and the affinity of the material interface. The final signal is formed with the contribution of the charged particles in solution that react forming the double layer capacitance [6–8]. There are numerous applications of ISFETs described in the literature ranging from sweat biomarker sensors, physiological measurements, to monitoring the enzymatic activity of polymerase to assist DNA sequencing [9–12].

Regardless of the impressive progress, these devices have triggered a serious limitation known to anyone using ISFETs, which is that they require continuous re-calibration. ISFETs show an instability that can be manifested in the drift of the threshold voltage or the output current used to transduce the acidity/basicity [13–15]. This effect limits the applications that require accurate monitoring of the pH during periods lasting hours as well as the miniaturisation of highly multiplexed devices. As a result of the drift, ISFETs require compensation strategies based on calculations or using reference devices that require extra resources.

At the origin of the drift, there is an irreversible chemical degradation of the dielectric barrier responsible for the change in capacitance. There are several explanations proposed for this effect that involve the migration of charges into the oxide materials [16–19] that decreases the dielectric constant of the affected region [20]. When possible, the re-calibrations are achieved using external reference buffers with known acidity. More sophisticated systems make use of an internal generation of acid that performs a titration curve [21,22]. Finally, there are models that propose to predict the degradation of the capacitance [18,20,23,24]. However, all of the above methods require experiment interruptions or can be sensitive to drastic changes in the ambient conditions. Most of the ongoing work focuses on the stability and material properties of different oxides to enhance the IS-FET performance but the operation of a FET device may be affected due to several process parameters that may make the results less reliable. Here, we present a detailed methodology to enhance the sensor reliability by aiming to get accurate values irrespective of the interacting oxide or FET operation.

In this paper, we propose a system for the absolute measurement of pH by simultaneously using two devices exposed to the same conditions and configured in a way that makes it possible to parameterize a synchronized response. We provide an analytical framework equivalent to an experimental mapping of the simultaneous current of two sensors that are used to describe the system and the methods that can determine the absolute pH irrespective of the experimental conditions. Our system improves the state-of-the-art by reducing the need of continuous calibrations and solving the problem of drift. The derivation of our analytical model is applied to protons, but it can be extended to correct the detection of any other ions that are selectively adsorbed and which suffer from the drift in the current.

The structure of this paper is as follows: the model to include the degradation region in the oxide capacitance is described in Section 2.1 followed by the equations to compute the depletion width. To describe experimental conditions, we applied the system to a high aspect ratio FinFET which enhances the sensitivity and reliability by a 3D gating while increasing the total surface area (Section 2.2). Section 3 outlines the main results and their discussion, including a meticulous analysis of the impact of degradation region on the depletion width and current as a function of the pH for ideal biosensors with two different oxides (Section 3.1) and non-ideal biosensors with SiO_2 (Section 3.3). Finally, conclusions are drawn in Section 4.

2. Methodology

2.1. Oxide Capacitance

The migration of ions from the electrolyte into the dielectric is observed experimentally as a decrease of the capacitance resulting from the irreversible chemical transformation of a layer of the original material. The ions diffuse down to an effective depth that in some cases can be calculated for given experimental conditions [20,25]. The degraded material experiences a decrease of the dielectric constant which in some materials like Al_2O_3 reaches values of 20% of the original one [20]. The typical penetration depths of ions account for several nanometers in the span of hours depending on experimental conditions that include the pH of the electrolyte, the temperature and the ionic strength. In steady-state conditions, the degradation often leads to a fast transition and a complete failure of the device when leakage currents appear between the electrolyte and the semiconductor channel.

To simulate the degradation, both the effective dielectric constant and penetration depth can be adjusted phenomenologically to match the capacitance with several combinations that can provide a successful description of the sensor behaviour before the avalanche of leakage currents makes the device fail. In our model, we have modelled the degradation with a reduction to an arbitrary dielectric constant with a value 20% of the original one. The degraded region is associated with a corresponding effective penetration depth of the ions of x that is used to parameterize the degradation. To determine the absolute pH, we will consider two ISFET devices with different thickness t_{ox1} and t_{ox2} of the dielectric barrier, subjected to the same experimental conditions and thus, with the same penetration of the effective degradation x on both of devices.

Figure 1a shows sections of the interface between the electrolyte and the silicon channel using our model for two cases corresponding to device couples of SiO$_2$ and HfO$_2$, respectively. The dielectric barrier in each device can be considered to be made of two materials in series. The first material in contact with the electrolyte accounts for the degraded region with the adopted effective dielectric constant 20% of the original material and the total effective thickness x corresponding to the penetration of the degradation. The second material has the dielectric properties of the original material (SiO$_2$ or HfO$_2$) with a total thickness of $(t_{ox1} - x)$ or $(t_{ox2} - x)$ for the first and the second device or each sensor couple that will be used to determine the pH. Underlying the non-degraded dielectric in each device, there is the silicon channel as shown in gray in the schemes of Figure 1. Each sensor of absolute pH could consist of more than two of these devices which would be redundant but could help in improving precision. However, for simplicity in this work, we have considered only the use of two devices to determine the absolute pH. We also consider that all the devices with SiO$_2$ or HfO$_2$ dielectric barriers will have the same configurations (silicon dimensions, doping, length, etc.) except for the oxide thickness (t_{ox1} = 5 nm and t_{ox2} = 10 nm). For practical reasons, in the simulations of this work, we have considered a maximum penetration of ions degrading the oxide of 3 nm. We have also considered in our configuration a common reference electrode for both devices.

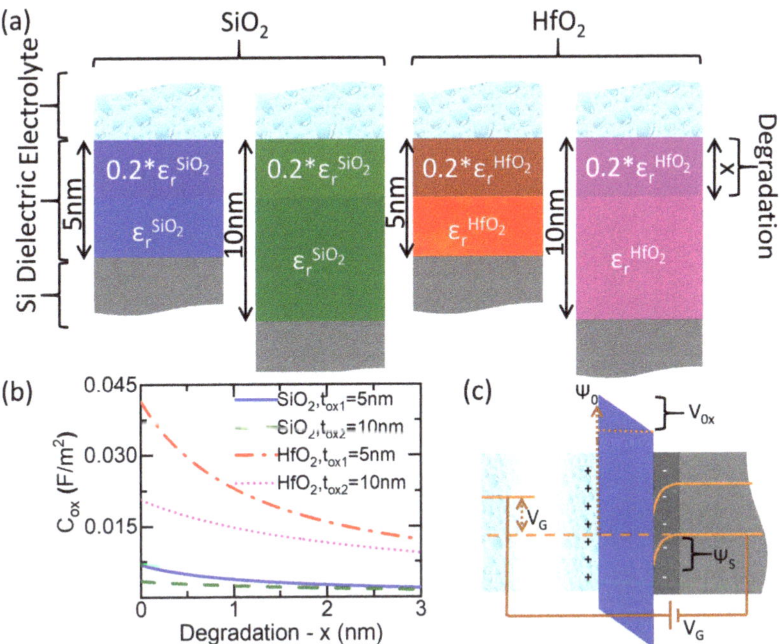

Figure 1. (**a**) Schematic representation that shows how the oxide capacitance is modelled considering a degraded region. Two different couple of devices are described in the diagrams including the combinations for two different oxide materials in the dielectric region. For each device couple, we have considered two oxide thicknesses: (blue) t_{ox1} = 5 nm and SiO$_2$, (green) t_{ox2} = 10 nm and SiO$_2$, (red) t_{ox1} = 5 nm and HfO$_2$, and (magenta) t_{ox2} = 10 nm and HfO$_2$. This color notation to identify each device has been kept the same throughout the whole paper. (**b**) Total capacitance vs. degradation using the penetration depth of the degrading charges as a parameter of the degradation. (**c**) Schematic of the energy band alignment along one interface in a generic ISFET sensor.

For each device, the total oxide capacitance (C_{ox}) is calculated as two capacitances in series including a capacitance without any degradation (C_{ox}^1) and another one with the degradation (C_{ox}^2):

$$\frac{1}{C_{ox}} = \frac{1}{C_{ox}^1} + \frac{1}{C_{ox}^2} = \frac{t_{ox} - x}{\varepsilon_r^1 \varepsilon_0} + \frac{x}{\varepsilon_r^2 \varepsilon_0} \tag{1}$$

where ε_r^1 is the relative dielectric constant of the original material, ε_r^2 is the relative dielectric constant of the degraded region material and ε_0 is the vacuum dielectric constant.

Accordingly, the total oxide capacitance for each sensor has been calculated from Equation (1):

$$C_{ox_i} = \frac{\varepsilon_r^1 \varepsilon_r^2 \varepsilon_0}{\varepsilon_r^1 x + \varepsilon_r^2 (t_{ox,1/2} - x)} \tag{2}$$

where the index i has been added to the total capacitance to determine ox_1 or ox_2 referring to the devices with the original dielectric of 5 nm or 10 nm, respectively.

Figure 1b shows the oxide capacitance calculated using Equation (2) as a function of the degraded region x, for the sensor interfaces with both thicknesses and materials. The observed behaviour corresponds to the decrease of the capacitance with increasing degradation depth x, which is equivalent to what observed in other works [25]. Regarding the total change in the oxide capacitance in the degradation range studied, it is much more pronounced for HfO_2 than for SiO_2. This is due to the higher dielectric constant ε_r of HfO_2 in comparison to the one of SiO_2 (23.4 and 3.9, respectively [26]). Comparing the devices within the same material, as expected, the total variation of the capacitance is more pronounced for the configurations with 5 nm oxide thickness with respect to the thicker oxides of 10 nm, as the degraded region represents a larger part of the total dielectric thickness. Overall, it can be concluded that the HfO_2 capacitance shows larger susceptibility to the degradation and the variations in the thickness due to the diffusion process when it is compared to the SiO_2.

2.2. Calculation of the Field Effect in the Semiconductor Channel

To calculate the effect of the adsorbed charges on the device's current, we consider the energy band diagram in the direction perpendicular to the surface of the oxide shown in Figure 1c. The model does not take into account the possible differences in the chemical potential between the semiconductor and the electrolyte; charges accumulated on the interface between the silicon and the dielectric barrier or phenomenon like the degradation of the reference electrode. When the semiconductor and the electrolyte are connected through a reference electrode and a gate voltage is applied between the two, it is possible to set the relation between the different potentials:

$$\Psi_0 = V_{ox} - V_G + \Psi_S \tag{3}$$

where Ψ_0 is the oxide-electrolyte interface potential, Ψ_S is the oxide-silicon interface potential, V_{ox} is the potential drop across the oxide and V_G is the external bias at the backgate.

The adsorption of protons in the oxide builds the potential Ψ_0, which is zero at the pH of point of zero charge (pH_{pzc}) where the adsorption and desorption processes are in equilibrium and which can be calculated from their proton affinities pK_a and pK_b ($pH_{pzc} = \frac{pK_a + pK_b}{2}$). This potential has a Nerstian response which has been well described in literature [6]:

$$\frac{\partial \Psi_0}{\partial pH} = -2.303 \frac{k_B T}{q} \alpha \tag{4}$$

where k_B, T, and q are the Boltzmann constant, the temperature, and the electron charge, respectively. The sensitivity parameter α depends on the chemical buffering capacity of the dielectric surface in contact with the electrolyte and the response of the ions in solution that will create the double layer capacitance. Often α is considered only in the linear range of the sensor, and in the ideal case it can be approximated by 1, showing the theoretical maximum variation (slope) of the surface potential with respect to the pH ($\frac{\partial \Psi_0}{\partial pH} = -2.3\frac{k_B T}{q}$).

To consider the chemical response of the biosensor (non-ideal sensor), α can be calculated making use of an iterative method with Ψ_0 making use of the dissociation model and Gouy-Chapman-Stern theory [6,7]. This more realistic assumption has been taken into account in our analytic model in Section 3.3. In this method, the sensitive parameter is calculated making use of the diffusion capacitance (C_{diff}), which depends on the electrolyte properties (considering the double layer and the Stern capacitances), and the intrinsic buffering capacitance of the dielectric (β_{diff}), which depends on the number of binding sites on the sensing surface N_s and their corresponding proton affinities pK_a and pK_b:

$$\alpha = \frac{1}{1 + \frac{2.303 k_B T C_{diff}}{q^2 \beta_{diff}}} \tag{5}$$

Ψ_0 in Equation (4) changes relative to the pH at pH_{pzc}, i.e. the acidity in the electrolyte at which the equilibrium of protonated and deprotonated species in the surface accounts for neutrality. The conduction band-bending at this pH_{pzc} depends only on the chemical equilibrium at the interface between the dielectric and the semiconductor interface, which is accounted by the flat band potential (V_{fb}). It is always possible normalization of the device current at pH_{pzc} to account for the band bending due to V_{fb}. At $pH \neq pH_{pzc}$, Ψ_0 will be equilibrated in the semiconductor channel by the mobile charge carriers that will migrate and generate a potential within the semiconductor (Ψ_S).

The term V_{ox} in Equation (3) accounts for the energy accumulated across the dielectric barrier. It can be expressed using the charge in the semiconductor side and considering a planar condenser:

$$V_{ox} = \frac{q N_A W_D}{C_{ox}} \tag{6}$$

where C_{ox} is the area capacitance of the dielectric barrier (typically a metal oxide) described in Section 2.1, q is the elementary charge, N_A is the density of dopants in the semiconductor and W_D is the region in the semiconductor channel depleted from carriers shown in Figure 1c in darker grey color. W_D can be derived solving the Poisson equation for Ψ_S with a planar configuration:

$$\frac{\partial^2 \Psi_S}{\partial x^2} = \frac{-q N_A}{\varepsilon_s \varepsilon_0}\bigg|_{x=0}^{x=W_D} \longrightarrow \Psi_S = \frac{q N_A W_D^2}{2\varepsilon_s \varepsilon_0} \tag{7}$$

where ε_s is relative dielectric constant of the semiconductor. We have replaced $\varepsilon_s \varepsilon_0 = \varepsilon_{Si}$ as p-type doped Silicon is used as a semiconductor channel for this work. Note that W_D changes the region populated with carriers and thus can modulate the conductivity of the FET channel. Combining Equations (3), (6) and (7), we have the following dependence:

$$\Psi_0 + V_G = \frac{q N_A W_D}{C_{ox}} + \frac{q N_A W_D^2}{2\varepsilon_{Si}} \tag{8}$$

In Equation (8), W_D changes with respect to the pH through the dependence of Ψ_0 with the acidity expressed in Equation (4) and so it is possible to get a final expression for W_D as a function of the pH:

$$W_D = -\frac{\varepsilon_{Si}}{C_{ox}} + \sqrt{\left(\frac{\varepsilon_{Si}}{C_{ox}}\right)^2 + 2\left(\frac{\varepsilon_{Si}}{q N_A}\right)(\Psi_0 + V_G)} \tag{9}$$

A drift in the current will be observed due to the dependence of W_D with the degradation of the different parameters. In particular, the parameters from Equation (9) which can be responsible for the drift are: (i) the changes in the dielectric material and thus in

C_{ox} due to the possible penetration of ions or modifications of the dielectric (Section 2.1); and (ii) the changes in the sensitivity (α) of the material, mainly due to the modifications in β_{diff} because of the degradation of the surface with absorbed molecules that change the number of sites (N_S) for the binding of protons. In this work, the sensitivity has been calculated assuming the ideal sensor (in order to equally compare the oxides) and making use of an iterative method with respect to Ψ_0 to consider the real sensor. Accordingly, we have focused on studying the impact of different penetration of ions in C_{ox}.

3. Results and Discussion

3.1. Impact of the Dielectric Degradation on the Depletion Width of Different Materials

Based on the analytical model described above, we have simulated the four devices detailed in Section 2.1 grouped in couples having two thicknesses (5 nm and 10 nm) for each material (SiO_2 or HfO_2). We have calculated the effect of the degradation in W_D as a function of the pH. As a first step, to simplify the study of the drift from other effects like the combination of the chemical affinity with the changes in the electrolyte, we have considered the case with ideal sensitivity ($\alpha = 1$).

We used Equation (9) to calculate the parameters of the semiconductor channel, considering a desirable dynamic range from 2 to 12. Thus, considering a p-doped semiconductor channel that is going to be depleted in acidic conditions, we calculated a bias external voltage V_G necessary to have full conductivity ($W_D = 0$ nm) at pH = 12, and calculated the value for both oxide materials. N_A was chosen to have a depletion region of $W_D = 100$ nm at pH = pH_{pzc} considering the devices with $t_{ox1} = 5$ nm.

Figure 2 shows W_D vs. pH for the interfaces described in figure 1 using the designated colour codes. A tone scale convention from darker to a lighter colour for increasing x has been added and will be maintained hereafter. As expected, W_D decreases with pH in all the devices as a result of the effect of the adsorbed protons. Comparing SiO_2 and HfO_2, the latter has a larger variation across the pH dynamic range due to the higher dielectric constant. The impact of the drift caused by the degradation on the pH determination by each of the devices is clearly observed in these graphics, as for a single depletion width, there are a broad number of possible pH values corresponding to different states of degradation in the material. For instance, if the constant $W_D = 60$ nm is considered (solid orange line in Figure 2), the pH uncertainty between the cases of no degradation ($x = 0$ nm) and the maximum degradation considered ($x = 3$ nm) are ΔpH = 3.15 and ΔpH = 3.30 for the SiO_2 dielectric with $t_{ox1} = 5$ nm and $t_{ox2} = 10$ nm, respectively; whereas the uncertainty is dramatically reduced to ΔpH = 0.30 and ΔpH = 0.31 for the HfO_2 dielectric with $t_{ox1} = 5$ nm and $t_{ox2} = 10$ nm, respectively. In both cases, ΔpH is slightly higher for the device with $t_{ox1} = 10$ nm.

On the other side, as considering Figure 1b, HfO_2 devices can offer a better pH resolution as the current range of pH values possible relative to the total current variation in the dynamic range, is much more restricted than for SiO_2. In addition, even if it is not taken into account by these simulations, the chemical stability of HfO_2 largely exceeds the one of SiO_2, and thus is less prone to ion penetration which makes that the degradation occurs in longer time periods. Elseways, SiO_2 has proportionally a larger variation of W_D as the degradation increases. In order to resolve the absolute pH, we intend to determine the degradation considering the current from a dual device composed of the two sensors with one of the two materials that we had calculated. In this sense, SiO_2 may have the advantage to determine an absolute pH as it provides proportionally larger current contrasts within a given pH range.

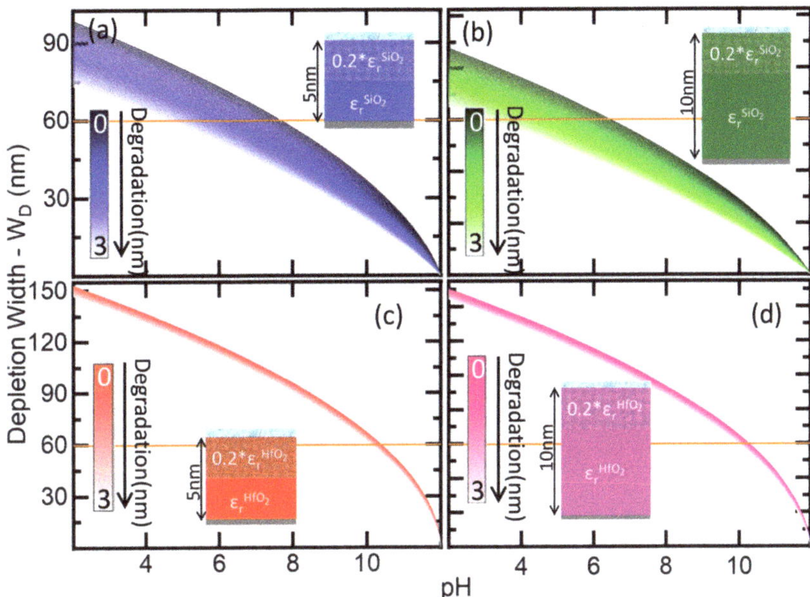

Figure 2. Depletion Width (W_D) as a function of the pH considering different degraded region in the oxide (x) from $x = 0$ nm (non-degraded oxide) to $x = 3$ nm for the two different oxides [(**a**)/(**b**) SiO_2 and (**c**)/(**d**) HfO_2] and two different ideal biosensors ($\alpha = 1$) which main difference is the oxide thickness [(**a**)/(**c**) $t_{ox1} = 5$ nm and (**b**)/(**d**) $t_{ox2} = 10$ nm]. The solid orange line represents the example of the variation of the pH range for a constant $W_D = 60$ nm. The calculated gate bias for the above simulations is 0.3825 V.

3.2. Determination of Absolute pH from Current Acquisition in FET Sensors

To illustrate the determination of pH in a case scenario, we used our model to calculate the response of a pair of sensors with the geometry of a high aspect ratio FinFET shown in Figure 3a. This ISFET geometry has been recently proposed by us as a robust and advanced design for a biosensor [27]. Similarly to single Silicon-Nanowire (SiNW), this geometry offers a three-dimensional direct gating which is advantageous with respect to typical planar devices or extended gates. Respect to the nanowires (NWs), the high aspect ratio FinFETs can also improve: (i) the reproducibility of the sensitivity for ion sensing (pH), (ii) the total signal, and (iii) the linearity of the current response. Moreover, high aspect ratio FinFETs have better linearity and a smaller footprint if compared to NW arrays. Due to the planar configuration of the conduction channel, the influence of small defects in pH sensing is localised and negligible for the sensor signal if compared to their influence in nanoscale SiNWs. For our work, we have chosen the device dimensions similar to the one shown in Figure 3a, where the width W was 200 nm, the height h was 2 μm and the length L was 10 μm.

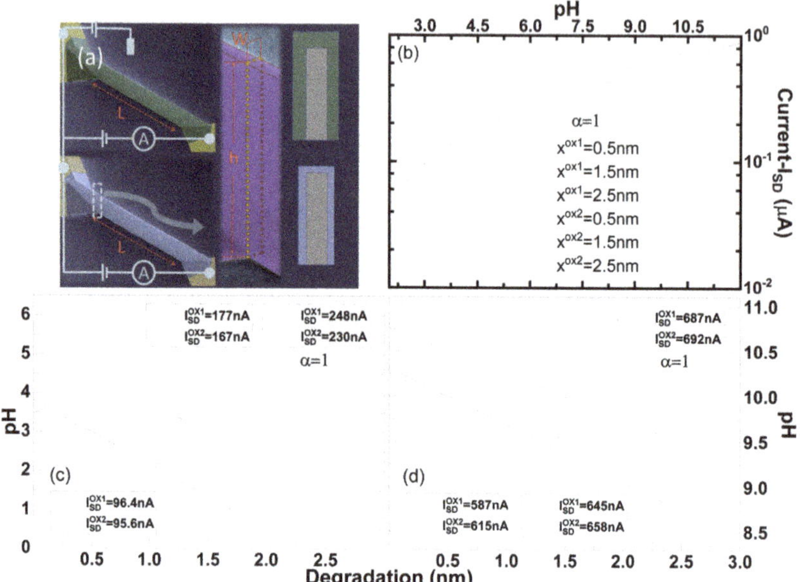

Figure 3. (**a**) SEM pictures from a typical FinFET device fabricated in LIST, schematically showing the electrical connections and the dimensions. In our work W, h and L have been chosen 200 nm, 2 µm and 10 µm, respectively. (**b**) Current (I_{SD}) as a function of the pH considering three degraded regions in the oxide ($x = 0.5$ nm, $x = 1.5$ nm, and $x = 2.5$ nm) for SiO$_2$ and two different ideal biosensors ($\alpha = 1$) having main difference is the oxide thickness ($t_{ox1} = 5$ nm and $t_{ox2} = 10$ nm). (**c**,**d**) Calculated pH as a function of the degradation (x) for SiO$_2$ and two different ideal biosensors ($\alpha = 1$) having main difference is the oxide thickness ($t_{ox1} = 5$ nm and $t_{ox2} = 10$ nm). The pH has been calculated considering the W$_D$ given by Equation (9) with three degraded regions in the oxide ($x = 0.5$ nm, $x = 1.5$ nm, and $x = 2.5$ nm) and an initial (**c**) pH=3 and (**d**) pH = 10. The solid orange line represents the constant initial pH. Drain bias equals 50 mV and calculated gate bias equals 0.5914 V are used for the simulation.

For a given FinFET, W_D can be related to the measured current depending on the geometry of the sensor considering that the size of the channel is diminished across the cross-sectional area by W_D in all the directions perpendicular to the surfaces in contact with the electrolyte, and then the total current (I_{SD}) can be calculated as:

$$I_{SD} = \sigma \frac{A}{L} V_{SD} \tag{10}$$

where σ, A, and L are the conductivity (material property), the cross-section, and the length of the silicon channel , respectively. At the point of zero charge, A coincides with the geometrical dimensions of the FinFET channel ($A = W \times h$) as pH increases [$A = (W - 2W_D) \times (h - W_D)$]. In this way, W_D is connected to the experimental data using the original geometrical cross-section and the actual resistance of the channel ($\rho = 1/\sigma$). Given the large aspect ratio, we have considered $h \gg W_D$ and thus we have approximated as $A = (W - 2W_D) \times (h)$. This possibility to neglect the depletion width in one direction, is indeed the origin of the higher linearity of the high aspect ratio FinFET respect to NWs described in our works [27,28]. Figure 3a shows two SEM pictures from different perspectives of a typical high aspect ratio FinFETs in which we have included schematics showing the electrical connections and the geometrical parameters W, h and L.

Figure 3b shows I_{SD} vs. pH for the pair of devices with silica dielectric at three different degradation points ($x = 0.5$ nm, $x = 1.5$ nm and $x = 2.5$ nm shown in darker to lighter colours and using solid blue lines for the thinner sensor and green dashed lines for the thicker sensor). We illustrate that at an arbitrary pH, the acidity can be unequivocally determined using the current values that intersect for example the orange lines in Figure 3b–d that mark a constant pH for values 3 and 10. For each of these pH values and for each state of degradation (x), Figure 3b provides a pair of current values that will be observed on the pair of sensors at the intersection with the indicated orange lines with each of the curves respective to each degradation x. For each sensor alone, there are several combinations of pH vs. degradation x that provide such currents values, but only at one point, a pair of currents converge with the same pH and degradation x. Figure 3d shows the equivalent situation for pH 10.

Both Figures 2 and 3 show that at more basic pH values, the differences in signal between devices of the same material becomes smaller, and thus discriminating the value of the currents for each state of degradation becomes more difficult depending on the values of the noise signal. By this effect, the determination of pH is also more affected by the noise signal at a more basic pH as the acidity has not acted on the surface potential that builds the depletion width W_D. This becomes also apparent comparing the range of pH variation for a given current in Figure 3c,d. The range of pH in the degradation span of our studies for each current is nearly three times larger for the pair of current values acquired at pH 3 (Figure 3c) than for the ones at pH 10 (Figure 3d), showing that the degradation can depend less in the measuring error in the first case. It is to be noted, that the current map calculations shown in Figure 3 using our simultaneous of current vs. pH, are equivalent to an experimental-mapping in a pair of devices with the same fabrication parameters except for oxide thickness, and where the simultaneous current response would be mapped during the degradation of the oxide. In such a case, we would bet a current map equivalent to Figure 3b. Given the broader response of SiO_2 to the degradation, it would require less precision on the determination of the current to obtain a match in the current response to a single pH compared to materials with less change with degradation, as for example the case of HfO_2. However, the lifetime and the variability of the sensor over time would still be beneficial for the material with higher chemical stability and dielectric constant.

3.3. Implementation of the Proton Affinity on the Sensor Response for Non-Linear Sensitivities

The simplification of ideal sensitivity $\alpha = 1$ predicts an excessive sensitivity and fails to describe the effect of saturation of proton adsorption that occurs at acidic concentrations. Materials like SiO_2 decrease in α due to the saturation of the silanol groups accepting protons. In this section, we are presenting results calculating the chemical response of the sensor to describe the best way to operate these devices To include the chemical sensitivity of the materials and the effects of the electrolyte, it is necessary to include the values of β_{diff} and C_{diff}. The buffering capacitance is calculated from the site-binding model:

$$\beta_{diff} = 2.303 \cdot a_{H_s} \cdot N_S \frac{K_b a_{H_s}^2 + 4 K_a K_b a_{H_s} + K_a K_b^2}{\left(K_a K_b + K_b a_{H_s} + a_{H_s}^2 \right)^2} \tag{11}$$

where a_{H_s} is the surface proton activity that depends on the bulk pH and on the surface potential. C_{diff} can be estimated with the Gouy-Chapman-Stern approximation. It is calculated as the Stern capacitance (C_{Stern}) in series with the double layer capacitance (C_{DL}). $C_{Stern} = 0.8\frac{F}{m^2}$ has been considered in this work in order to consider a realistic behaviour of an ISFET [6]. C_{DL} is calculated by:

$$C_{DL} = Q_o cosh\left(\frac{q\Psi_0}{2k_B T}\right) = \sqrt{\frac{2\varepsilon_W I_0 N_{avo} q^2}{k_b T}} cosh\left(\frac{q\Psi_0}{2k_B T}\right) \tag{12}$$

where ε_W is the electrolyte permittivity, I_0 is the ion concentration in mol/L, and N_{avo} is the Avogadro constant. As β_{diff} and C_{diff} depend on the surface potential α and Ψ_0 are computed in a self-consistent loop described above.

Figure 4 shows the relationship between α, Ψ_0 and pH. The model is able to reproduce the saturation of the surface potential observed in our previous experiments [27–29]. At pH = 2, which is the point of zero charge for SiO_2, α has its minimum value. The origin of the surface potential Ψ_0 is also set equal to 0 at pH 2. Figure 4 shows that with increasing of the pH, the value of α becomes close to 1.0 (α must have value between 0 and 1) and the surface potential Ψ_0 increases to a higher negative value. As the acidity is increased, the decrease in α results in the saturation of change in Ψ_0. It is also to be noted, that contrary to what is assumed in most cases, the behaviour of Ψ_0 is not linear through the pH range, and that has singularities due to the interplay of proton affinities with the double-layer capacitance.

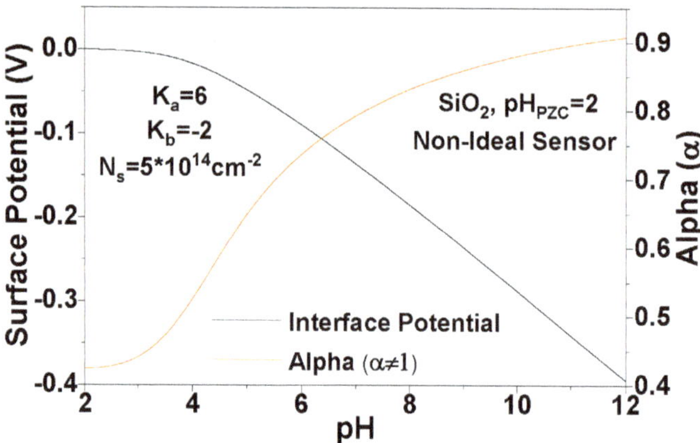

Figure 4. Surface potential (Ψ_0) and sensitivity parameter (α) as a function of the pH calculated using the iterative method for a non-ideal sensor considering only SiO_2.

Figure 5 shows I_{SD} vs. pH for a device couple of SiO_2 with corrected α. As in the case of the ideal sensitivity, the current values in both devices converge to the maximum at basic conditions due to the vanishing W_D. Contrary, when the pH is very acidic (pH = 2 or 3), there is a larger drift of the current values with x. The effect of the drift is even larger for the device with t_{ox1} in comparison to the t_{ox2}. This is expected due to the larger proportion of degraded material in the device with oxide thickness equals to 5 nm. Another interesting point is that both devices have almost identical current profile for all pH values at maximum degradation of 3 nm (I_{SD} vs. pH curves with lighter colours). Hence, it seems that once the degraded region of the oxide dominates the contribution of the capacitance. The effects of the saturation of the sensitivity α are also observed in the acidic range for both currents simulated in Figure 5 as the variation of I_{SD} vs. pH decreases as the pH becomes more acidic. This loss of sensitivity affects also the determination of pH, as for a given noise signal, more pH values will fall within the range of error. However, this is a property of the material observed in the saturation of the surface potential in Figure 4 which cannot be resolved with a different operation mode.

The lines of constant I_{SD} at 20% of the total conductance (I_{SD} = 0.2 µA) are indicated as horizontal orange lines in Figure 5. It can be noticed, that the pH uncertainty associated to that measurement is much greater for the device with thinner oxide.

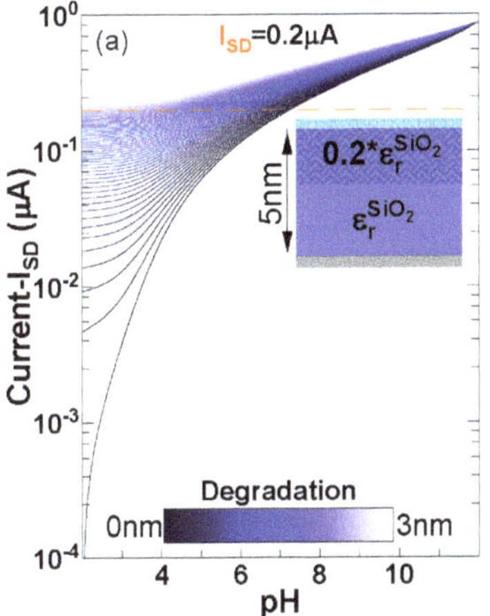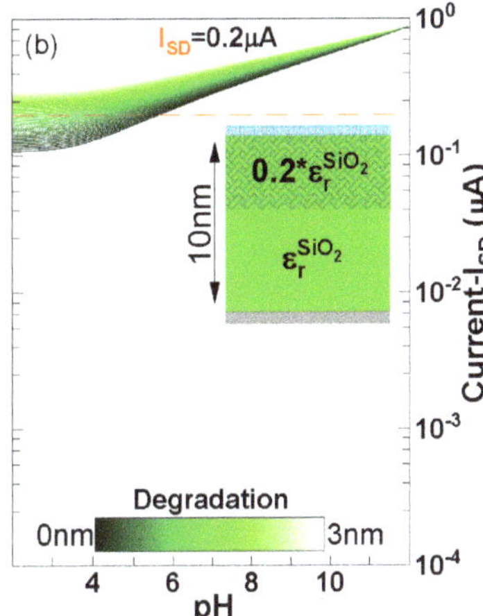

Figure 5. Current (I_{SD}) as a function of the pH considering different degraded region in the oxide (x) from x = 0 nm (non-degraded oxide) to x = 3 nm for SiO_2 and two different non-ideal biosensors (α is self-consistently computed using the iterative method as shown in Figure 4) which main difference is the oxide thickness ((**a**) t_{ox1} = 5 nm and (**b**) t_{ox2} = 10 nm). The solid orange line represents the example of the variation of the pH range for a constant I_{SD} = 0.2 µA. Drain bias equals 50 mV and calculated gate bias equals 0.3825 V are used for the simulation.

3.4. Optimisation of pH Determination Using a I_{SD} Follower in One of the Sensors

The current response obtained in Figure 5a,b can be used to reproduce the plan of action described at the end of Section 3.2 to obtain the absolute pH. However, using a constant gate voltage is detrimental to the accuracy at more basic pH values due to the similar values between currents at different x because of the small values of W_D. The traditional method to measure the acidity follows the surface potential Ψ_0 by compensating with a voltage bias applied between the channel and the reference electrode to maintain a constant current, usually closed to one obtained with the threshold voltage of the transistor (maximum W_D), but not too low as to increase the signal to noise ratio.

Figure 6a,b show the calculation of the gate voltage correction to maintain a current of 0.2 µA (equivalent to a W_D of 80 % of the width of the sensor) as a function of pH and for all the states of degradation within the range of our study for the devices with thinner and thicker dielectrics, respectively. The different curves of V_G vs. pH for each state of degradation are parallel to each other, showing an opposite behaviour to Ψ_0 (shown in Figure 4) to compensate the charge accumulated due to the pH.

In order to use Figure 6a,b as a map of values to determine the pH, we have to take into consideration that there is only a common reference electrode in the system. Consequently, only one of the devices can be kept at a constant current I_{DS}. Here, we have arbitrarily chosen to maintain constant the device with the smaller oxide thickness, and use the map in Figure 6a corresponding to a particular pH and state of degradation, while using the obtained values of V_G and parameter of degradation x to calculate the current that corresponds in the second device. Figure 6c,d show the possible values of pH vs. degradation that could be obtained at the values mapped for pH 3 and 10, respectively, for both sensors shown in blue and green for the 5 nm and 10 nm sensors, respectively. We have extracted the values of V_G and current of the second sensor obtained at the levels of degradation of 0.5 nm, 1.5 nm and 2.5 nm. It can be observed that equally to the method of the current, for each pair of devices, there is a single point that determines the pH and the parameter of degradation. Comparing the slopes obtained in the pH determination using the current output Figure 3c,d with the ones obtained with the mapping of V_G and the current of the second sensor in Figure 6c,d, we can notice that the later have a steeper slope.

This signals also the better determination as determine range of degradation corresponds to a shorter range of pH with the best precision acquired in Figure 6c of the determination only with the current. Thus, using current as the mapping parameter for an unknown variable "oxide degradation" and controlled variable "operating bias", we were able to accurately determine the pH value with the help of two similar devices with different oxide thickness.

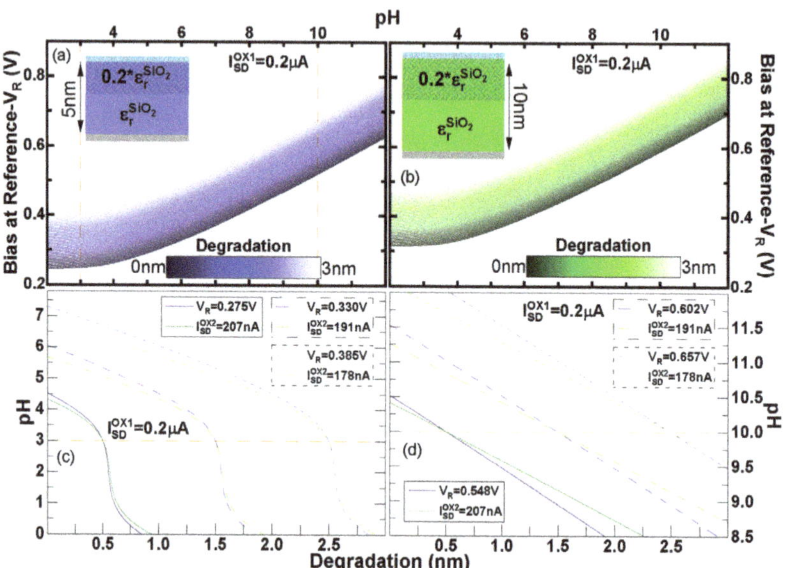

Figure 6. (**a**,**b**) External bias at the backgate (V_G) as a function of the pH calculated using the Equation (3) for a constant I_{SD} = 0.2 μA considering different degraded region in the oxide (x) from x = 0 nm (non-degraded oxide) to x = 3 nm for SiO$_2$ and two different non-ideal biosensors which main difference is the oxide thickness ((**a**) t_{ox1} = 5 nm and (**b**) t_{ox2} = 10 nm). (**c**,**d**) Calculated pH value as a function of the oxide degradation (x) for SiO$_2$ [having different external bias (V_G) and current (I_{SD})] and two different non-ideal biosensors which main difference is the oxide thickness (t_{ox1} = 5 nm and t_{ox2} = 10 nm). The V_G has been calculated using Equation (3) considering the W_D given by Equation (9) with three degraded regions in the oxide (x = 0.5 nm, x = 1.5 nm, and x = 2.5 nm) and an initial (**c**) pH = 3 and (**d**) pH = 10. The solid orange line represents the constant V_G in which the curves for both devices cross. In all the figures (**a**–**d**), α is self-consistently computed using the iterative method as shown in Figure 4.

4. Conclusions

In this paper, we have developed an analytical model to calculate the surface potential and the current response of ISFETs to pH. We have implemented the effect of the degradation at the dielectric barrier that induces the current drift. The derived model used a capacitance representing the degraded region which is adjusted with a phenomenological effective dielectric constant and depth connected in series with the capacitance of the rest of non-degraded material with the original properties. We calculated the response of the degradation of the capacitance for two materials, SiO$_2$ and HfO$_2$ as examples of low and high dielectric constants, respectively. The relative effect of the degradation is higher for materials with lower dielectric strength. Further, without any correction, the materials with a higher dielectric constant have less uncertainty of the measured pH.

Using the modification of the capacitance with degradation, we propose a method to determine the absolute pH using a mapping of dual sensor response. In our paper, we have used a mapping with calculations equivalent to a mapping that would be produced experimentally with reproducible devices. To simulate the effects of the chemical response of the materials, we have implemented the site-binding model interacting with stern and double layer capacitances. This model does not take into account the modification of binding sites at the interface of the dielectric and the electrolyte. We have shown that using a common reference electrode at constant voltage, the current values are less accurate to determine the pH at basic pH where there is less action of the acid and less depleted region in the semiconductor. This effect can be partially corrected using the voltage of the reference electrode as a current follower for one of the devices. However, in the case of materials like SiO_2, the effect of site-binding saturation at acidic pH also causes a decrease in the sensitivity, which affects also the possibility to determine the absolute pH.

In summary, we have shown a method to determine the absolute pH using dual measurements from two sensors, which can be a breakthrough for the applications of ISFET in physiological monitoring, or quantification of ion activity.

Author Contributions: Conceptualization, C.P.G., D.L.; methodology, C.P.G., C.M.-B., N.K. and V.P.G.; software, C.M.-B. and N.K.; validation, R.P.S.D., D.L. and I.T.; formal analysis, C.P.G., C.M.-B., D.L., N.K. and V.P.G.; visualization, C.P.G., C.M.-B., N.K. and R.P.S.D.; supervision, D.L., C.P.G. and V.P.G. All authors have read and agreed to the published version of the manuscript

Funding: This project has received funding from the European Union's Horizon 2020 research and innovation program under grant agreement No 862539-Electromed-FET OPEN.

Institutional Review Board Statement: Not applicable.

Informed Consent Statement: Not applicable.

Conflicts of Interest: The authors declare no conflict of interest.

References

1. Bergveld, P. Development of an ion-sensitive solid-state device for neurophysiological measurements. *IEEE Trans. Biomed. Eng.* **1970**, *BME-17*, 70–71. [CrossRef]
2. Ihantola, H.; Moll, J.L. Design theory of surface field-effect transistor. *Solid State Electron.* **1964**, *7*, 423–430. [CrossRef]
3. Garrett, C.G.B.; Brattain, W.H. Physical Theory of semiconductor surfaces. *Phys. Rev.* **1955**, *99*, 376–387. [CrossRef]
4. Rigante, S.; Scarbolo, P.; Wipf, M.; Stoop, R.L.; Bedner, K.; Buitrago, E.; Bazigos, A.; Bouvet, D.; Calame, M.; Schönenberger, C.; et al. Sensing with Advanced Computing Technology: Fin Field-Effect Transistors with High-k Gate Stack on Bulk Silicon. *ACS Nano* **2015**, *9*, 4872–4881. [CrossRef]
5. Teng, N.Y.; Wu, Y.T.; Wang, R.X.; Lin, C.T. Sensing Characteristic Enhancement of CMOS-Based ISFETs With Three-Dimensional Extended- Gate Architecture. *IEEE Sens. J.* **2021**, *21*, 8831–8838. [CrossRef]
6. Van Hal, R., Eijkel, J.; Bergveld, P. A general model to describe the electrostatic potential at electrolyte oxide interfaces. *Adv. Colloid Interface Sci.* **1996**, *69*, 31–62. [CrossRef]
7. Nair, P.R.; Alam, M.A. Screening-limited Response of Nanobiosensors. *Nano Lett.* **2008**, *8*, 1281–1285. [CrossRef]
8. Kulkarni, G.S.; Zhong, Z. Detection beyond the Debye Screening Length in a High-Frequency Nanoelectronic Biosensor. *Nano Lett.* **2012**, *12*, 719–723. [CrossRef]
9. Zhang, J.; Rupakula, M.; Bellando, F.; Garcia Cordero, E.; Longo, J.; Wildhaber, F.; Herment, G.; Guérin, H.; Ionescu, A.M. Sweat Biomarker Sensor Incorporating Picowatt, Three Dimensional Extended Metal Gate Ion Sensitive Field Effect Transistors. *ACS Sens.* **2019**, *4*, 2039–2047. [CrossRef]
10. Sakata, T. Biologically Coupled Gate Field-Effect Transistors Meet in Vitro Diagnostics. *ACS Omega* **2019**, *4*, 11852–11862. [CrossRef]
11. Hashim, U.; Chong, S.W.; Liu, W.W. Fabrication of Silicon Nitride Ion Sensitive Field-Effect Transistor for pH Measurement and DNA Immobilization/Hybridization. *J. Nanomater.* **2013**, *2013*, 542737. [CrossRef]
12. Romele, P.; Gkoupidenis, P.; Koutsouras, D.A.; Lieberth, K.; Kovács-Vajna, Z.M.; Blom, P.W.M.; Torricelli, F. Multiscale real time and high sensitivity ion detection with complementary organic electrochemical transistors amplifier. *Nat. Commun.* **2020**, *11*, 3743. [CrossRef] [PubMed]
13. Kim, S.; Kwon, D.W.; Kim, S.; Lee, R.; Kim, T.H.; Mo, H.S.; Kim, D.H.; Park, B.G. Analysis of current drift on p-channel pH-Sensitive SiNW ISFET by capacitance measurement. *Curr. Appl. Phys.* **2018**, *18*, S68–S74. [CrossRef]
14. Jakobson, C.; Feinsod, M.; Nemirovsky, Y. Low frequency noise and drift in Ion Sensitive Field Effect Transistors. *Sens. Actuators B Chem.* **2000**, *68*, 134–139. [CrossRef]
15. Jamasb, S. Current-Mode Signal Enhancement in the Ion-Selective Field Effect Transistor (ISFET) in the Presence of Drift and Hysteresis. *IEEE Sens. J.* **2021**, *21*, 4705–4712. [CrossRef]

16. Esashi, M.; Matsuo, T. Integrated micro multi ion sensor using field effect of semiconductor. *IEEE Trans. Biomed. Eng.* **1978**, 184–192. [CrossRef]
17. Matsuo, T.; Esashi, M. Methods of ISFET fabrication. *Sens. Actuators* **1981**, *1*, 77–96. [CrossRef]
18. Bousse, L.; Bergveld, P. The role of buried OH sites in the response mechanism of inorganic-gate pH-sensitive ISFETs. *Sens. Actuators* **1984**, *6*, 65–78. [CrossRef]
19. Buck, R.P. Kinetics and drift of gate voltages for electrolyte-bathed chemically sensitive semiconductor devices. *IEEE Trans. Electron Devices* **1982**, *29*, 108–115. [CrossRef]
20. Elyasi, A.; Fouladian, M.; Jamasb, S. Counteracting threshold-voltage drift in ion-selective field effect transistors (ISFETs) using threshold-setting ion implantation. *IEEE J. Electron Devices Soc.* **2018**, *6*, 747–754. [CrossRef]
21. Böhm, S.; Timmer, B.; Olthuis, W.; Bergveld, P. A closed-loop controlled electrochemically actuated micro-dosing system. *J. Micromech. Microeng.* **2000**, *10*, 498. [CrossRef]
22. Briggs, E.M.; Sandoval, S.; Erten, A.; Takeshita, Y.; Kummel, A.C.; Martz, T.R. Solid state sensor for simultaneous measurement of total alkalinity and pH of seawater. *ACS Sens.* **2017**, *2*, 1302–1309. [CrossRef]
23. Sinha, S.; Sahu, N.; Bhardwaj, R.; Ahuja, H.; Sharma, R.; Mukhiya, R.; Shekhar, C. Modeling and simulation of temporal and temperature drift for the development of an accurate ISFET SPICE macromodel. *J. Comput. Electron.* **2020**, *19*, 367–386. [CrossRef]
24. Bhardwaj, R.; Sinha, S.; Sahu, N.; Majumder, S.; Narang, P.; Mukhiya, R. Modeling and simulation of temperature drift for ISFET-based pH sensor and its compensation through machine learning techniques. *Int. J. Circuit Theory Appl.* **2019**, *47*, 954–970. [CrossRef]
25. Jamasb, S.; Collins, S.; Smith, R.L. A physical model for drift in pH ISFETs. *Sens. Actuators B Chem.* **1998**, *49*, 146–155. [CrossRef]
26. Schlom, D.G.; Guha, S.; Datta, S. Gate Oxides Beyond SiO2. *MRS Bull.* **2008**, *33*, 1017–1025. [CrossRef]
27. Rollo, S.; Rani, D.; Leturcq, R.; Olthuis, W.; Garc, P.; Group, C. A high aspect ratio Fin-Ion Sensitive Field Effect Transistor: compromises towards better electrochemical bio-sensing. *Nano Lett.* **2019**, *19*, 2741–3386. [CrossRef]
28. Rollo, S.; Rani, D.; Olthuis, W.; García, C.P. High performance Fin-FET electrochemical sensor with high-k dielectric materials. *Sens. Actuators B Chem.* **2020**, *303*, 127215. [CrossRef]
29. Rani, D.; Rollo, S.; Olthuis, W.; Krishnamoorthy, S.; Pascual García, C. Combining Chemical Functionalization and FinFET Geometry for Field Effect Sensors as Accessible Technology to Optimize pH Sensing. *Chemosensors* **2021**, *9*, 20. [CrossRef]

Article

Realization of a PEDOT:PSS/Graphene Oxide On-Chip Pseudo-Reference Electrode for Integrated ISFETs

Marcel Tintelott, Tom Kremers, Sven Ingebrandt, Vivek Pachauri and Xuan Thang Vu *

Institute of Materials in Electrical Engineering 1, RWTH Aachen University, Sommerfeldstr. 24, 52074 Aachen, Germany; tintelott@iwe1.rwth-aachen.de (M.T.); tom.kremers@rwth-aachen.de (T.K.); ingebrandt@iwe1.rwth-aachen.de (S.I.); pachauri@iwe1.rwth-aachen.de (V.P.)
* Correspondence: vu@iwe1.rwth-aachen.de; Tel.: +49-241-80-27816

Abstract: A stable reference electrode (RE) plays a crucial role in the performance of an ion-sensitive field-effect transistor (ISFET) for bio/chemical sensing applications. There is a strong demand for the miniaturization of the RE for integrated sensor systems such as lab-on-a-chip (LoC) or point-of-care (PoC) applications. Out of several approaches presented so far to integrate an on-chip electrode, there exist critical limitations such as the effect of analyte composition on the electrode potential and drifts during the measurements. In this paper, we present a micro-scale solid-state pseudo-reference electrode (pRE) based on poly(3,4-ethylene dioxythiophene): poly(styrene sulfonic acid) (PEDOT:PSS) coated with graphene oxide (GO) to deploy with an ion-sensitive field-effect transistor (ISFET)-based sensor platform. The PEDOT:PSS was electropolymerized from its monomer on a micro size gold (Au) electrode and, subsequently, a thin GO layer was deposited on top. The stability of the electrical potential and the cross-sensitivity to the ionic strength of the electrolyte were investigated. The presented pRE exhibits a highly stable open circuit potential (OCP) for up to 10 h with a minimal drift of ~0.65 mV/h and low cross-sensitivity to the ionic strength of the electrolyte. pH measurements were performed using silicon nanowire field-effect transistors (SiNW-FETs), using the developed pRE to ensure good gating performance of electrolyte-gated FETs. The impact of ionic strength was investigated by measuring the transfer characteristic of a SiNW-FET in two electrolytes with different ionic strengths (1 mM and 100 mM) but the same pH. The performance of the PEDOT:PSS/GO electrode is similar to a commercial electrochemical Ag/AgCl reference electrode.

Keywords: PEDOT:PSS; 2D materials; biosensor; stability; gate electrode; diffusion barrier

Citation: Tintelott, M.; Kremers, T.; Ingebrandt, S.; Pachauri, V.; Vu, X.T. Realization of a PEDOT:PSS/ Graphene Oxide On-Chip Pseudo-Reference Electrode for Integrated ISFETs. *Sensors* **2022**, *22*, 2999. https://doi.org/10.3390/s22082999

Academic Editor: Hai-Feng (Frank) Ji

Received: 25 December 2021
Accepted: 11 April 2022
Published: 14 April 2022

Publisher's Note: MDPI stays neutral with regard to jurisdictional claims in published maps and institutional affiliations.

1. Introduction

In recent years, field-effect transistors (FETs) have become one of the most promising sensor platforms for the electronic detection of biomolecules. Besides the well-known ion-sensitive field-effect transistor (ISFET) [1] and its nanoscale counterpart, the silicon nanowire field-effect transistor (SiNW-FET) [2], several new materials, such as graphene [3], reduced graphene oxide [4,5], carbon nanotubes [6], poly(3,4-ethylene dioxythiophene): poly(styrene sulfonic acid) (PEDOT:PSS) [7], or zinc oxide nanowires [8], have been investigated for FET-based (bio)sensing applications. Even though the working principle of these devices is based on different physical phenomena, a stable reference electrode (RE) is mandatory for reliable and reproducible measurements. Commonly, bulky and often fragile electrochemical silver/silver chloride (Ag/AgCl) REs are used to provide a stable gate potential to operate an ISFET for the detection of biomolecules or other analytes. An electrochemical Ag/AgCl RE requires a chloride solution at a given concentration (e.g., 3 M KCl), surrounding the Ag/AgCl wire in a container, and an ion-conductive membrane allowing electrical contact between the electrode and the electrolyte solution. Due to the need for a membrane and an electrolyte solution, the miniaturization of an electrochemical RE remains challenging [9–11].

With the advancement of nanoscale fabrication methods, the transduction area of FET-based sensors has shrunk down to a few tens of nanometers [12,13]. While the sensors and microfluidic systems are becoming smaller, the average sizes of the mandatory RE have remained virtually the same [9]. Apart from the miniaturization on its own, miniaturized and integrated micro-scale REs are expected to eliminate current limitations, such as the use of large sample sizes, microfluidic integration towards Lab-on-a-Chip (LoC) systems, or better portability of (bio)sensor systems [12,14]. An on-chip micro-scale RE should ideally be an entirely solid-state building-block to prevent electrolyte leakages and exert a potential independent of the electrolyte or analyte [9]. Several concepts of on-chip pREs exist, such as Ag/AgCl redox systems or those utilizing the catalytic properties of platinum or iridium oxide. Such pREs, however, either exhibit low potential stability or show high pH or ionic cross-sensitivity [10,15–17].

Conductive polymers have been used in various applications, especially for biological and biochemical sensing applications. PEDOT:PSS is one of the most studied and promising conductive polymers. It is a polymer–polyelectrolyte complex that offers both ion and electron conductivity with semiconducting and redox-active charge conduction properties [18,19], making it a popular material, often used in organic electrochemical transistors (OECTs) [20], light-emitting diodes (LEDs) [21], biohybrid synapse [22], neural probes [23], ion-selective electrodes [24], and ion-pumps [25]. As a material, PEDOT:PSS has shown the ability of coupling between ionic and electronic species [26,27]. The electronic conduction of the π-conjugated PEDOT:PSS is based on weakly bound electrons that can move along a molecule through delocalized π-orbitals and between different molecules if a sufficient π-π overlap is present. However, delocalization and overlap are limited by the structural disorder in the material. In this case, thermally activated hops can describe the nature of the electrical conductivity of PEDOT:PSS. Different models, such as ion hopping, solvated/vehicle, and Grotthuss mechanisms, can describe the ionic current [26,27]. The ionic-electronic interaction is based on either electrostatic ion-electron coupling or direct electron transfer [26,27]. In addition to its electrical properties, the selective coating of PEDOT:PSS (e.g., using electropolymerization) onto metallic substrates allows rapid, low-cost, and high throughput fabrication. Conductive polymers have been used as pREs in various applications, such as electrochemical impedance spectroscopy (EIS) [28,29], ISFETs [30], or OECTs [31]. Due to its ion conductivity, a high cross-sensitivity to the ion concentration of the electrolyte can be expected [32]. Therefore, an ion diffusion barrier that does not degrade the electrode performance but eliminates the cross-sensitivity to ions would be of great interest to increase the stability of polymeric pREs. Here, graphene and graphene derivatives (e.g., graphene oxide (GO)) have proven to be excellent materials to prevent diffusion due to their pinhole-free layers and close interlayer distance packing [33–36]. However, the large-scale deposition of graphene is still challenging [37]. GO, by contrast, allows solution-based processing, which is compatible with standard cleanroom processes and, therefore, enables high throughput and low-cost fabrication [4,38].

In this article, we present a solid-state pRE based on PEDOT:PSS coated with GO as an ion diffusion barrier, which exhibited a highly stable long-term potential and reduced cross-sensitivity to the ionic strength of the electrolyte. Furthermore, we used SiNW-FETs to perform well-known pH experiments. Here, a SiNW-FET gated with the presented pRE exhibited a similar performance as the ones gated with a commercial electrochemical Ag/AgCl RE. To evaluate the cross-sensitivity to changes in ion concentration, measurements were carried out at the same pH but with different ion concentrations. A GO-coated PEDOT:PSS electrode showed a significantly reduced cross-sensitivity to the ionic strength of the electrolyte compared to a bare PEDOT:PSS electrode. Furthermore, we could show that the quality of the GO layer on top of the polymeric electrode has a huge impact on the reliability of SiNW-FETs. The performance of the PEDOT:PSS/GO electrode is similar to a commercial electrochemical Ag/AgCl reference electrode.

2. Materials and Methods

2.1. Materials

Phosphate-buffered saline (PBS) was prepared by dissolving pH buffer capsules (Sigma-Aldrich Chemie GmbH, Taufkirchen, Germany) in deionized (DI) water. Phosphate buffer solutions (pH 7, 100 mM) were prepared by dissolving sodium phosphate dibasic dihydrate and sodium phosphate monobasic monohydrate (Sigma-Aldrich Chemie GmbH, Taufkirchen, Germany) in DI water. The 1 mM phosphate buffer was prepared by the dilution of a 100 mM phosphate buffer using DI water. The pH was measured using a HI5522 pH-meter (Hanna Instruments Deutschland GmbH, Vöhringen, Germany). A leak-free Ag/AgCl double junction RE (DRIREF-2SH, World Precision Instruments, Sarasota, FL, USA) was used as a reference. The GO solution was synthesized using low-temperature exfoliation as described before [39]. The chemical exfoliation was performed using the improved Hummers method.

2.2. Electrode Fabrication

The deposition of PEDOT:PSS was carried out by the electropolymerization of a mixture of 3,4-ethylene dioxythiophene (EDOT) and PSS. Both solutions were dissolved in ultra-pure DI water with a concentration of 20 mM, respectively. An EG&G Model 283 Potentiostat/Galvanostat (Princeton Applied Research, Oak Ridge, TN, USA) was used for depositions of PEDOT:PSS on Ti/Au microelectrodes. Defined charge depositions were carried out potentiostatically at 1 V vs. Ag/AgCl. The potentiostat terminated the deposition when a predetermined total charge was transferred. This deposition method was used to obtain reproducible film characteristics for depositions using the same parameters. Assuming that the ohmic current between the counter electrode (CE) and the working electrode (WE) was negligible in contrast to the current flow caused by the electropolymerization at the electrode, multiple depositions with the same defined charge and electrode area should lead to the same amount of monomers reacting and therefore to the same film thickness. The GO coating of the electrodes was performed using the drop-casting technique. An amount of 10 mL of the GO solution obtained from an exfoliation process developed earlier [4] was drop-casted on the electrode area and dried for 5 min at 50 °C. Accordingly, the choice of charge is dependent on the electrode size and the desired film thickness. Table 1 provides an overview of the four different pREs used in this study.

Table 1. Nomenclature of the investigated pREs.

Electrode Name	Electrode Composition
pRE 1	Au electrode coated with GO
pRE 2	Au electrode coated with PEDOT:PSS (termination charge of 100 µC)
pRE 3	Au electrode coated with PEDOT:PSS (termination charge of 10 µC), additional GO coating
pRE 4	Au electrode coated with PEDOT:PSS (termination charge of 700 µC), additional GO coating

2.3. SiNW-FET Fabrication

The SiNW-FETs were fabricated based on the "top-down" approach using a mix & match process involving electron beam lithography (EBL) and photolithography on a 4-inch wafer scale. Briefly, a silicon-on-insulator (SOI) wafer (Soitec, Bernin, France) with a 70 nm top silicon layer and 145 nm buried-oxide (BOX) was thinned down to ~50 nm by thermal oxidation to define the resulting height of the nanowires. The resulting oxide was used as a hard mask for wet chemical patterning of the top silicon layer [40]. The nanowires and the drain and source regions were defined by EBL and optical lithography using mix & match resist (AR-N7520.11 new, Allresist GmbH, Strausberg, Germany). After the resist development, a CHF_3 dry etching process was carried out to selectively transfer the resist structures onto the silicon oxide hard mask. Tetramethylammonium hydroxide (TMAH) solution (25%) was used to etch the top silicon layer, selectively [41]. An ion implantation process was carried out with a dose of 5×10^{15} atoms/cm^2, implantation energy of 8 keV, and a 7° tilt to form the drain and source regions. A combination of dry oxidation and silicon oxide deposition by high-quality plasma-enhanced chemical vapor deposition (PECVD) was used for the passivation of the drain and source feed lines [42]. The passivation was then etched away on the gate area and the drain and source contact area. A high-quality (dry oxidation process) silicon oxide (~7 nm) was grown on the SiNW gate areas acting as a gate dielectric layer. On-chip temperature sensors and pREs were fabricated using optical lithography combined with a layer stack of chromium, platinum, and titanium lift-off process. A low-temperature oxidation process was carried out to oxidize the top layer of titanium to passivate the temperature sensors and the contact line of the electrode. The TiO$_2$ on top of the electrode was removed by optical lithography and wet etching in buffered HF (BOE 71). A layer stack of aluminum (150 nm), titanium (10 nm), and Au (100 nm) was deposited by electron beam evaporation on the source and drain contact areas to form reliable ohmic contacts with the SiNWs. Before the metal evaporation, an HF-dip was performed to remove native SiO$_2$ from the source and drain contact pads. Finally, the wafer was annealed at 350 °C for 10 min in forming gas (N$_2$/H$_2$) to create ohmic contacts. Further descriptions of the SiNW-FET fabrication process can be found in a previous publication [43].

2.4. Impedance Measurements

The PEDOT:PSS and PEDOT:PSS/GO films were initially investigated using EIS. The electrode under test was connected to the WE while an Ag/AgCl pellet served as the CE, and an electrochemical Ag/AgCl RE was connected to the RE port. A Novocontrol Technologies Alpha-A High-Performance Frequency Analyzer (Novocontrol Technologies GmbH & Co. KG, Montabaur, Germany) was used to measure the impedance spectra. The spectra were obtained in the frequency range between 0.1 Hz and 1 MHz, with an applied voltage amplitude of 10 mV in 1× PBS (pH 7.4) as the electrolyte.

2.5. OCP Measurements

Open Circuit Potential (OCP) was recorded for different electrodes in phosphate buffer solution (1 mM, pH 7) using a Potentiostat/Galvanostat Model 283 (EG&G Instruments, Princeton Applied Research, Oak Ridge, TN, USA). A 2-electrode setup was used to measure the OCP of the electrode under test versus a leak-free Ag/AgCl double junction RE (DRIREF-2SH, World Precision Instruments, Inc., Sarasota, FL, USA).

2.6. Electrical Measurements

The characterization of the SiNW-FETs was performed using a Keithley 4200A-SCS (Keithley Instruments, Solon, OH, USA). A drain-source voltage of –0.1 V was applied between the drain and source terminals of the SiNW-FET. The gate voltage was applied to the RE. In general, reported gate potentials refer to the potential, which is applied to the RE, and not the effective voltage at the transistor gate. Characterizations were carried out with an Ag/AgCl electrode, GO-coated Au electrodes (pRE 1), PEDOT:PSS-coated Au electrodes (pRE 2), and PEDOT:PSS/GO-coated Au electrodes (pRE 3 and pRE 4). The threshold voltage has been extracted using the transconductance extrapolation method (see Figure S1).

3. Results and Discussion

3.1. Electrode Preparation and Characterization

A 1 cm × 1.5 cm chip with a circularly patterned Au electrode, as shown in Figure 1a, was used to study the RE materials. The electrode had a diameter of 500 μm and thus an area of 0.169 mm^2. The fabrication involved the sputter deposition of a 30 nm thick titanium adhesion layer, 220 nm Au, and 50 nm titanium as a protective layer. Afterward, the chip surface was coated with a 3.5 μm thick Parylene C layer. A standard photolithography process was followed by a dry etching process to define the electrode and contact pad areas. This electrode design was chosen to match the electrode design of the fabricated SiNW-FET chip used in this work. The 1 cm × 1 cm SiNW-FET chip consisted of 16 individually addressable SiNW-FETs (~120 nm top width and 6 μm length (compare Figure 1f), monolithically integrated temperature sensors, and on-chip electrodes (Figure 1e).

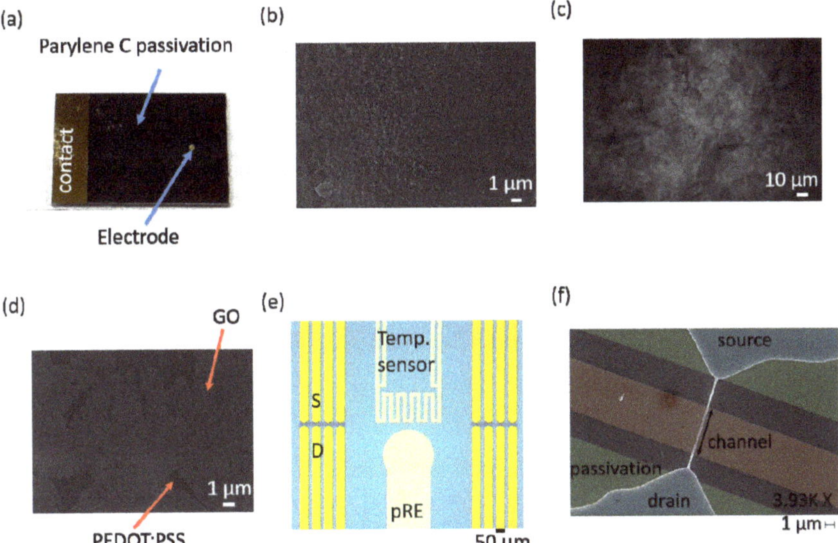

Figure 1. Photograph of the used electrode test structure (**a**). An image of a PEDOT:PSS electrode surface (**b**). SEM image of a PEDOT:PSS electrode coated with GO (**c**). High-resolution image of a GO-coated PEDOT:PSS electrode showing micro holes in the GO film (**d**) Microscopy image showing the SiNW arrays, an integrated temperature sensor, and an on-chip pRE (**e**). SEM image of a single SiNW-FET (**f**).

The PEDOT:PSS was deposited on the designed electrode by electropolymerization of EDOT and PSS in DI water. A former study showed that our charge terminated deposition (combining a surface cleaning step and the actual electropolymerization process) results in a highly reproducible electrode coatings [29]. Before the electropolymerization process, the top Ti layer was etched using ammonium hydroxide-hydrogen peroxide solution to obtain a clean Au surface [44]. A charge terminated electropolymerization process was performed to coat the electrode surface with PEDOT:PSS. Scanning electron microscope (SEM) images were taken after each fabrication step to evaluate the respective process (Figure 1b–d). As shown in Figure 1b, the electropolymerization process resulted in a continuous PEDOT:PSS film. Figure 1c shows an SEM image of a PEDOT:PSS electrode coated with GO. The drop-casting of GO resulted in a continuous coating of the electrode surface (Figure 1c); however, several micro holes in the GO film were observed (Figure 1d).

After each deposition, EIS measurements were performed on samples with different PEDOT:PSS deposition charges in PBS (pH 7.4) to determine the optimal coating parameters. As shown in Figure S2 of the supplementary material, a bare Au electrode exhibited the highest impedance compared to all electrodes coated with PEDOT:PSS. A higher termination charge for the electropolymerization of PEDOT:PSS resulted in a lower impedance, as shown in Figure S2, for termination charges of 10 μC, 1000 μC, and 10,000 μC, respectively. This test was performed to evaluate the electrode performance in dependency on the deposition charge. Even though the electrode coated with a deposition charge of 10,000 μC exhibited the lowest electrode impedance, this high deposition charge is not suitable for pREs. Firstly, the deposition of PEDOT:PSS took several hours, which was not beneficial for high throughput production. Secondly, such a lengthy deposition could induce significant variations in the electrode performance due to current flows, which do not originate from the electropolymerization process itself. The electropolymerization process could be accelerated by using a larger counter electrode. However, using a larger counter electrode resulted in the delamination of the PEDOT:PSS film (see Figure S3). Therefore, electrodes coated with a termination charge of 10 μC and 700 μC were identified for further investigations. Figure 2 shows the electrical impedance spectra of the electrode before and after the deposition of PEDOT:PSS and PEDOT:PSS coated with GO. Here, it can be seen that the deposition of PEDOT:PSS reduced the electrode impedance compared to a bare Au electrode as described before. An additional coating with GO further reduced the electrode impedance in the capacitive regime and slightly increased the impedance in the resistive regime. Furthermore, the percentage impedance change between 10 Hz and 100 kHz has been investigated. Here, a bare Au electrode exhibited an impedance change of ~36,000%, a PEDOT:PSS-coated electrode a change of ~3000%, and a PEDOT:PSS/GO electrode an impedance change of only 272%.

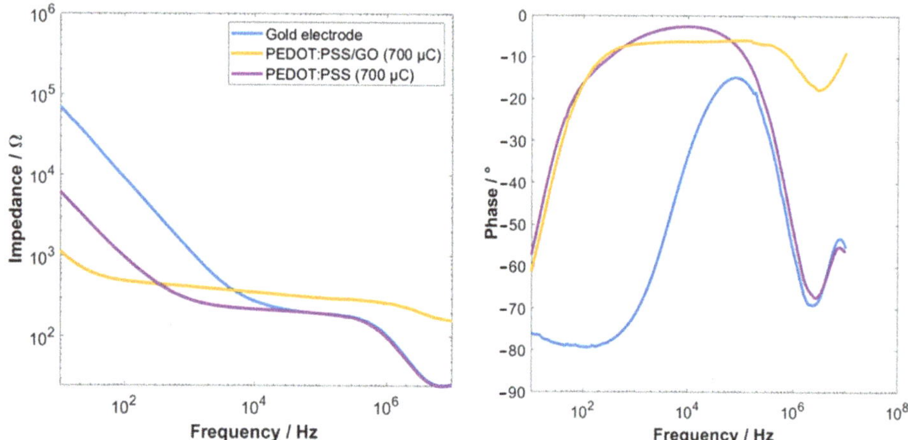

Figure 2. Bode plots of electrochemical impedance spectra (**left**) and phase (**right**) of an Au electrode, a PEDOT:PSS-coated Au electrode, and a GO-coated PEDOT:PSS electrode.

PEDOT:PSS-coated electrodes (termination charge of 10 μC and 700 μC) were coated with GO, using the drop-casting technique. To evaluate the potential stability of these electrodes, OCP measurements were performed in phosphate buffer (pH 7, 1 mM ion concentration). The measurements were performed in a 3D-printed fluidic chamber with insertion slots for the electrode under test and the RE (see Figure S4). The obtained OCPs of four different types of electrodes (PEDOT:PSS (pRE 2), an Au electrode coated with GO (pRE 1), and PEDOT:PSS coated with GO (pRE 3 + pRE 4)), which was recorded for 10 h, are presented in Figure 3. As shown in Figure 3, pRE 2 and pRE 1 exhibited an unstable OCP throughout the measurement with a significant drift. pRE 2 showed a high drifting rate for the first hour and a lower but continuous drifting rate with an overall OCP change of approximately 70 mV. Figure S9 shows additional transient OCP measurements for ~50 min and around 20 h, proving the drifting behavior of PEDOT:PSS-based electrodes. pRE 1 exhibited an OCP change of 10 mV within the first few minutes and remained unstable within the next 7 h, with a total drift of approximately 70 mV. pRE 3 showed an unstable OCP for the first hour and exhibited a lower drifting rate for the next 9 h. Superior results were observed for pRE 4. This electrode exhibited a minimal change in its OCP during the 10 h recording and did not show any significant OCP change in the first minutes after the immersion into the electrolyte. An inset of the first ~2 h can be found in the supporting information (Figure S5), which elucidates the differences in the OCP stability during the first minutes. Three additional OCP recordings for the pRE 4 are shown in the supporting information (Figure S6), proving the stable OCP for short-term measurements (20 min) and long-term measurements (3 h). Furthermore, two additional measurements are shown in the supporting information. Figure S7 shows an OCP recording over 3 days. Here, the electrode (pRE 4) exhibited a stable potential with a minimal drift for 10 h and a significant drift afterwards. As shown in Figure S8, the pRE 4 had the potential for minimal drifting. Here, the electrode exhibited a drift of 0.65 mV/h over the first 10 h and a slightly higher drift for the next 4.5 h. Table 2 shows the change in OCP at different times during the long-term recording shown in Figure 3. The percentage change in the OCP of the different pREs shows that pRE 1 and pRE 3 exhibited a highly unstable behavior within the first minutes. Furthermore, pRE 1 and pRE 2 showed a significant (larger than 39% within 10 h) OCP change. Compared to the starting point, the change in OCP of pRE 4 was always less than 10%.

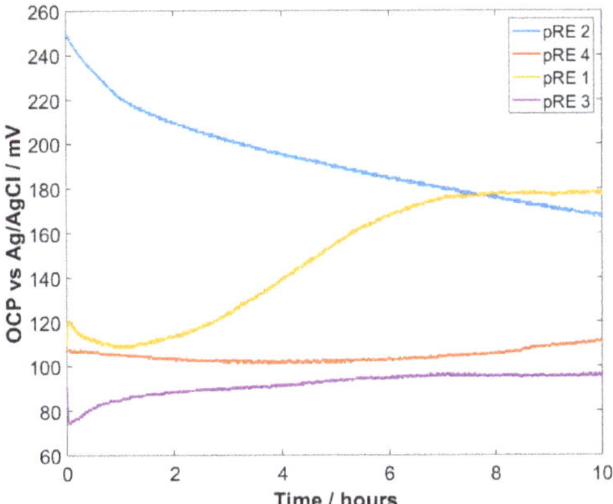

Figure 3. OCP measurements of four different electrodes against an electrochemical Ag/AgCl RE. pRE 4 exhibits the lowest drift, while the other electrodes exhibit an unstable OCP, especially within the first hour.

Table 2. OCP changes at different points of the long-term recording shown in Figure 3.

Electrode Name	1 min	10 min	1 h	5 h	10 h
pRE 1	8.4%	6%	−0.9%	29%	39%
pRE 2	−0.4%	−2.8%	−12.6%	−30%	−48%
pRE 3	−15%	−21%	8.2%	2%	4%
pRE 4	−0.9%	−0.9%	−2.86%	−6.9%	2.7%

To investigate the impact of changes in ion strength on the OPC of a PEDOT:PSS (pRE 2) and a PEDOT:PSS/GO electrode (pRE 4), the OCP measurements were first performed in 1 mM phosphate buffer (pH 7). During the recording of the OCPs, a phosphate buffer with higher ionic strength (100 mM, pH 7) was added to increase the ionic strength of the electrolyte. As shown in Figure 4, the OCP of the PEDOT:PSS electrode (pRE 2) showed a significant response to the addition of the high-concentration buffer. The OCP exhibited a highly unstable potential for almost 1 min after adding the 100 mM buffer and showed an overall potential change of approximately 25 mV. In comparison, the OCP of the GO-coated electrode (pRE 4) showed only a shallow change in its OCP of 1–2 mV.

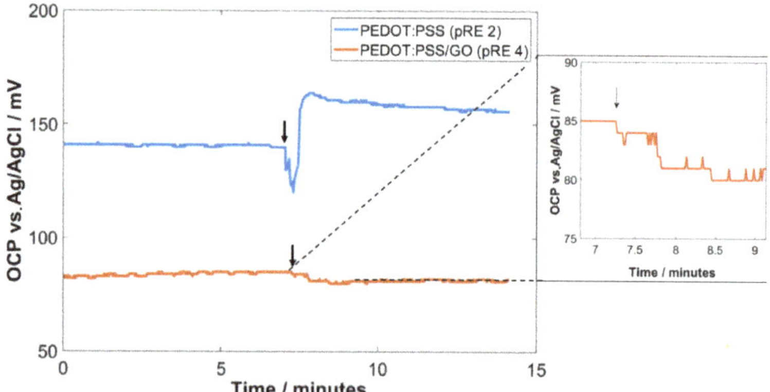

Figure 4. Impact of the addition of higher ionic strength droplets on the OCP of pRE 2 and pRE 4. The black arrows indicate the addition of high ionic strength solution.

The PEDOT:PSS/GO pRE 4 exhibited a highly stable OCP with a shallow drifting rate of 7 mV over 3 h. Compared to the pRE 2, the electrode potential does not need a specific time to become stable and exhibited a much lower drifting. This stable OCP behavior makes the electrode highly suitable for integrated FET biosensing applications. Furthermore, we could prove that the GO layer on top of the PEDOT:PSS thin film hinders the incorporation of ions into the polymer thin film. Therefore, the electrode potential is not dependent on the ion concentration of the surrounding electrolyte. This feature is of high importance when it comes to diagnostics with clinical samples because the ionic strength of these samples can differ from each other.

3.2. Sensing Performance

To evaluate the ability of PEDOT:PSS-based electrodes as solid-state pREs, pH measurements were performed with SiNW-FETs, using the coated electrodes as the gate electrode. The transfer characteristics of the devices were measured with three different pH solutions (pH 4, pH 7, and pH 10) with different ionic strengths. The measurements were repeated four times per pH value (see Figure 5a). Besides using the PEDOT:PSS-based electrode, the pH sensitivity of the SiNW was also characterized using a commercial electrochemical Ag/AgCl electrode as a comparison. The transfer characteristics were obtained by sweeping the gate-source voltage from 0 V to -2 V at a constant drain-source voltage of -0.1 V or -0.5 V. Figure 5b shows the resulting threshold voltage change due to changes in pH using different kinds of pREs. All electrodes exhibited a larger threshold voltage change due to changing the pH from pH 7 to pH 10 compared to changing the pH from pH 4 to pH 7. Non-linear behavior of the pH response was observed due to the non-functionalized SiO_2 surface of the gate oxide layer [45]. Here, the SiNW-FETs gated with our PEDOT:PSS/GO (pRE 4) exhibited only a slightly higher threshold voltage change compared to the devices gated with an electrochemical Ag/AgCl RE. A thinner PEDOT:PSS electrode coated with GO (pRE 3) led to lower threshold voltage changes, while the device showed the highest pH response using pRE 2. Changing the electrolyte from pH 4 to pH 7 resulted in a tiny and unpredictable change in threshold voltage when pRE 2 was used. Due to the significant difference in ionic strength between these two electrolytes, the large variation can be attributed to the remaining cross-sensitivity to ions. In addition, the threshold voltage change due to changes in pH may be superimposed with the remaining ion sensitivity of the pREs. The SiNW-FET exhibited a larger change in threshold voltage when using both PEDOT:PSS/GO electrodes compared to an Ag/AgCl RE. Overall, the SiNW-FET showed much higher reliability when an Ag/AgCl electrode was used compared to all other electrodes. A remaining cross-sensitivity to ions can explain the higher standard deviation of the polymeric pREs. As shown in Figure 1d, the GO film on top of the PEDOT:PSS had some micro-scale holes, which may allow ion diffusion into the PEDOT:PSS layer (illustrated in Figure 5c). In addition, pH measurements for pRE 1 are shown in Figure S10. The SiNW-FET exhibited a clear signal change due to changes in pH (95 mV due to changing the pH from 7 to 10). However, due to the lack of OCP stability, drifting during real-time measurements is expected. To verify this hypothesis, real-time measurements were performed using pRE 1 and pRE 4. As shown in Figure S11, a SiNW-FET gated with pRE 1 exhibits an unstable drain current over time, while a SiNW-FET gated with pRE 4 exhibits a stable drain current after a short drift.

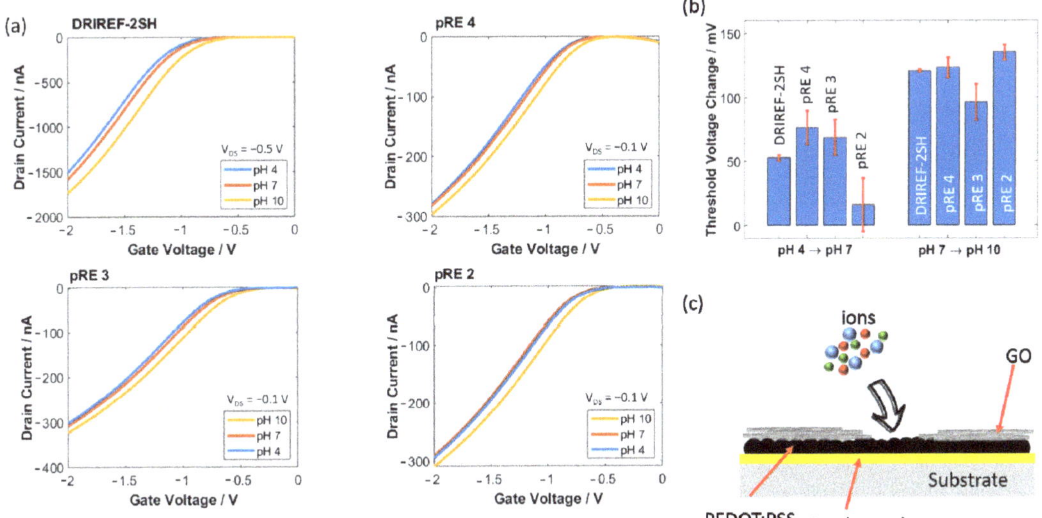

Figure 5. Transfer characteristics at different pH of a SiNW-FET gated with different gate electrodes (a). Threshold voltage change of a SiNW-FET due to changes in pH (b). Schematic illustration of the remaining cross-sensitivity and the reason for the relatively high standard deviation of our pRE (c).

The transfer characteristic of the SiNW-FETs was measured using two phosphate buffer solutions with the same pH (pH 7) and different ionic strengths (1 mM and 100 mM) to evaluate the ionic strength cross-sensitivity of the electrodes. The measurements were carried out for pRE 2 and pRE 4. The change in threshold voltage due to changes in ionic strength for the two electrodes is shown in Figure S12. It can be seen that the gating with a PEDOT:PSS electrode (pRE 2) resulted in a significant threshold shift of approximately 130 mV due to changing the ionic strength from 1 mM to 100 mM. In comparison, a SiNW-FET gated with a GO-coated PEDOT:PSS electrode (pRE 4) exhibited a lower threshold voltage shift of around 50 mV. These results match the above-mentioned OCP measurements. For both results, slight differences in pH need to be taken into account because the pH value of both solutions differed by around 0.2 pH. The remaining shift in threshold voltage identifies a remaining ion sensitivity of the GO-coated pRE. Here, the ion concentration differs by a factor of 100. The difference in ionic strength of the pH solutions used for the measurements shown in Figure 5 was less than a factor of 10. Therefore, the change in threshold voltage (compare Figure 5b) is mainly based on changes in pH with a minor but significant impact of the ion concentration of the electrolyte and a possible OCP drift of the pRE (as shown by the high standard deviation).

A variation between single measurements using the presented pREs was observed, which may limit the use of pREs for sensing applications. This variation can be attributed to micro-scale holes inside the GO film and different immersion times into the electrolyte solution. To overcome the variability between different measurements under the same conditions, the GO coating of pRE 4 was further improved. Here, multiple small droplets were drop-casted on top of the PEDOT:PSS electrode to achieve a higher GO coverage. As shown in Figure 6a, the optimized drop-casting of the GO film results in a dense film without micro-scale holes in comparison to the non-optimized drop-casting (Figure 1d). A SiNW-FET gated with the optimized pRE 4 exhibits a distinguishable change in threshold voltage against electrolyte pH with an extremely reduced standard deviation (see Figure 6b and the inset). Changing the pH from 4 to 7 resulted in a threshold voltage change of 52 ± 3 mV (Ag/AgCl 53 ± 2 mV), and a change in pH from 7 to 10 resulted in a threshold voltage change of 119 ± 2 mV (Ag/AgCl 121 ± 1 mV) (see Figure 6c). These results show that the optimization in the GO film quality has great potential for overcoming the existing limitations of metallic and polymeric-based pREs (e.g., cross-sensitivity to the ionic strength of the electrolyte). Furthermore, due to the comparison of Figure 1d (non-optimized GO coating) and Figure 6a (optimized GO coating) it can be argued that micro-scale holes inside the GO film are the main reason behind the high standard deviation of pH measurements using a non-optimized GO-coated pRE.

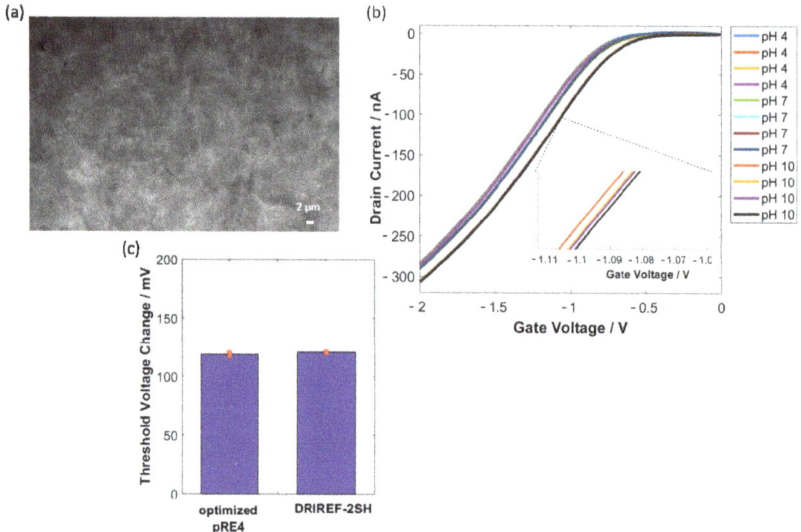

Figure 6. SEM image of the optimized GO coating (**a**). Multiple transfer characteristic measurements at different pH (**b**). Comparison in threshold voltage change due to changing the pH from 7 to 10 for an optimized pRE4 and a commercial Ag/AgCl RE (**c**).

Finally, we compared the potential drifting of our pRE 4 with several pRE approaches presented in the state-of-the-art literature. The comparison is shown in Table 3. Duarte-Cuevara et al. reported pREs based on metal electrodes coated with polypyrrole (PPy) [30]. Here, they have shown that the drifting of electrode potential can be reduced from 23.2 mV/h (only Pt) to 0.75 mV/h due to the coating of a Pt electrode with PPy. According to their finding, the metal has a huge impact on the stability of polymeric pRE. Using Au as an electrode material resulted in a ~2.9 times higher drift compared to a Pt electrode (both coated with PPy). Furthermore, they investigated the drifting of an electrochemical Ag/AgCl reference electrode that was found to be 0.6 mV/h. In comparison, our pRE approach (pRE 4) exhibited a drifting of 0.65 mV/h, which is comparable with the drifting of an Ag/AgCl RE [30]. Furthermore, we compared the achieved pH sensitivity using pRE 4 with other reported pRE concepts. Here, our findings show that the use of a GO-coated PEDOT:PSS electrode exhibited a similar pH sensitivity compared to an ISFET (with hafnium oxide dielectric) gated with a Pt electrode coated with PPy. However, it is noteworthy that the pH sensitivity of different devices should be compared carefully. For instance, the type and quality of the gate oxide are important for the pH sensitivity of such a device [46]. Additional data and information about the possibility of on-chip integration can be found in Table 3 to compare different pRE approaches. The achieved pH sensitivity of 39.7 mV/pH is comparable to the former results of our group [47].

Table 3. Comparison of different pRE approaches in terms of drifting and achieved pH sensitivity of ISFETs gated with pREs.

pRE Concept	OPC Drift	V_{th} Change of pRE Gated ISFETs	Possibility of On-Chip Integration	Refs.
Ag/AgCl reference elelctrode	0.6 mV/h	54.9 mV/pH (hafnium oxide)	yes	[30]
Pt	23.2 mV/h	5.4 mV/pH (hafnium oxide)	yes	[30]
Pt + PPy	0.75 mV/h	44.2 mV/pH (hafnium oxide)	yes	[30]
Au + PPy	2.17 mV/h	-	yes	[30]
Palladium + PPy	0.92 mV/h	-	yes	[30]
Inkjet-printed pRE	4.16 mV/h	-	yes	[48]
Activated Carbon	0.8 mV/day	-	no	[49]
Ag/AgCl screen-printed	0.2 mV/h	-	yes	[50]
Ag/AgCl	0.2 mV/h	-	yes	[51]
PEDOT:PSS/GO	0.65 mV/h	39.7 mV/pH (silicon oxide)	yes	This work

4. Conclusions and Outlook

We have demonstrated that the coating of PEDOT:PSS electrodes with GO resulted in a much better pRE performance compared to bare PEDOT:PSS electrodes. The stability of the OCP was investigated by long-term measurements. The PEDOT:PSS/GO electrode exhibited a shallow drifting rate of 5 mV over the first 6 h and a slight drifting for the next 4 h and exhibited an overall constant electrode potential in most of the measurements. However, the pRE 4 configuration showed the potential of a low-drifting gate electrode with a minimal drift of 0.65 mV/h, which is comparable with state-of-the-art pREs. Transient OCP measurements were carried out by adding a 100 mM phosphate buffer (10% of the initial volume) to a 1 mM phosphate buffer of the same pH to evaluate the cross-sensitivity of the ionic strength of an electrolyte to the OCP of the electrode. A significant change in the OCP was observed for the PEDOT:PSS electrode, while the GO-coated PEDOT:PSS electrode showed a relatively stable behavior. The slight change in the OCP of GO-coated PEDOT:PSS electrodes can be attributed to micro-scale holes inside the GO film. Together with an electrochemical Ag/AgCl RE, the PEDOT:PSS and the GO-coated PEDOT:PSS electrodes were used with SiNW–FETs to characterize the pH sensitivity of the SiNW-FET for solutions with different ionic strength. The pH response of the SiNW-FET using the GO-coated PEDOT:PSS electrodes resulted in a similar pH response to the commercial electrochemical RE, while the SiNW-FET gated with a PEDOT:PSS electrode exhibited a partly unpredictable pH response. A change in the ionic strength led to a change in the threshold voltage of the SiNW-FET; however, the GO-coated PEDOT:PSS electrode showed an almost three times lower change compared to the PEDOT:PSS electrode. Furthermore, we have shown that the optimization in drop-casting of the GO resulted in a highly reliable and reproducible pH response. A SiNW-FET gated with a GO-coated PEDOT:PSS electrode exhibited the same pH response as one gated with a commercial Ag/AgCl RE.

In conclusion, a combination of GO with PEDOT:PSS by initial electropolymerization of PEDOT:PSS on a metal electrode and a subsequent coating of the electrode with GO has improved the performance of the electrode by lowering drifting of the OPC and eliminating the interference of the ionic strength to the OCP, a crucial characteristic of an RE. We assume that the GO has a function to stop the diffusion of ions in the electrolyte to the PEDOT:PSS layer underneath while maintaining the insensitivity to the pH of the PEDOT:PSS layer. One major drawback of the presented pRE is the remaining, but much lower, drifting and cross-sensitivity to the ionic strength of an electrolyte solution. The optimization of the GO coating further reduced the cross-sensitivity to the ionic strength of the analyte. A SiNW-FET gated with an optimized GO coating exhibited a similar threshold voltage change (ΔV_{th} 119 $\pm$ 2 mV) due to changing the analyte pH from 7 to 10 as one gated with an Ag/AgCl RE (ΔV_{th} 121 $\pm$ 1 mV). In comparison to the non-optimized GO coating, much higher reliability could be achieved due to optimization of the GO coating. With the help of SEM images, we could show that an optimized GO coating resulted in a continuous film without micro holes. Therefore, it can be concluded that the quality of the GO coating is highly influencing the performance of the pRE. The work presented here establishes the great potential of combining polymeric electrodes with ion diffusion barriers. In future work, we plan to utilize more controllable processes (e.g., spin-coating) to further improve the fabrication of a reliable on-chip pRE. In addition to the coating of the ion diffusion barrier, the impact of the metal underneath the polymer can be investigated to further improve the pRE performance.

Supplementary Materials: The following supporting information can be downloaded at: https://www.mdpi.com/article/10.3390/s22082999/s1, Figure S1: V_{th} extraction method; Figure S2: Additional EIS data of PEDOT:PSS coated electrodes; Figure S3: Impact of deposition speed on the PEDOT:PSS film quality; Figure S4: 3D printed fluidic chambers; Figure S5: Inset of OCP measurements; Figure S6: Reproducibility of the OCP; Figure S7: Long-term OCP measurement; Figure S8: Low-drift OCP measurement; Figure S9: Additional OCP measurements of a PEDOT:PSS electrode; Figure S10: pH sensitivity of an SiNW-FET gated with pRE 1; Figure S11: Drain current stability using differen pREs; Figure S12: threshold voltage change due to changes in ionic strength.

Author Contributions: Conceptualization, M.T. and X.T.V.; SiNW-FET fabrication, M.T.; methodology, M.T., T.K. and X.T.V.; formal analysis, M.T.; writing—original draft, M.T. and X.T.V.; supervision, V.P., S.I. and X.T.V.; funding acquisition, S.I. and V.P. All authors have read and agreed to the published version of the manuscript.

Funding: This research was funded by the Deutsche Forschungsgemeinschaft (DFG), grant number Nr. 391107823.

Data Availability Statement: All relevant data generated or analyzed during this study are included in this published article and its supplementary information files.

Acknowledgments: The authors thank the DFG for funding under project no. 391107823 and 40055779. The authors would like to thank Joachim Knoch, Stefan Scholz, Birger Berghoff, Noel Wilck, and Jochen Heiss for assistance and valuable discussions during the fabrication of the SiNW-FETs. The authors would like to thank Linda Wetzel for linguistic editing of the manuscript.

Conflicts of Interest: The authors declare no conflict of interest.

References

1. Bergveld, P. Development of an Ion-Sensitive Solid-State Device for Neurophysiological Measurements. *IEEE Trans. Biomed. Eng.* **1970**, *17*, 70–71. [CrossRef] [PubMed]
2. Pachauri, V.; Ingebrandt, S. Biologically sensitive field-effect transistors: From ISFETs to NanoFETs. *Essays Biochem.* **2016**, *60*, 81–90. [PubMed]
3. Ohno, Y.; Maehashi, K.; Matsumoto, K. Label-free biosensors based on aptamer-modified graphene field-effect transistors. *J. Am. Chem. Soc.* **2010**, *132*, 18012–18013. [CrossRef] [PubMed]
4. Lu, X.L.; Munief, W.M.; Heib, F.; Schmitt, M.; Britz, A.; Grandthyl, S.; Muller, F.; Neurohr, J.U.; Jacobs, K.; Benia, H.M.; et al. Front-End-of-Line Integration of Graphene Oxide for Graphene-Based Electrical Platforms. *Adv. Mater. Technol.* **2018**, *3*, 14. [CrossRef]

5. Figueroa-Miranda, G.; Liang, Y.; Suranglikar, M.; Stadler, M.; Samane, N.; Tintelott, M.; Lo, Y.; Tanner, J.A.; Vu, X.T.; Knoch, J. Delineating charge and capacitance transduction in system-integrated graphene-based BioFETs used as aptasensors for malaria detection. *Biosens. Bioelectron.* **2022**, *208*, 114219. [CrossRef] [PubMed]

6. Balasubramanian, K.; Burghard, M. Biosensors based on carbon nanotubes. *Anal. Bioanal. Chem.* **2006**, *385*, 452–468. [CrossRef]

7. Torsi, L.; Magliulo, M.; Manoli, K.; Palazzo, G.J.C.S.R. Organic field-effect transistor sensors: A tutorial review. *Chem. Soc. Rev.* **2013**, *42*, 8612–8628. [CrossRef]

8. Pachauri, V.; Vlandas, A.; Kern, K.; Balasubramanian, K.J.S. Site-Specific Self-Assembled Liquid-Gated ZnO Nanowire Transistors for Sensing Applications. *Small* **2010**, *6*, 589–594. [CrossRef]

9. Kaisti, M. Detection principles of biological and chemical FET sensors. *Biosens Bioelectron* **2017**, *98*, 437–448. [CrossRef]

10. Simonis, A.; Dawgul, M.; Luth, H.; Schoning, M.J. Miniaturised reference electrodes for field-effect sensors compatible to silicon chip technology. *Electrochim. Acta* **2005**, *51*, 930–937. [CrossRef]

11. Yee, S.; Jin, H.; Lam, L.K.C. Miniature liquid junction reference electrode with micromachined silicon cavity. *Sens. Actuators* **1988**, *15*, 337–345. [CrossRef]

12. Tintelott, M.; Pachauri, V.; Ingebrandt, S.; Vu, X.T. Process variability in top-down fabrication of silicon nanowire-based biosensor arrays. *Sensors* **2021**, *21*, 5153. [CrossRef] [PubMed]

13. Hu, Q.; Chen, S.; Solomon, P.; Zhang, Z. Ion sensing with single charge resolution using sub–10-nm electrical double layer–gated silicon nanowire transistors. *Sci. Adv.* **2021**, *7*, eabj6711. [CrossRef] [PubMed]

14. Bergveld, P. Thirty years of ISFETOLOGY—What happened in the past 30 years and what may happen in the next 30 years. *Sens. Actuators B Chem.* **2003**, *88*, 1–20. [CrossRef]

15. Lewenstam, A. *Handbook of Reference Electrodes*; Inzelt, G., Lewenstam, A., Scholz, F., Eds.; Springer: Berlin/Heidelberg, Germany, 2013; pp. 1–344. ISBN 978-3-642-36187-6.

16. Yang, H.; Kang, S.K.; Choi, C.A.; Kim, H.; Shin, D.-H.; Kim, Y.S.; Kim, Y.T. An iridium oxide reference electrode for use in microfabricated biosensors and biochips. *Lab Chip* **2004**, *4*, 42–46. [CrossRef] [PubMed]

17. Shinwari, M.W.; Zhitomirsky, D.; Deen, I.A.; Selvaganapathy, P.R.; Deen, M.J.; Landheer, D. Microfabricated Reference Electrodes and their Biosensing Applications. *Sensors* **2010**, *10*, 1679–1715. [CrossRef]

18. Petsagkourakis, I.; Kim, N.; Tybrandt, K.; Zozoulenko, I.; Crispin, X. Poly(3,4-ethylenedioxythiophene): Chemical Synthesis, Transport Properties, and Thermoelectric Devices. *Adv. Electron. Mater.* **2019**, *5*, 1800918. [CrossRef]

19. Park, H.-S.; Ko, S.-J.; Park, J.-S.; Kim, J.Y.; Song, H.-K. Redox-active charge carriers of conducting polymers as a tuner of conductivity and its potential window. *Sci. Rep.* **2013**, *3*, 2454. [CrossRef]

20. Hempel, F.; Law, J.K.Y.; Nguyen, T.C.; Munief, W.; Lu, X.L.; Pachauri, V.; Susloparova, A.; Vu, X.T.; Ingebrandt, S. PEDOT:PSS organic electrochemical transistor arrays for extracellular electrophysiological sensing of cardiac cells. *Biosens. Bioelectron.* **2017**, *93*, 132–138. [CrossRef]

21. Reineke, S.; Thomschke, M.; Lüssem, B.; Leo, K. White organic light-emitting diodes: Status and perspective. *Rev. Mod. Phys.* **2013**, *85*, 1245. [CrossRef]

22. Keene, S.T.; Lubrano, C.; Kazemzadeh, S.; Melianas, A.; Tuchman, Y.; Polino, G.; Scognamiglio, P.; Cinà, L.; Salleo, A.; van de Burgt, Y.; et al. A biohybrid synapse with neurotransmitter-mediated plasticity. *Nat. Mater.* **2020**, *19*, 969–973. [CrossRef] [PubMed]

23. Schander, A.; Stemmann, H.; Tolstosheeva, E.; Roese, R.; Biefeld, V.; Kempen, L.; Kreiter, A.; Lang, W. Design and fabrication of novel multi-channel floating neural probes for intracortical chronic recording. *Sens. Actuators A Phys.* **2016**, *247*, 125–135. [CrossRef]

24. Bobacka, J. Potential Stability of All-Solid-State Ion-Selective Electrodes Using Conducting Polymers as Ion-to-Electron Transducers. *Anal. Chem.* **1999**, *71*, 4932–4937. [CrossRef] [PubMed]

25. Isaksson, J.; Kjäll, P.; Nilsson, D.; Robinson, N.; Berggren, M.; Richter-Dahlfors, A. Electronic control of Ca^{2+} signalling in neuronal cells using an organic electronic ion pump. *Nat. Mater.* **2007**, *6*, 673. [CrossRef] [PubMed]

26. Berggren, M.; Malliaras, G.G. How conducting polymer electrodes operate. *Science* **2019**, *364*, 233–234. [CrossRef]

27. Paulsen, B.D.; Tybrandt, K.; Stavrinidou, E.; Rivnay, J. Organic mixed ionic–electronic conductors. *Nat. Mater.* **2019**, *19*, 13–26. [CrossRef]

28. Heine, V.; Kremers, T.; Menzel, N.; Schnakenberg, U.; Elling, L. Electrochemical Impedance Spectroscopy Biosensor Enabling Kinetic Monitoring of Fucosyltransferase Activity. *ACS Sens.* **2021**, *6*, 1003–1011. [CrossRef]

29. Kremers, T.; Tintelott, M.; Pachauri, V.; Vu, X.T.; Ingebrandt, S.; Schnakenberg, U. Microelectrode Combinations of Gold and Polypyrrole Enable Highly Stable Two-electrode Electrochemical Impedance Spectroscopy Measurements under Turbulent Flow Conditions. *Electroanalysis* **2021**, *33*, 197–207. [CrossRef]

30. Duarte-Guevara, C.; Swaminathan, V.V.; Burgess, M.; Reddy, B.; Salm, E.; Liu, Y.S.; Rodriguez-Lopez, J.; Bashir, R. On-chip metal/polypyrrole quasi-reference electrodes for robust ISFET operation. *Analyst* **2015**, *140*, 3630–3641. [CrossRef]

31. Han, S.; Polyravas, A.G.; Wustoni, S.; Inal, S.; Malliaras, G.G. Integration of organic electrochemical transistors with implantable probes. *Adv. Mater. Technol.* **2021**, *6*, 2100763. [CrossRef]

32. Hempel, F.W. Organic electrochemical transistors based on PEDOT: PSS for the sensing of cellular signals from confluent cell layers down to single cells. *Mater. Sci.* **2019**, *180*, 113101.

33. Leenaerts, O.; Partoens, B.; Peeters, F. Graphene: A perfect nanoballoon. *Appl. Phys. Lett.* **2008**, *93*, 193107. [CrossRef]

34. Bong, J.H.; Yoon, S.J.; Yoon, A.; Hwang, W.S.; Cho, B.J. Ultrathin graphene and graphene oxide layers as a diffusion barrier for advanced Cu metallization. *Appl. Phys. Lett.* **2015**, *106*, 063112. [CrossRef]
35. Yoo, B.M.; Shin, H.J.; Yoon, H.W.; Park, H.B. Graphene and graphene oxide and their uses in barrier polymers. *J. Appl. Polym. Sci.* **2014**, *131*, 39628. [CrossRef]
36. Sung, S.J.; Park, J.; Cho, Y.S.; Gihm, S.H.; Yang, S.J.; Park, C.R. Enhanced gas barrier property of stacking-controlled reduced graphene oxide films for encapsulation of polymer solar cells. *Carbon* **2019**, *150*, 275–283. [CrossRef]
37. Ren, W.; Cheng, H.-M. The global growth of graphene. *Nat. Nanotechnol.* **2014**, *9*, 726–730. [CrossRef]
38. Yamaguchi, H.; Granstrom, J.; Nie, W.; Sojoudi, H.; Fujita, T.; Voiry, D.; Chen, M.; Gupta, G.; Mohite, A.D.; Graham, S. Reduced graphene oxide thin films as ultrabarriers for organic electronics. *Adv. Energy Mater.* **2014**, *4*, 1300986. [CrossRef]
39. Lu, X. Reduced Graphene Oxide Biosensors for Prostate Cancer Biomarker Detection. Ph.D. Thesis, Justus-Liebig-Universität Gießen, Gießen, Germany, 2018.
40. Vu, X.; GhoshMoulick, R.; Eschermann, J.; Stockmann, R.; Offenhäusser, A.; Ingebrandt, S. Fabrication and application of silicon nanowire transistor arrays for biomolecular detection. *Sens. Actuators B Chem.* **2010**, *144*, 354–360. [CrossRef]
41. Schnakenberg, U.; Benecke, W.; Lange, P. TMAHW etchants for silicon micromachining. In Proceedings of the TRANSDUCERS'91: 1991 International Conference on Solid-State Sensors and Actuators. Digest of Technical Papers, San Francisco, CA, USA, 24–27 June 1991; pp. 815–818.
42. Klos, J.; Sun, B.; Beyer, J.; Kindel, S.; Hellmich, L.; Knoch, J.; Schreiber, L. Spin Qubits Confined to a Silicon Nano-Ridge. *Appl. Sci.* **2019**, *9*, 3823. [CrossRef]
43. Tintelott, M.; Ingebrandt, S.; Pachauri, V.; Vu, X.T. Lab-on-a-chip based silicon nanowire sensor system for the precise study of chemical reaction-diffusion networks. In Proceedings of the MikroSystemTechnik Congress 2021, Ludwigsburg, Germany, 8–10 November 2021; pp. 1–4.
44. Lazar, J.; Schnelting, C.; Slavcheva, E.; Schnakenberg, U. Hampering of the stability of gold electrodes by ferri-/ferrocyanide redox couple electrolytes during electrochemical impedance spectroscopy. *Anal. Chem.* **2016**, *88*, 682–687. [CrossRef]
45. Cui, Y.; Wei, Q.; Park, H.; Lieber, C.M. Nanowire nanosensors for highly sensitive and selective detection of biological and chemical species. *Science* **2001**, *293*, 1289–1292. [CrossRef] [PubMed]
46. Fu, W.; Nef, C.; Knopfmacher, O.; Tarasov, A.; Weiss, M.; Calame, M.; Schönenberger, C. Graphene transistors are insensitive to pH changes in solution. *Nano Lett.* **2011**, *11*, 3597–3600. [CrossRef] [PubMed]
47. Vu, X.T.; Eschermann, J.F.; Stockmann, R.; GhoshMoulick, R.; Offenhäusser, A.; Ingebrandt, S. Top-down processed silicon nanowire transistor arrays for biosensing. *Phys. Status Solidi A* **2009**, *206*, 426–434. [CrossRef]
48. Papamatthaiou, S.; Zupancic, U.; Kalha, C.; Regoutz, A.; Estrela, P.; Moschou, D. Ultra stable, inkjet-printed pseudo reference electrodes for lab-on-chip integrated electrochemical biosensors. *Sci. Rep.* **2020**, *10*, 17152. [CrossRef]
49. Abbas, Y.; Olthuis, W.; van den Berg, A. Activated carbon as a pseudo-reference electrode for electrochemical measurement inside concrete. *Constr. Build. Mater.* **2015**, *100*, 194–200. [CrossRef]
50. Ying, K.S.; Heng, L.Y.; Hassan, N.I.; Hasbullah, S.A. A New and All-Solid-State Potentiometric Aluminium Ion Sensor for Water Analysis. *Sensors* **2020**, *20*, 6898. [CrossRef]
51. Tymecki, Ł.; Zwierkowska, E.; Koncki, R. Screen-printed reference electrodes for potentiometric measurements. *Anal. Chim. Acta* **2004**, *526*, 3–11. [CrossRef]

Article

High Spatial Resolution Ion Imaging by Focused Electron-Beam Excitation with Nanometric Thin Sensor Substrate

Kiyohisa Nii [1], **Wataru Inami** [2] **and Yoshimasa Kawata** [1,2,*]

[1] Graduate School of Medical Photonics, Shizuoka University, 3-5-1 Johoku, Naka, Hamamatsu 432-8011, Japan; nii.kiyohisa.17@shizuoka.ac.jp

[2] Research Institute of Electronics, Shizuoka University, 3-5-1 Johoku, Naka, Hamamatsu 432-8011, Japan; inami.wataru@shizuoka.ac.jp

* Correspondence: kawata@eng.shizuoka.ac.jp; Tel.: +81-53-478-1069

Abstract: We developed a high spatially-resolved ion-imaging system using focused electron beam excitation. In this system, we designed a nanometric thin sensor substrate to improve spatial resolution. The principle of pH measurement is similar to that of a light-addressable potentiometric sensor (LAPS), however, here the focused electron beam is used as an excitation carrier instead of light. A Nernstian-like pH response with a pH sensitivity of 53.83 mV/pH and linearity of 96.15% was obtained. The spatial resolution of the imaging system was evaluated by applying a photoresist to the sensing surface of the ion-sensor substrate. A spatial resolution of 216 nm was obtained. We achieved a substantially higher spatial resolution than that reported in the LAPS systems.

Keywords: light-addressable potentiometric sensor; high spatial resolution; electron-beam-induced current; thin sensor substrate; chemical imaging system; electron-beam addressable potentiometri sensor

Citation: Nii, K.; Inami, W.; Kawata, Y. High Spatial Resolution Ion Imaging by Focused Electron-Beam Excitation with Nanometric Thin Sensor Substrate. *Sensors* **2022**, *22*, 1112. https://doi.org/10.3390/s22031112

Academic Editors: Michael J. Schöning and Sven Ingebrandt

Received: 24 December 2021
Accepted: 27 January 2022
Published: 1 February 2022

Publisher's Note: MDPI stays neutral with regard to jurisdictional claims in published maps and institutional affiliations.

1. Introduction

An ion sensor is a device used to measure the concentration of a target ion. It is used in various applications, such as water quality surveys, blood chemistry, and the adjustment of a cell culture medium; pH is an important parameter in chemical measurement, and several ion sensors for detecting pH have been developed to date. In addition, miniature ion sensors, ion-sensitive field-effect transistors (ISFETs), have been developed, which further expand the application range of ion sensors [1,2].

Recently, ion sensors that can image the two-dimensional distribution of ions have been developed. Two-dimensional ion imaging provides visualization and dynamic analysis of processes such as electrolysis and corrosion [3]. It is also possible to obtain the distribution of ion concentrations in the vicinity of cells and measure the metabolic activity of living cells [4–10].

An ISFET array, which is a two-dimensional array of ISFETs, can be used for ion imaging. The ISFET array sensor achieves real-time ion imaging with a frame rate of 6100 fps [11]. CMOS ion-sensitive field-effect transistor (ISFET) arrays with column offset compensation have been proposed for long-term bacterial metabolism monitoring [12]. In addition, the graphene field-effect transistor arrays are proposed for real-time, high resolution, simultaneous measurement of multiple ionic species [13]. In ISFETs, the gate-insulating layer is directly immersed in the solution without using metal as the gate electrode of the metal oxide semiconductor field-effect transistor (MOSFET), and the ion concentration is measured from the change in drain current generated by the interface potential between the solution and the gate-insulating layer. The spatial resolution is limited by cell size, and can only be achieved at 9.22 μm × 7.56 μm with a chip size of 1024 × 1024 array [14].

A charge transfer-type ion-image sensor is capable of real-time, high-sensitivity measurement of ions. The concentration images of four kinds of ions (H^+, K^+, Na^+, and Ca^{2+}) could be obtained simultaneously through the CCD multi-ion-image sensor [15]. A proton image sensor was inserted into the brain to detect the changes of pH in the brain caused by any visual stimulation [16]. The pH value was converted into an electrical charge. The charges were then transferred to the output circuit for reading. The spatial resolution of a charge-transfer-type ion-image sensor was determined by the pixel size. A spatial resolution of 3.75 µm with 1.3 megapixels has been achieved in such a sensor [17].

In the LAPS, the ion concentration distribution can be imaged by scanning two-dimensionally with a laser beam. An analog micromirror was adopted for the raster scan of the sensor substrate, which enables high-resolution pH imaging with a frame rate of 8 fps [18]. The pH changes in the hippocampal formation of rats were obtained using an all-in-one pH probe [8]. The flat-band voltage shift due to the ions of the semiconductor in an EIS structure was measured with light. The spatial resolution of the LAPS depends on the spot size of the laser beam used for excitation, the thickness of the ion sensor substrate, and the diffusion of the carriers [19–23]. A spatial resolution of 0.8 µm was achieved by using the SOS substrate and the two-photon excitation method [24].

In this study, we developed an electron-beam-addressable potentiometric sensor (EAPS) to improve the spatial resolution of the ion sensor. In the EAPS, the light used in the LAPS is replaced by a focused electron beam [25,26]. The electron beam can realize a spot size of several nanometers. In addition, the silicon layer of the sensor substrate was thinned to suppress the diffusion of minority carriers. The pH measurement capability and spatial resolution were evaluated using an EAPS system.

2. Materials and Methods

2.1. Principle of Operation of an Electron-Beam-Addressable Potentiometric Sensor

2.1.1. Method for Measuring the Distribution of the Ion Concentration

A schematic diagram of the electron-beam-excited ion-imaging system developed in this study is depicted in Figure 1. This system consists of an ion sensor substrate, analyte solution, reference electrode, power supply, ammeter, and electron beam for excitation. Insulating layers of SiO_2 and Si_3N_4 were present on the surface of the sensor substrate. The Si_3N_4 layer in contact with the analyte solution acted as an ion-sensitive membrane. In addition, electrodes are attached to the rear surface of the sensor substrate through an ohmic contact. The Si_3N_4 on the surface of the ion sensor substrate forms a silanol group (Si-OH), and its state changes depending on the concentration of hydrogen ions in the analysis solution [27] as follows:

$$SiOH_2^+ \leftrightarrow SiOH + H^+, \; SiOH \leftrightarrow SiO^- + H^+ \tag{1}$$

Consequently, the potential on the surface of the substrate changes according to the degree of ionic binding and dissociation. At the time of measurement, a power supply was used to create a depletion state by applying a voltage between the reference electrode and the electrode on the rear surface. When the potential on the substrate surface changes owing to the binding and dissociation of the ions, the thickness of the depletion layer formed at the substrate–insulating layer interface of the ion sensor substrate also changes. The thickness of the depletion layer was calculated from the amount of alternating current flowing through the circuit after irradiation using an electron beam. The thickness of the depletion layer depends on the ion concentration which can be calculated by measuring the thickness of the depletion layer.

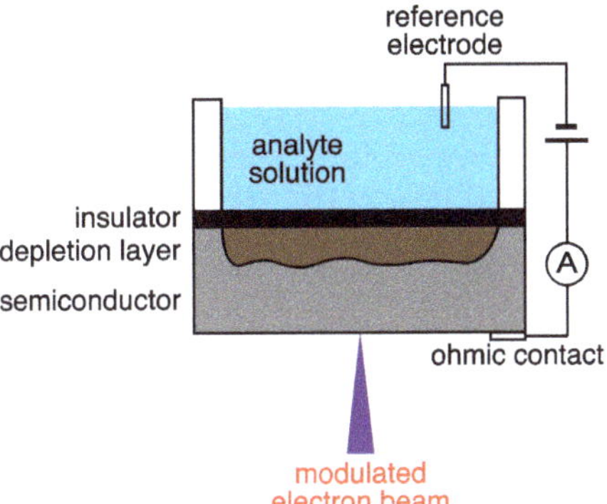

Figure 1. Schematic diagram of the electron-beam-excited ion-imaging system.

2.1.2. Measurement of the Depletion Layer Width with a Circuit Model and Electron Beam Irradiation

A band diagram showing the alternating current generated by electron beam irradiation is presented in Figure 2a. When irradiated with an electron beam, electron-hole pairs are generated inside the semiconductor substrate that diffuses into the depletion layer. Electrons and holes are separated by an electric field in the depletion layer, resulting in a transient current. When the electron beam irradiation ceases, the excess holes in the depletion layer are removed by recombination. As a result, a transient current flows in the opposite direction [28,29].

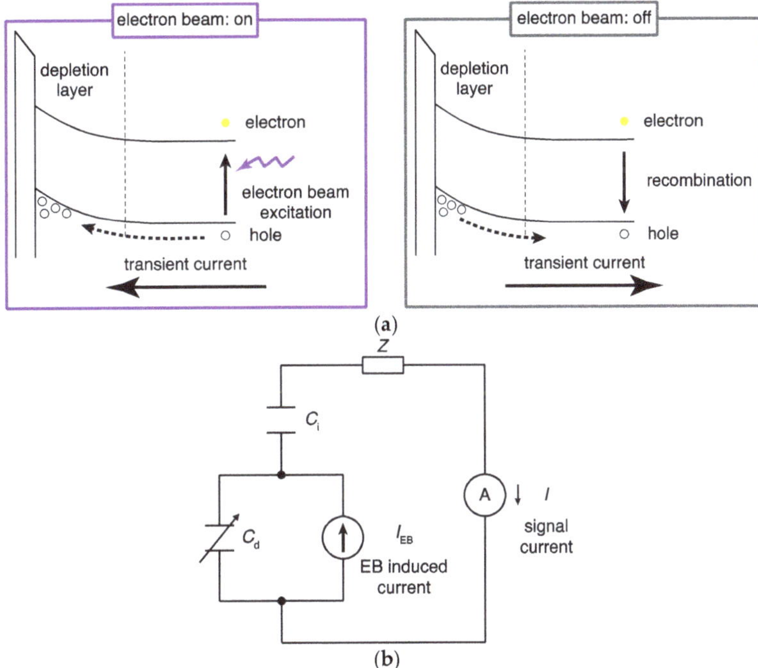

Figure 2. (**a**) Energy band diagrams explaining the generation of transient currents in an electron-beam-addressable potentiometric sensor after the electron beam is turned on and off. (**b**) Circuit model of an electron-beam-addressable potentiometric sensor.

The transient current generated through repeated irradiation by the modulated electron beam is represented by the alternating current (I_{EB}) of the circuit model shown in Figure 2b. The alternating current (I_{EB}) is divided according to the capacitances (C_d) of the depletion layer and (C_i) of the insulating layer in the area irradiated with the electron beam, and the alternating current flowing through the latter is measured externally as signal I. The series impedance Z includes the impedance of the solution, the resistance of the semiconductor, the contact resistance, and the input impedance of the transimpedance amplifier. If the series impedance Z is negligibly small, the alternating current I is ex-pressed using the following equation [28]:

$$I = I_{EB} \frac{C_i}{C_i + C_d} \tag{2}$$

2.1.3. Relationship between the Ion Sensor Substrate and Spatial Resolution

The LAPS constitutes a system that uses light to excite electrons to measure the ion concentration. In the system developed in this study, a high spatial resolution is achieved by adapting the light used for excitation of the electron beam in the LAPS, such that it can form a smaller spot diameter. Spatial resolution is one of the most important indicators of the performance of chemical sensors. In the LAPS, the spatial resolution is determined by the range of diffusion of the charge carriers in the semiconductor layer if the light beam is sufficiently focused [20,22,30]. When irradiating an electron beam, the spatial resolution is defined according to the range of scattering of the electron beam and the diffusion of charge carriers. Under irradiation, the electron beam enters the silicon, generating electron-hole pairs that diffuse into the depletion layer. Even for a small spot diameter, a high spatial resolution cannot be achieved if the film is too thick. The spatial resolution can be improved by increasing the doping concentration of impurities or by reducing the thickness of the ion sensor substrate. It has been reported that a reduction in the thickness of the ion sensor substrate is a better method for improving the spatial resolution [19]. In our device, we prepared a window-structured ion sensor substrate by thinning the silicon-on-insulator (SOI) substrate through etching to achieve ion imaging with high spatial resolution.

2.2. Ion-Imaging System Based on Electron Beam Excitation

A schematic diagram of the ion-imaging system based on electron beam excitation is shown in Figure 3a. The electron beam emitted from the electron gun was converged by an electrostatic lens and irradiated on the ion sensor substrate. It is also possible to irradiate the sample while scanning the electron beam with a scan coil.

To irradiate the electron beam, it is necessary to install the rear surface of the ion sensor substrate under vacuum conditions. Therefore, as shown in Figure 3a, the inside of the electron microscope barrel is maintained under vacuum by installing an O-ring on the lower surface of the holder onto which the ion sensor substrate is attached. In the ion-sensing unit, only the rear surface of the ion sensor substrate was under vacuum. Therefore, even though the measurement system uses electron beam irradiation, it is possible to culture living cells on the surface of the ion sensor substrate and perform measurements in vivo.

The irradiation current and acceleration voltage were controlled by a control unit attached to an inverted scanning electron microscope (MINI-EOC, APCO Ltd., Tokyo, Japan). However, it was not possible to irradiate the electron beam while switching on and off with the control unit alone. Therefore, a function generator (AFG3021C, Tektronix, Beaverton, OR, USA) was connected to the blank part of the electron microscope (as shown in Figure 3a). By inputting an on/off electric signal, the electron beam can be irradiated while being modulated by the status of the signal.

An image of the manufactured electron-beam-excited ion-imaging system is shown in Figure 3b. The part shown in the yellow frame is the electron microscope, and the part shown in the red frame is the ion-measuring part. The measurement was performed by irradiating the ion sensor substrate fixed onto the holder with an electron beam emitted from an inverted scanning electron microscope.

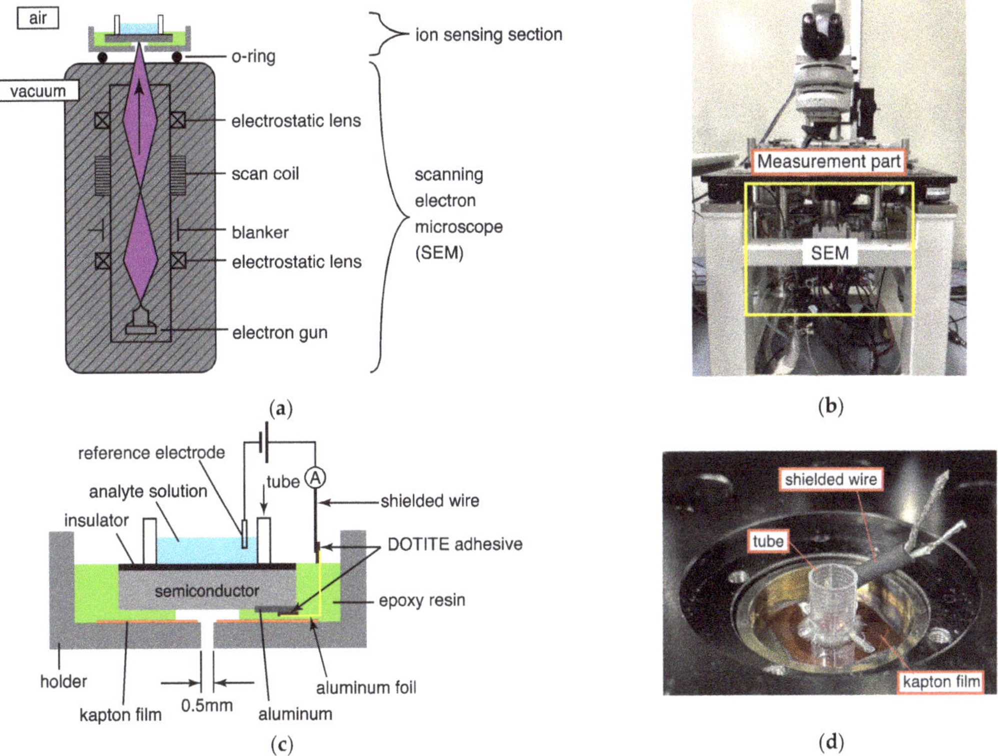

Figure 3. (**a**) Schematic diagram of the electron-beam-excited ion-imaging system, (**b**) image of the manufactured electron-beam-excited ion-imaging system, (**c**) enlarged view of the measuring section, and (**d**) image of the actual measuring section.

An enlarged view of the measurement section is presented in Figure 3c. The aluminum electrode attached to the rear surface of the ion sensor substrate was glued with the conductive adhesive DOTITE (D-500, FUJIKURA KASEI Co., Ltd., Tokyo, Japan), and the electrode was extended using aluminum foil. The substrate with the extended electrodes was glued with epoxy resin (Araldite, Huntsman, The Woodlands, TX, USA) to a holder with a hole diameter of 0.5 mm for passing the electron beam. The Kapton film (DuPont de Nemours, Inc., Wilmington, DE, USA) was sandwiched between the substrate and holder to prevent electrical continuity between these components. The electrode was further extended from the aluminum foil with the help of the DOTITE adhesive using a shielded wire. An image of the actual measurement section is presented in Figure 3d. A cylinder intended to contain the analyte solution was attached to the epoxy resin. At the time of measurement, the analyte solution was placed in the cylinder, and a reference electrode was inserted. The measurement was performed by applying a voltage between the reference electrode and the electrode extending from the aluminum foil attached to the rear of the ion sensor substrate.

2.3. Fabrication of the Thin Ion Sensor Substrate

We formed a SiO_2 and Si_3N_4 layer on an SOI substrate and fabricated a thin-film window-structured ion-sensor substrate on the rear surface by etching. A schematic diagram of the configuration of the window-structured ion-sensor substrate is shown in Figure 4a. A SiO_2 layer of 12 nm was formed on the surface of a p-type SOI wafer (10–20 $\Omega\cdot$cm, Si = 50 nm) by thermal oxidation method, and a 50 nm Si_3N_4 layer was deposited via the low-pressure chemical vapor deposition (LP-CVD) method. The Si_3N_4 layer acted as an ion-sensitive membrane. The dimensions of the ion sensor substrate were 5 mm × 5 mm, and those of the window part were 100 μm × 100 μm. The prepared substrate was washed with piranha solution (H_2SO_4 (UN1830 H_2SO_4, FUJIFILM Wako Pure Chemical Corporation, Osaka, Japan): H_2O_2 (35% H_2O_2, Hirota chemical Industry, Co., Ltd., Tokyo, Japan) = 4:1) for 15 min. To form ohmic contacts on the rear surface of the substrate, the natural oxide film was removed by immersing it in a 1% HF (49% HF, Hirota chemical Industry, Co., Ltd., Tokyo, Japan) solution for 30 s. Subsequently, aluminum was attached as an electrode through vacuum vapor deposition. The thickness of the aluminum electrode was 30 nm. Images of the front and rear surfaces of the manufactured substrate are shown in Figure 4b. The front surface was flat, and the rear surface was dented because only the part used for imaging was thinned. Aluminum was deposited on the recessed window structure.

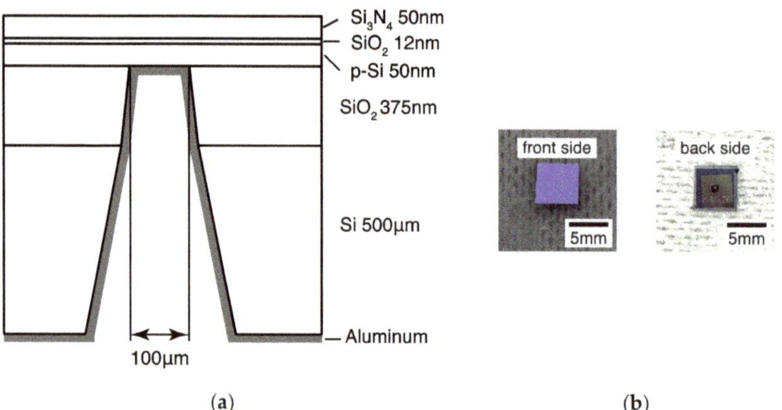

(a)　　　　(b)

Figure 4. (**a**) Schematic diagram of the window-structure ion-sensor substrate configuration and (**b**) image of the front and rear surfaces of the manufactured substrate.

3. Results and Discussion

3.1. Measurement Results for Solutions with Various Values of pH

The pH dependence of the bias voltage–current characteristics of the electron-beam-induced current was measured. The bias voltage–current characteristics acquired by the developed device are shown in Figure 5. The acceleration voltage of the electron beam was 5 keV, the irradiation current was 3.1 nA, and the modulation frequency was 820 Hz. pH standard solutions (pH standard solution, Horiba, Kyoto, Japan) with pH values of 4.01, 6.86, and 9.18 were used as the measurement solution. A water-based reference electrode (Ag/AgCl) (RE-1B, BAS, Tokyo, Japan) was used as the reference electrode. A dual-output power supply (E3620A, Keysight Technologies, CA, USA) was used as the power supply, and a current input preamplifier (LI-76, NF Corporation, Kanagawa, Japan) was used as the I-V amplifier.

At all three pH values, the bias voltage increased with a concomitant increase in the current through the circuit. This observation is in accordance with Equation (2). On applying a reverse bias voltage to the substrate, the depletion layer becomes thicker. This reduces the capacitance C_d of the depletion layer, and the capacitance of the insulating layer C_i and the alternating current I_{EB} are constant during measurement. Therefore, from Equation (2), the alternating current I measured by the external circuit increases. When a voltage of 0.1 V is applied, the resulting current flowing through the circuit differs for different values of pH. This is in accordance with Equation (1). The processes of ionic binding and dissociation, represented in Equation (1), depend on the hydrogen ion concentration of the standard pH solution. As a result, the potential on the surface of the substrate changed, and the thickness of the depletion layer also changed. When the voltage applied from the outside is constant, the AC current I measured by the external circuit also varies with the pH because the thickness of the depletion layer varies. This result clarifies how ion concentration can be measured. The bias voltage corresponding to the inflection point in each of the curves in Figure 5 was calculated to calculate the pH sensitivity and linearity of the prepared ion-sensor substrate. The relationship between the bias voltage and pH, as represented by the inflection points obtained by calculation, is shown in Figure 6.

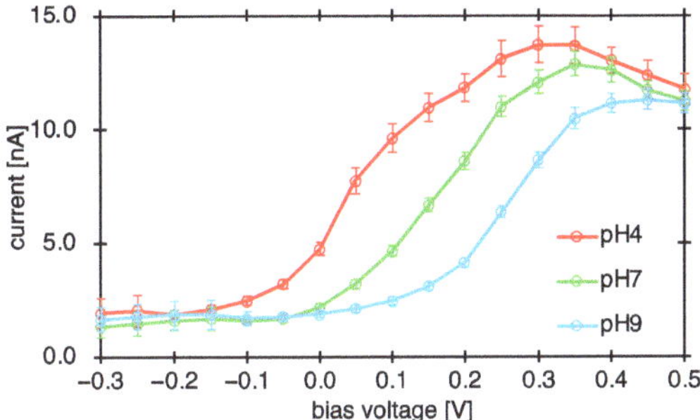

Figure 5. Bias voltage-current characteristics of the electron-beam-induced current.

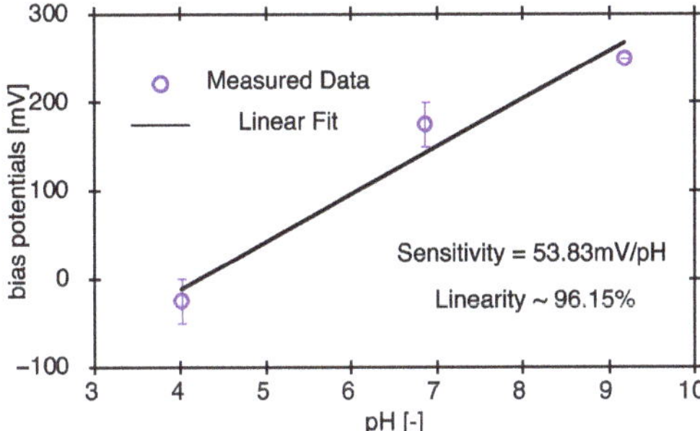

Figure 6. Sensitivity and linearity of the window-structure ion sensor.

The resulting pH sensitivity and linearity were calculated to be 53.83 mV/pH and 96.15%, respectively. These values are in good agreement with the ideal Nernstian values (59.16 mV/pH at 25 °C).

3.2. Evaluation of the Spatial Resolution

To evaluate the spatial resolution, an ion sensor substrate with half the surface covered with a photoresist was manufactured through a photolithography process [31–33]. The manufacturing procedure included the following steps: (1) coating a photoresist layer, (2) ultraviolet exposure, and (3) developing the image. A positive photoresist, OFPR800 (Tokyo Ohka Kogyo Co., Ltd., Kanagawa, Japan), was formed via spin coating using a spin coater (ACT-220DII, Active, Saitama, Japan). The film was formed at a rotation speed of 2600 rpm with a formation time of 16 s. Subsequently, using a manual mask aligner (MJB4, SUSS MicroTec, Garching, Germany), irradiation with ultraviolet rays for 3 s ensured that the resist remained on half of the window part of the substrate. The substrate, which was exposed to ultraviolet rays, was immersed in a developing solution (NMD-3, Tokyo Ohka Kogyo Co., Ltd., Kanagawa, Japan) for 40 s. After development, it was rinsed by immersion in pure water for 30 s. After development, the photoresist was baked and hardened by baking it for 300 s on a hot plate heated to 135 °C. After baking, the knife-edge was confirmed using a semiconductor/flat-panel-display (FPD) inspection microscope (MX51, OLYMPUS, Tokyo, Japan). An image of the ion-sensor substrate after photolithography is shown in Figure 7a. An edge exists at the center of the sensor. The left side was covered with a photoresist. The Si_3N_4 layer was exposed to the right side. The thickness of the photoresist at the knife-edge was measured as 1.35 µm using a profilometer (Alpha-step IQ, KLA-Tencor, Milpitas, CA, USA).

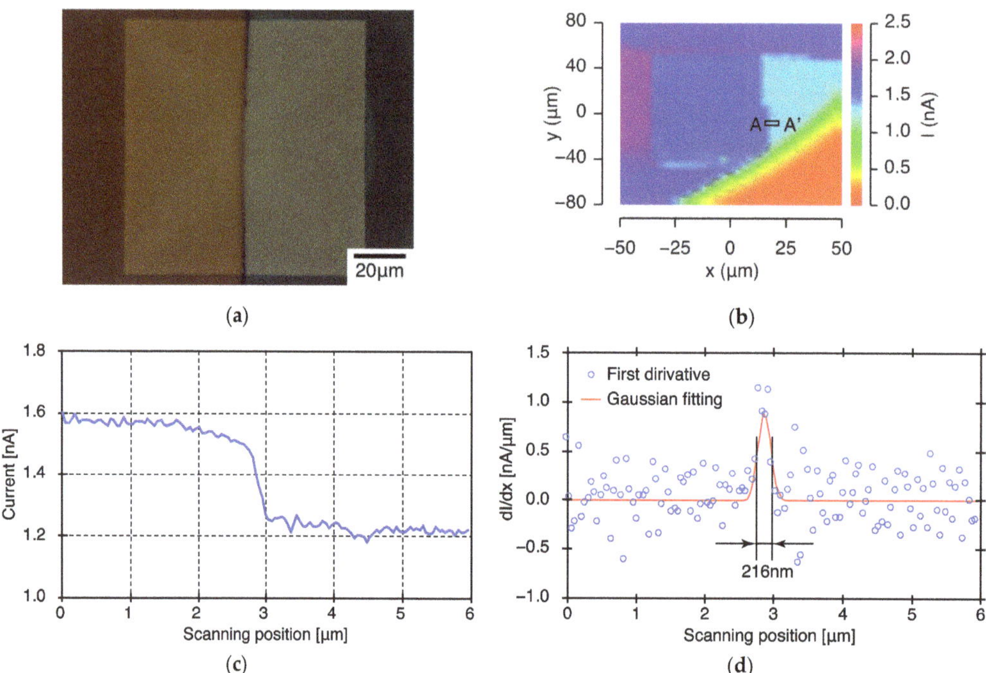

Figure 7. (**a**) Optical microscope image of the ion sensor substrate after photolithography. The photoresist is applied to the left half. (**b**) Electron-beam-induced current image, (**c**) intensity line-profile, and (**d**) first derivative of the curve shown in (**c**).

An electron-beam-induced current image obtained using a photoresist-coated ion-sensor substrate is presented in Figure 7b. The electron-beam-induced current image was obtained from the current flowing through the circuit by scanning the substrate two-dimensionally with an electron beam. The acceleration voltage of the electron beam was 5 keV, the irradiation current was 1.0 nA, and the modulation frequency was 1 kHz. A standard pH solution (pH 6.86) was used as the measurement solution. The bias voltage was set at 0 V. The number of pixels was 32×32 and the dwell time was 0.02 s. The dark blue part of the figure corresponds to the photoresist-coated part, and the light blue part corresponds to the non-photoresist-coated part. The red part on the lower right corresponds to the holder attached to the ion-sensor substrate. The electron-beam-induced current image clearly shows that the current in the part where the photoresist is applied differs from that in the part where the photoresist is not applied. In addition, an electron-beam-induced current image was acquired in the region of the black rectangle A-A' in Figure 7b, corresponding to the edge part. The number of pixels was 128×1. The intensity line profile of the acquired electron-beam-induced current image is shown in Figure 7c. The spatial resolution was determined from the FWHM of the first derivative of the intensity line profile. This method produced a spatial resolution of 216 nm, as indicated by the vertical lines in Figure 7d.

To estimate the extent of scattering of the electron beam irradiated onto the ion sensor, the scattering state of the electron beam at the thickness of the substrate used was calculated using the Monte Carlo simulation method. The relationship with the FWHM was investigated by calculating the scattering range of the electron beam that reached the depletion layer formed at the interface between the Si and SiO_2 layers. The free software package CASINO (Monte Carlo simulation of electroN trajectory in sOlids) was used for the simulation [34].

The results of the electron beam scattering simulations are shown in Figure 8a. The composition of the ion-sensor substrate was Si_3N_4(50 nm)/SiO_2(12 nm)/Si(50 nm)/Al(30 nm) and water (H_2O) was placed on the sample side. The electron beam irradiation conditions were calculated under the assumption that the acceleration voltage was 5 keV, the number of calculated electrons was 1000, and the diameter of the electron beam was 20 nm.

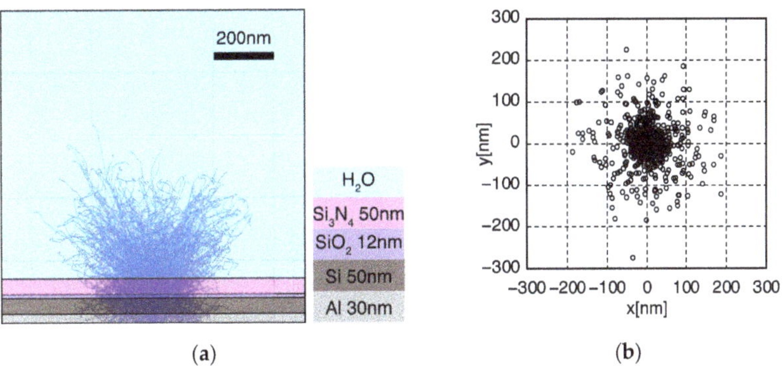

Figure 8. (**a**) Monte Carlo simulation results of electron beam scattering on the ion-sensor substrate and (**b**) electron beam passage position at the Si/SiO_2 interface.

The spread of the electron beams in the depletion layer was calculated from the Monte Carlo simulation results of the electron beam scattering. The electron beam passage position at the Si/SiO_2 interface is shown in Figure 8b. The origin was located at the position of the electron beam irradiation. When an electron beam with an acceleration voltage of 5 keV was irradiated, the scattering range of the electron beam at the Si/SiO_2 interface became 276 nm at the maximum displacement from the origin. The FWHM was calculated from the result of the electron beam scattering simulation in the same way as the spatial resolution evaluation by the first-derivative FWHM method in the experiment. The FWHM was found to be 64 nm. This value was given only by electron scattering for the excitation of carriers, and did not include the diffusion and drift effect of the excited carriers.

In the system developed in this study, electron-hole pairs were generated along with the scattering of electron beams. These charged particles diffused into the depletion layer. Consequently, the spatial resolution was negatively impacted. In addition, carriers trapped on the semiconductor surface by the electric field of the depletion layer can diffuse parallel to the surface [19]. It is considered that such diffusion caused the carriers to spread laterally; the spatial resolution obtained in the experiment was lower than the FWHM obtained in the Monte Carlo simulation.

From the simulation results depicted in Figure 8a, it can be seen that the scattered electron beams reached the water region. If imaging is performed under these conditions, the cells are damaged by the electron beam. A higher spatial resolution may be achieved by suppressing the scattering range of electron beams. Therefore, it is necessary to perform imaging under conditions that do not damage the cells, while suppressing the scattering range of electron beams by lowering the acceleration voltage.

4. Conclusions

A window-structure ion-sensor substrate was prepared with the aim of increasing the spatial resolution of the electron-beam-addressable potentiometric sensor. The pH values of the solutions were measured using the substrate. A Nernstian-like pH response with a pH sensitivity of 53.83 mV/pH and linearity of 96.15% was obtained. In addition, the spatial resolution was evaluated by applying a photoresist to the sensing surface of the ion-sensor substrate. The spatial resolution was 216 nm. From the scattering range of the electron beam by Monte Carlo simulation, the FWHM was 64 nm. In this system, electron-hole pairs were generated along with the scattering of electron beams, and these pairs diffused into the depletion layer, negatively impacting the spatial resolution. In addition, carriers trapped on the semiconductor surface owing to the electric field of the depletion layer could diffuse parallel to the surface. It is considered that such diffusion caused the carriers to spread laterally; consequently, the spatial resolution obtained in the experiment was lower than the FWHM obtained in the Monte Carlo simulation. A high spatial resolution was achieved by thinning the substrate.

As reported for LAPS systems, it is considered that higher spatial resolution could also be achieved in our system by suppressing the diffusion in the lateral direction by increasing the doping concentration of impurities. However, most minority carriers generated by excitation are lost by recombination with the majority carriers in the background, causing a decrease in the signal-to-noise ratio. Thus, a higher spatial resolution is achieved at the expense of the signal-to-noise ratio. Therefore, it is necessary to adjust the doping concentration of the impurities by considering the signal-to-noise ratio.

To achieve a high spatial resolution, it is necessary to increase the doping concentration of impurities and to use higher acceleration voltage to focus the electron beam. However, when an electron beam is irradiated with a high acceleration voltage to sensor substrate, the electron beam reaches the insulating layers SiO_2 and Si_3N_4, and the sensor chip is charged. The charging of the sensor chip affects the pH measurement. Therefore, it is necessary to investigate the effect of the charging on the substrate to the pH measurement. When the film thickness is fixed, damage to the sample can be eliminated, and the loss of spatial resolution can be suppressed by using an acceleration voltage that does not reach the SiO_2 layer. In the future, we aim to further improve the spatial resolution of our sensor by using an acceleration voltage that does not reach the SiO_2 layer.

Author Contributions: Conceptualization, W.I. and Y.K.; methodology, W.I. and Y.K.; software, W.I.; formal analysis, K.N.; investigation, K.N. and W.I.; writing—original draft preparation, K.N.; writing—review and editing, K.N., W.I. and Y.K.; visualization, K.N.; supervision, Y.K.; project administration, Y.K.; funding acquisition, Y.K. All authors have read and agreed to the published version of the manuscript.

Funding: This research was funded by the JST ASTEP Program (grant number JPMJTS1516).

Institutional Review Board Statement: Not applicable.

Informed Consent Statement: Not applicable.

Data Availability Statement: Not applicable.

Conflicts of Interest: The authors declare no conflict of interest.

References

1. Bergveld, P. Development of an Ion-Sensitive Solid-State Device for Neurophysiological Measurements. *IEEE Trans. Biomed. Eng.* **1970**, *BME-17*, 70–71. [CrossRef] [PubMed]
2. Matsuo, T.; Wise, K.D. An Integrated Field-Effect Electrode for Biopotential Recording. *IEEE Trans. Biomed. Eng.* **1974**, *BME-21*, 485–487. [CrossRef]
3. Nose, K.; Miyamoto, K.; Yoshinobu, T. Estimation of Potential Distribution during Crevice Corrosion through Analysis of I–V Curves Obtained by LAPS. *Sensors* **2020**, *20*, 2873. [CrossRef] [PubMed]
4. Kono, A.; Sakurai, T.; Hattori, T.; Okumura, K.; Ishida, M.; Sawada, K. Label Free Bio Image Sensor for Real Time Monitoring of Potassium Ion Released from Hippocampal Slices. *Sens. Actuators B* **2014**, *201*, 439–443. [CrossRef]

5. Zhang, D.-W.; Wu, F.; Wang, J.; Watkinson, M.; Krause, S. Image Detection of Yeast Saccharomyces Cerevisiae by Light-Addressable Potentiometric Sensors (LAPS). *Electrochem. Commun.* **2016**, *72*, 41–45. [CrossRef]
6. Wu, F.; Zhou, B.; Wang, J.; Zhong, M.; Das, A.; Watkinson, M.; Hing, K.; Zhang, D.-W.; Krause, S. Photoelectrochemical Imaging System for the Mapping of Cell Surface Charges. *Anal. Chem.* **2019**, *91*, 5896–5903. [CrossRef]
7. Zhou, B.; Das, A.; Kappers, M.J.; Oliver, R.A.; Humphreys, C.J.; Krause, S. InGaN as a Substrate for AC Photoelectrochemical Imaging. *Sensors* **2019**, *19*, 4386. [CrossRef]
8. Guo, Y.; Werner, C.F.; Handa, S.; Wang, M.; Ohshiro, T.; Mushiake, H.; Yoshinobu, T. Miniature Multiplexed Label-Free pH Probe in Vivo. *Biosens. Bioelectron.* **2021**, *174*, 112870. [CrossRef]
9. Doi, H.; Parajuli, B.; Horio, T.; Shigetomi, E.; Shinozaki, Y.; Noda, T.; Takahashi, K.; Hattori, T.; Koizumi, S.; Sawada, K. Development of a Label-Free ATP Image Sensor for Analyzing Spatiotemporal Patterns of ATP Release from Biological Tissues. *Sens. Actuators B* **2021**, *335*, 129686. [CrossRef]
10. Wagner, T.; Molina, R.; Yoshinobu, T.; Kloock, J.P.; Biselli, M.; Canzoneri, M.; Schnitzler, T.; Schöning, M.J. Handheld Multi-Channel LAPS Device as a Transducer Platform for Possible Biological and Chemical Multi-Sensor Applications. *Electrochim. Acta* **2007**, *53*, 305–311. [CrossRef]
11. Zeng, J.; Kuang, L.; Cacho-Soblechero, M.; Georgiou, P. An Ultra-High Frame Rate Ion Imaging Platform Using ISFET Arrays With Real-Time Compression. *IEEE Trans. Biomed. Circuits Syst.* **2021**, *15*, 820–833. [CrossRef] [PubMed]
12. Duan, M.; Zhong, X.; Xu, J.; Lee, Y.-K.; Bermak, A. A High Offset Distribution Tolerance High Resolution ISFET Array with Auto-Compensation for Long-Term Bacterial Metabolism Monitoring. *IEEE Trans. Biomed. Circuits Syst.* **2020**, *14*, 463–476. [CrossRef] [PubMed]
13. Fakih, I.; Durnan, O.; Mahvash, F.; Napal, I.; Centeno, A.; Zurutuza, A.; Yargeau, V.; Szkopek, T. Selective Ion Sensing with High Resolution Large Area Graphene Field Effect Transistor Arrays. *Nat. Commun.* **2020**, *11*, 3226. [CrossRef] [PubMed]
14. Nakazato, K. An Integrated ISFET Sensor Array. *Sensors* **2009**, *9*, 8831–8851. [CrossRef] [PubMed]
15. Hattori, T.; Dasai, F.; Sato, H.; Kato, R.; Sawada, K. CCD Multi-Ion Image Sensor with Four 128×128 Pixels Array. *Sensors* **2019**, *19*, 1582. [CrossRef]
16. Horiuchi, H.; Agetsuma, M.; Ishida, J.; Nakamura, Y.; Lawrence Cheung, D.; Nanasaki, S.; Kimura, Y.; Iwata, T.; Takahashi, K.; Sawada, K.; et al. CMOS-Based Bio-Image Sensor Spatially Resolves Neural Activity-Dependent Proton Dynamics in the Living Brain. *Nat. Commun.* **2020**, *11*, 712. [CrossRef]
17. Edo, Y.; Tamai, Y.; Yamazaki, S.; Inoue, Y.; Kanazawa, Y.; Nakashima, Y.; Yoshida, T.; Arakawa, T.; Saitoh, S.; Maegawa, M.; et al. 1.3 Mega Pixels CCD pH Imaging Sensor with 3.75 μm Spatial Resolution. In Proceedings of the 2015 IEEE International Electron Devices Meeting (IEDM), Washington, DC, USA, 7–9 December 2015. [CrossRef]
18. Zhou, B.; Das, A.; Zhong, M.; Guo, Q.; Zhang, D.-W.; Hing, K.A.; Sobrido, A.J.; Titirici, M.-M.; Krause, S. Photoelectrochemical Imaging System with High Spatiotemporal Resolution for Visualizing Dynamic Cellular Responses. *Biosens. Bioelectron.* **2021**, *180*, 113121. [CrossRef]
19. George, M.; Parak, W.J.; Gerhardt, I.; Moritz, W.; Kaesen, F.; Geiger, H.; Eisele, I.; Gaub, H.E. Investigation of the Spatial Resolution of the Light-Addressable Potentiometric Sensor. *Sens. Actuators A* **2000**, *86*, 187–196. [CrossRef]
20. Parak, W.J.; Hofmann, U.G.; Gaub, H.E.; Owicki, J.C. Lateral Resolution of Light-Addressable Potentiometric Sensors: An Experimental and Theoretical Investigation. *Sens. Actuators A* **1997**, *63*, 47–57. [CrossRef]
21. Wu, F.; Campos, I.; Zhang, D.-W.; Krause, S. Biological Imaging Using Light-Addressable Potentiometric Sensors and Scanning Photo-Induced Impedance Microscopy. *Proc. R. Soc. A* **2017**, *473*. [CrossRef]
22. Yoshinobu, T.; Miyamoto, K.; Wagner, T.; Schöning, M.J. Recent Developments of Chemical Imaging Sensor Systems Based on the Principle of the Light-Addressable Potentiometric Sensor. *Sens. Actuators B* **2015**, *207*, 926–932. [CrossRef]
23. Yang, C.-M.; Chen, C.-H.; Akuli, N.; Yen, T.-H.; Lai, C.-S. A Revised Manuscript Submitted to Sensors and Actuators B: Chemical Illumination Modification from an LED to a Laser to Improve the Spatial Resolution of IGZO Thin Film Light-Addressable Potentiometric Sensors in pH Detections. *Sens. Actuators B* **2021**, *329*, 128953. [CrossRef]
24. Chen, L.; Zhou, Y.; Jiang, S.; Kunze, J.; Schmuki, P.; Krause, S. High Resolution LAPS and SPIM. *Electrochem. Commun.* **2010**, *12*, 758–760. [CrossRef]
25. Shibano, S.; Kawata, Y.; Inami, W. Evaluation of pH Measurement Using Electron-Beam-Induced Current Detection. *Phys. Status Solidi A* **2021**, *218*, 2100147. [CrossRef]
26. Inami, W.; Nii, K.; Shibano, S.; Tomita, H.; Kawata, Y. Improvement of the Spatial Resolution of Ion Imaging System Using Thinned Sensor Substrate. In Proceedings of the Biomedical Imaging and Sensing Conference 2020, Yokohama, Japan, 20–24 April 2020; Matoba, O., Awatsuji, Y., Luo, Y., Yatagai, T., Aizu, Y., Eds.; SPIE: New York, NY, USA, 15 June 2020; p. 115210.
27. Yates, D.E.; Levine, S.; Healy, T.W. Site-Binding Model of the Electrical Double Layer at the Oxide/Water Interface. *J. Chem. Soc. Faraday Trans. 1* **1974**, *70*, 1807–1818. [CrossRef]
28. Yoshinobu, T.; Schöning, M.J. Light-Addressable Potentiometric Sensors for Cell Monitoring and Biosensing. *Curr. Opin. Electrochem.* **2021**, *28*, 100727. [CrossRef]
29. Yoshinobu, T.; Miyamoto, K.; Werner, C.F.; Poghossian, A.; Wagner, T.; Schöning, M.J. Light-Addressable Potentiometric Sensors for Quantitative Spatial Imaging of Chemical Species. *Annu. Rev. Anal. Chem.* **2017**, *10*, 225–246. [CrossRef]
30. Sartore, M.; Adami, M.; Nicolini, C.; Bousse, L.; Mostarshed, S.; Hafeman, D. Minority Carrier Diffusion Length Effects on Light-Addressable Potentiometric Sensor (LAPS) Devices. *Sens. Actuators A* **1992**, *32*, 431–436. [CrossRef]

31. Nakao, M.; Yoshinobu, T.; Iwasaki, H. Improvement of Spatial Resolution of a Laser-Scanning pH-Imaging Sensor. *Jpn. J. Appl. Phys.* **1994**, *33*, L394–L397. [CrossRef]
32. Truong, H.A.; Werner, C.F.; Miyamoto, K.; Yoshinobu, T. A Partially Etched Structure of Light-Addressable Potentiometric Sensor for High-Spatial-Resolution and High-Speed Chemical Imaging. *Phys. Status Solidi A* **2018**, *215*, 1700964. [CrossRef]
33. Moritz, W.; Gerhardt, I.; Roden, D.; Xu, M.; Krause, S. Photocurrent Measurements for Laterally Resolved Interface Characterization. *Fresenius J. Anal. Chem.* **2000**, *367*, 329–333. [CrossRef] [PubMed]
34. Hovington, P.; Drouin, D.; Gauvin, R. CASINO: A New Monte Carlo Code in C Language for Electron Beam Interaction -Part I: Description of the Program. *Scanning* **2006**, *19*, 1–14. [CrossRef]

sensors

MDPI

Communication

Efficient Illumination for a Light-Addressable Potentiometric Sensor

Tatsuo Yoshinobu [1,*] and Ko-ichiro Miyamoto [2]

[1] Department of Biomedical Engineering, Tohoku University, Sendai 980-8579, Japan
[2] Department of Electronic Engineering, Tohoku University, Sendai 980-8579, Japan; koichiro.miyamoto.d2@tohoku.ac.jp
*   Correspondence: tatsuo.yoshinobu.a1@tohoku.ac.jp

Abstract: A light-addressable potentiometric sensor (LAPS) is a chemical sensor that is based on the field effect in an electrolyte–insulator–semiconductor structure. It requires modulated illumination for generating an AC photocurrent signal that responds to the activity of target ions on the sensor surface. Although high-power illumination generates a large signal, which is advantageous in terms of the signal-to-noise ratio, excess light power can also be harmful to the sample and the measurement. In this study, we tested different waveforms of modulated illuminations to find an efficient illumination for a LAPS that can enlarge the signal as much as possible for the same input light power. The results showed that a square wave with a low duty ratio was more efficient than a sine wave by a factor of about two.

Keywords: light-addressable potentiometric sensor; LAPS; pH sensor; field-effect device; photocurrent; modulated illumination; lock-in detection; waveform; square wave; duty ratio

Citation: Yoshinobu, T.; Miyamoto, K.-i. Efficient Illumination for a Light-Addressable Potentiometric Sensor. *Sensors* **2022**, *22*, 4541. https://doi.org/10.3390/s22124541

Academic Editor: Arunas Ramanavicius

Received: 2 May 2022
Accepted: 15 June 2022
Published: 16 June 2022

1. Introduction

A light-addressable potentiometric sensor (LAPS) [1,2] is a chemical sensor that is based on a semiconductor, with a surface that can be flexibly modified with various sensing materials, such as ionophores, enzymes, aptamers, receptors, and cells, to render it a versatile platform for the electrochemical sensing and imaging [3,4] of both inorganic and organic chemical species. Its potential application range is wide, ranging from materials science to biology and medicine, and researchers have recently devoted substantial efforts to developing cell-based sensors for biomedical applications [5,6].

A LAPS has a field-effect structure [7,8] similar to that of an ion-sensitive field-effect transistor (ISFET) [9]. In both devices, the distribution of the charge carriers at the insulator–semiconductor interface varies by the field effect in response to the activity of target ions on the sensor surface. A variation in the channel conductance of an ISFET is detected in the form of a drain current, while a variation in the width of the depletion layer of a LAPS is detected in the form of a photocurrent.

To read out the change in the depletion layer, the LAPS sensor plate must be illuminated by a light beam with photon energy that is larger than the energy bandgap of the semiconductor. In most cases, the sensor plate is illuminated from the back surface to avoid the absorption or scattering of light by the sample on the front surface. Electron–hole pairs are generated by the absorption of light in the vicinity of the back surface, and they diffuse towards the insulator–semiconductor interface [10]. The electrons and holes are separated by the electric field inside the depletion layer, which functions as a current source [11]. Because the DC current is blocked by the insulator, the light beam is modulated to generate an AC photocurrent signal, and its amplitude correlates with the activity of target ions.

For a high-precision measurement, the signal should be as large as possible [12]. An increase in the input light power is a direct approach to obtaining a large signal from a LAPS, but excess light power is not only wasteful but also harmful to the measurement.

As an extreme case, if the photon energy is so high as to cause ionization/deionization in the insulator or at the insulator–semiconductor interface, as in the case of a vacuum ultraviolet light, it can alter the properties of the field-effect structure, including the flat-band condition [13]. Even in the case of visible light, most of the light power is eventually converted into heat inside the semiconductor layer, which raises the temperature of the sensor plate. Both the charge-carrier properties of the semiconductor and the Nernst potential that is built up at the solution–insulator interface are responsive to temperature change, which may result in the drift in the slope sensitivity. In addition, a higher intensity of illumination not only increases the minority carriers that contribute to the photocurrent signal, but also raises the concentration of the majority carriers in the background. A device simulation of a LAPS revealed that this effect reduces the thickness of the depletion layer and lowers the spatial resolution of chemical imaging by a LAPS [14]. Finally, when a LAPS is applied to an in vivo measurement (for example, in the brain of an animal [15]), the injection of energy in any form into the body must be minimized as a safety measure, as well as to avoid its potential influence on the living organism.

In this study, we tested different waveforms of illuminations to find an efficient illumination for a LAPS to maximize the photocurrent signal that is generated by the same light power or, equivalently, to minimize the light power that generates the same photocurrent signal.

2. Materials and Methods

The setup for the LAPS measurement that we used in this study is shown in Figure 1a. The LAPS sensor plate was composed of n-type Si, with a resistivity of 1–10 Ωcm, a size of 35 mm $\times$ 35 mm, and a thickness of 200 μm. We formed a 50 nm thick thermal oxide and deposited a 50 nm thick Si_3N_4, in this order, onto the front surface, and evaporated an Ohmic rear-side contact near the edge of the back surface.

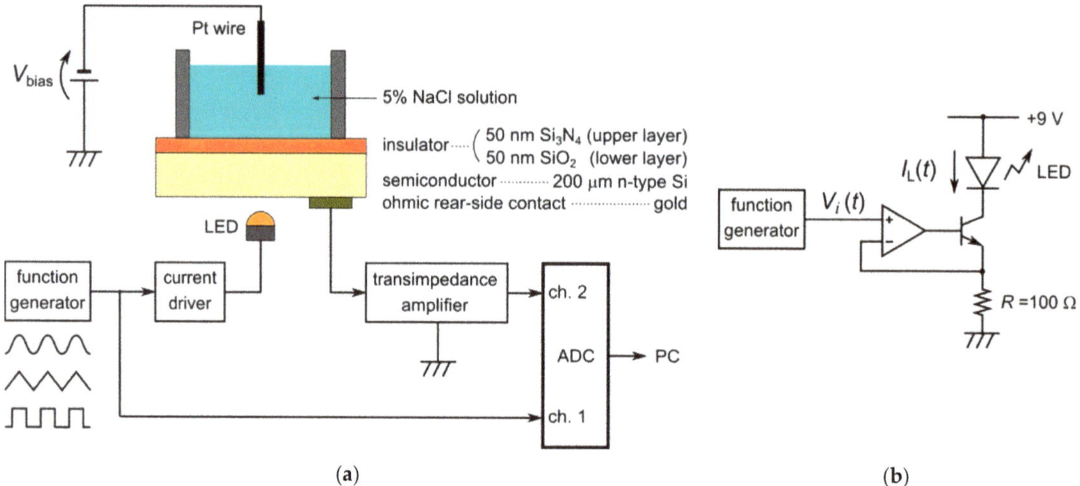

Figure 1. (**a**) Measurement setup for a LAPS. (**b**) LED current driver circuit.

Figure 1b shows the LED current driver circuit we used in this study. The input voltage ($V_i(t)$) controlled the LED current ($I_L(t)$), which is given by:

$$I_L(t) = \frac{\beta}{1+\beta} \cdot \frac{V_i(t)}{R} \approx \frac{V_i(t)}{R},$$ (1)

where β ($\gg 1$) is the common-emitter current gain of the bipolar junction transistor. We can calculate the light power ($P_L(t)$) emitted by the LED as a product of the photon energy and the number of photons emitted in a unit of time:

$$P_L(t) = \frac{hc}{\lambda} \cdot \eta \cdot \frac{I_L(t)}{q}, \tag{2}$$

where h is Planck's constant, c is the speed of light in a vacuum, λ is the wavelength of the light, η is the quantum yield of the LED, and q is the elementary charge. The light power is therefore proportional to the LED current ($I_L(t)$). In this study, we placed a 5 mm round-shaped amber LED (C503B-AAN-CY0B0251, Cree LED, Durham, NC, USA) with $\lambda = 591$ nm in proximity to the back surface of the sensor plate, and we supplied the input voltage ($V_i(t)$) by a digital function generator (DF1906, NF Corporation, Yokohama, Japan), which generated various shapes of periodic functions with a specified frequency.

The transimpedance amplifier virtually grounded the sensor plate, and we applied a fixed bias voltage of -1.5 V across the field-effect structure, at which the vicinity of the insulator–semiconductor interface was in the inversion state. To minimize the influence of the frequency characteristics of the measurement circuit, we directly applied the bias voltage to a Pt wire dipped in 5% NaCl solution on the sensor surface by a DC voltage source instead of using an electrochemical potentiostat.

A wideband transimpedance amplifier (SA-604F2, NF corporation) amplified and converted the photocurrent signal into voltage, with a gain of 10^7 V/A, and we set the cut-off frequency of the built-in low-pass filter at 30 kHz, which was sufficiently higher than the modulation frequencies used in this study. We digitally sampled the amplified signal, together with the input voltage ($V_i(t)$), at a sampling frequency of $f_s = 100$ kHz, by a 16-bit analog-to-digital converter of the multifunction I/O device (USB-6341, National Instruments), and we recorded it with a PC using a program written with LabVIEW (National Instruments, Austin, TX, USA).

The photocurrent signal ($I_{sig}(t)$) is essentially a periodic function with the same period (T) as that of the $I_L(t)$. To determine the amplitude of the $I_{sig}(t)$, we used the principle of dual-phase lock-in detection, which extracts only the component that corresponds to the reference frequency. Lock-in detection is not only advantageous for the reduction in noise, but it is also indispensable in cases where more than one light beam modulated at different frequencies is employed to simultaneously address a plurality of locations on the sensor plate for high-speed measurement [16,17].

When we consider the Fourier series expansion of $I_{sig}(t)$:

$$I_{sig}(t) = \frac{a_0}{2} + \sum_{n=1}^{\infty} \left(a_n \cos \frac{2\pi n t}{T} + b_n \sin \frac{2\pi n t}{T} \right), \tag{3}$$

$$a_n = \frac{2}{T} \int_0^T I_{sig}(t) \cos \frac{2\pi n t}{T} dt, \tag{4}$$

$$b_n = \frac{2}{T} \int_0^T I_{sig}(t) \sin \frac{2\pi n t}{T} dt, \tag{5}$$

the amplitude of the fundamental frequency of $I_{sig}(t)$ (hereafter called A_{sig}) is given by:

$$A_{sig} = \sqrt{a_1^2 + b_1^2}. \tag{6}$$

Our goal, therefore, was to maximize the value of $A_{sig}/\overline{I_L}$. Here, $\overline{I_L}$ is the average LED current, which is proportional to the average input light power.

For the calculation of A_{sig} from the experimentally obtained photocurrent signal, we always used 200 cycles of digitally sampled data ($I_{sig,1}$, $I_{sig,2}$, $\cdots I_{sig,N}$), where the number of samples was $N = 200 f_s T$. We then numerically calculated the values of a_1 and b_1 as:

$$a_1 = \frac{2}{N} \sum_{k=1}^{N} I_{sig,k} \, \cos \frac{2\pi k}{f_s T},$$ (7)

$$b_1 = \frac{2}{N} \sum_{k=1}^{N} I_{sig,k} \, \sin \frac{2\pi k}{f_s T},$$ (8)

from which we obtained A_{sig} by Equation (6).

3. Results and Discussion

First, we investigated the effect of the shape of $I_L(t)$ on the value of A_{sig}. We tested three different waveforms, namely, a sine wave, a triangle wave, and a square wave, which are plotted in blue in Figure 2. All three waveforms had the same fundamental frequency (1 kHz) and the same average LED current ($\overline{I_L}$ of 2 mA). The maximum and the minimum current values were 4 and 0 mA, respectively. The resultant photocurrent signals ($I_{sig}(t)$) for each waveform are plotted in red. Note that the photocurrent signal ($I_{sig}(t)$) has no DC component because the DC current is blocked by the insulator layer.

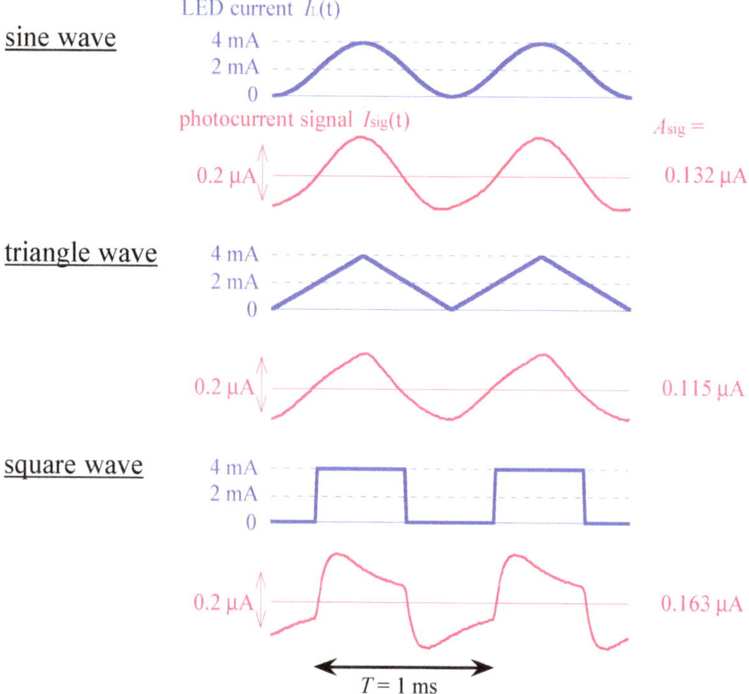

Figure 2. Waveforms of photocurrent signal ($I_{sig}(t)$) (plotted in red) in response to different waveforms (sine, triangle, and square) of $I_L(t)$ (plotted in blue) with the same frequency (1 kHz) and the same average LED current (2 mA). The values of the amplitude of fundamental frequency (A_{sig}) are also shown. All the waveforms shown in this figure are averages from over 100 cycles recorded.

The magnitude of $I_{sig}(t)$ was approximately four orders smaller than that of $I_L(t)$. This ratio is mostly determined by the decay factor ($\exp(-d/L)$), where d is the thickness of the semiconductor layer, and L is the diffusion length of the minority carriers (holes in the present case) [10,11]. The small value of this factor suggested that most of the photo carriers generated at the back surface of the sensor plate were lost by recombination in the course of diffusion across the sensor plate. The values of A_{sig} were 0.132, 0.115, and 0.163 µA for the sine, triangle, and square waves, respectively. The square wave resulted in the largest value of A_{sig} among these three waveforms, for the same value of $\overline{I_{\mathrm{L}}}$.

This result can be understood by considering the amplitude of the fundamental frequency of the $I_{\mathrm{L}}(t)$ (hereafter called A_{L}). We can calculate the values of the A_{L} for the sine, triangle, and square waves as follows:

$$\text{sine wave } A_{\mathrm{L}} = \frac{2}{T}\cdot\int_0^T \overline{I_{\mathrm{L}}}\cdot\sin\frac{2\pi t}{T}\cdot\sin\frac{2\pi t}{T}dt = \overline{I_{\mathrm{L}}}, \tag{9}$$

$$\text{triangle wave } A_{\mathrm{L}} = \frac{2}{T}\cdot 4\int_0^{\frac{T}{4}} \frac{\overline{I_{\mathrm{L}}}\cdot t}{T/4}\cdot\sin\frac{2\pi t}{T}dt = \frac{8}{\pi^2}\cdot\overline{I_{\mathrm{L}}} \approx 0.811\cdot\overline{I_{\mathrm{L}}}, \tag{10}$$

$$\text{square wave } A_{\mathrm{L}} = \frac{2}{T}\cdot\int_0^{\frac{T}{2}} 2\cdot\overline{I_{\mathrm{L}}}\cdot\sin\frac{2\pi t}{T}dt = \frac{4}{\pi}\cdot\overline{I_{\mathrm{L}}} \approx 1.27\cdot\overline{I_{\mathrm{L}}}. \tag{11}$$

The ratios among them were in good agreement with the ratios among the experimentally obtained values of A_{sig} ($0.115/0.132 = 0.871$ and $0.163/0.132 = 1.23$), which suggested that A_{sig} was primarily determined by A_{L} despite the nonlinear distortion of waveforms.

From a practical point of view, a square wave of $I_{\mathrm{L}}(t)$ is much easier to generate than a sine or a triangle wave, as it requires only one bit of output from a digital counter circuit to periodically switch the LED current on and off. This advantage increases in the case where a large number of light beams must be simultaneously controlled. An array of digital counters can be implemented, for example, in a single chip of a field-programmable gate array to output square waves [18].

We could further increase the value of A_{L} by changing the duty ratio of a square wave. Figure 3 shows a square wave with a duty ratio (D) and an average LED current ($\overline{I_{\mathrm{L}}}$):

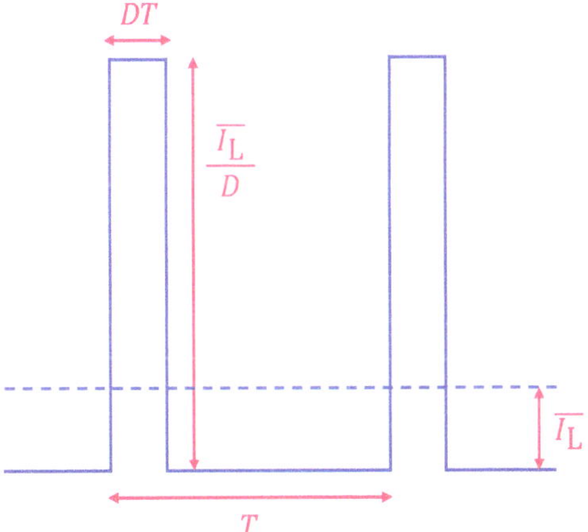

Figure 3. Square wave of $I_{\mathrm{L}}(t)$ with period (T), duty ratio (D), and an average LED current ($\overline{I_{\mathrm{L}}}$). Peak pulse height is $\overline{I_{\mathrm{L}}}/D$.

The value of A_{L}, in this case, is calculated as follows:

$$\text{square wave (duty ratio } (D)) \quad A_{\mathrm{L}} = \frac{2}{T}\cdot 2\int_0^{\frac{DT}{2}} \frac{\overline{I_{\mathrm{L}}}}{D}\cdot\cos\frac{2\pi t}{T}dt = \frac{2\sin D\pi}{D\pi}\cdot\overline{I_{\mathrm{L}}}. \tag{12}$$

The factor $2 \sin D\pi / D\pi$ coincides with the value $4/\pi$ in Equation (11) at $D = 0.5$, and asymptotically approaches 2 in the limit of $D \to 0$, which is the case of a periodic delta function. The peak height of the pulse becomes larger as D becomes smaller, but it is limited in practice by the absolute maximum current of the LED. In this study, we reduced the value of D to 0.20, while always maintaining the value of $\overline{I_\mathrm{L}}$ constant at 2 mA.

Figure 4a shows the experimentally obtained photocurrent signals ($I_\mathrm{sig}(t)$) for $D = 0.20$, 0.25, 0.32, and 0.50. As D becomes smaller, the pulses become narrower and taller.

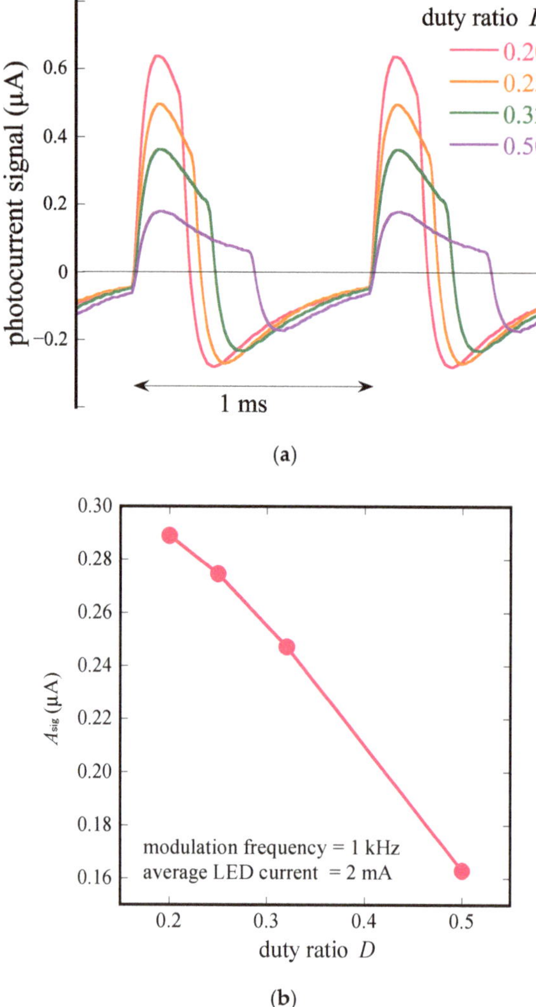

Figure 4. (**a**) Waveforms of $I_\mathrm{sig}(t)$ for different duty ratios (0.20, 0.25, 0.32, and 0.50) of square waves of $I_\mathrm{L}(t)$ with the same frequency (1 kHz) and the same average LED current (2 mA). All waveforms shown were averaged over 100 cycles. (**b**) Amplitude of fundamental frequency of photocurrent signal (A_sig), plotted as a function of duty ratio (D).

In Figure 4b, we plot the amplitude of the fundamental frequency of the photocurrent signal (A_sig) as a function of the duty ratio (D). As we expected, the amplitude (A_sig) increased as D reduced. The value of A_sig at $D = 0.20$ was 0.289 μA, which was slightly larger than twice the value for a sine wave (0.132 µA). A further reduction in D would result in even taller pulses, which, however, does not contribute to a substantial increase in A_{sig}. However, a higher peak value demands more allowance for both the output current of the LED current driver and the input range of the trans-impedance amplifier. Therefore, from a practical point of view, a duty ratio of 0.20 is an appropriate compromise.

Finally, Figure 5 compares the values of A_{sig} we obtained with a sine wave and a square wave ($D = 0.2$) of illumination in a typical frequency range of a LAPS, 100 to 5000 Hz. The overall shape of the frequency dependence is typical for a conventional LAPS sensor plate; the photocurrent had a peak in the kHz region and decayed at both lower and higher frequencies [10,11,19]. Except for the lowest frequency, the photocurrent generated by a square wave ($D = 0.2$) of illumination was always larger than a sine wave by a factor of about two. This result showed that the correct choice of the modulation frequency, as well as the waveform, results in a much higher photocurrent signal, which is advantageous for high-precision measurement with a LAPS sensor plate.

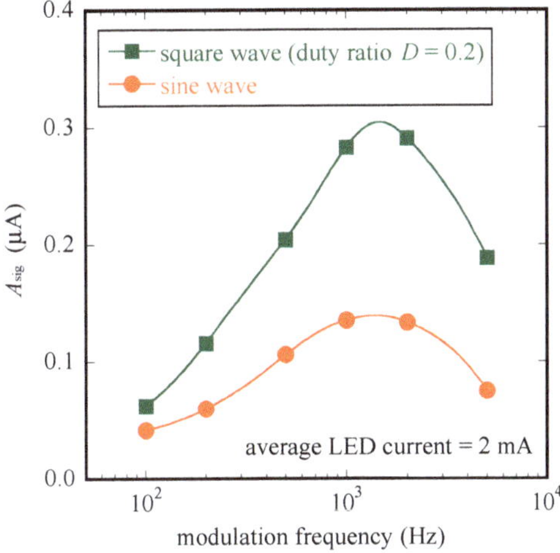

Figure 5. Comparison of values of the amplitude of fundamental frequency (A_{sig}) obtained with a sine and a square wave with a low duty ratio ($D = 0.2$) in a frequency range from 100 to 5000 Hz.

4. Conclusions

In this study, we tested different waveforms of illumination in the search for an efficient illumination for a LAPS that maximizes the photocurrent signal for the same input light power. We found that a square wave with a low duty ratio could generate a larger photocurrent signal than a sine wave by a factor of about two throughout the typical frequency range of a LAPS sensor plate. The correct choice of the modulation frequency, as well as the waveform, is important to maximize the efficiency of the signal generation and to obtain a higher signal-to-noise ratio in LAPS measurement.

Author Contributions: Conceptualization, T.Y.; investigation, T.Y. and K.-i.M.; writing—original draft preparation, T.Y.; writing—review and editing, K.-i.M.; project administration, T.Y. All authors have read and agreed to the published version of the manuscript.

Funding: This research was funded by JSPS KAKENHI, grant number: 22H02100.

References

1. Hafeman, D.G.; Parce, J.W.; McConnell, H.M. Light-addressable potentiometric sensor for biochemical systems. *Science* **1988**, *240*, 1182–1185. [CrossRef] [PubMed]
2. Owicki, J.C.; Bousse, L.J.; Hafeman, D.G.; Kirk, G.L.; Olson, J.D.; Wada, H.G.; Parce, J.W. The light-addressable potentiometric sensor–principles and biological applications. *Annu. Rev. Biophys. Biomol. Struct.* **1994**, *23*, 87–114. [CrossRef] [PubMed]
3. Yoshinobu, T.; Miyamoto, K.; Werner, C.F.; Poghossian, A.; Wagner, T.; Schöning, M.J. Light-addressable potentiometric sensors for quantitative spatial imaging of chemical species. *Annu. Rev. Anal. Chem.* **2017**, *10*, 225–246. [CrossRef] [PubMed]
4. Liang, T.; Qiu, Y.; Gan, Y.; Sun, J.; Zhou, S.; Wan, H.; Wang, P. Recent developments of high-resolution chemical imaging systems based on light-addressable potentiometric sensors (LAPSs). *Sensors* **2019**, *19*, 4294. [CrossRef] [PubMed]
5. Yoshinobu, T.; Schöning, M.J. Light-addressable potentiometric sensors for cell monitoring and biosensing. *Curr. Opin. Electrochem.* **2021**, *28*, 100727. [CrossRef]
6. Özsoylu, D.; Wagner, T.; Schöning, M.J. Electrochemical cell-based biosensors for biomedical applications. *Curr. Top. Med. Chem.* **2022**, *22*, 713–733. [CrossRef] [PubMed]
7. Poghossian, A.; Schöning, M.J. Recent progress in silicon-based biologically sensitive field-effect devices. *Curr. Opin. Electrochem.* **2021**, *29*, 100811. [CrossRef]
8. Wu, C.; Zhu, P.; Liu, Y.; Du, L.; Wang, P. Field-effect sensors using biomaterials for chemical sensing. *Sensors* **2021**, *21*, 7874. [CrossRef] [PubMed]
9. Bergveld, P. Development of an ion-sensitive solid-state device for neurophysiological measurements. *IEEE Trans. Biomed. Eng.* **1970**, *17*, 70–71. [CrossRef] [PubMed]
10. Sartore, M.; Adami, M.; Nicolini, C.; Bousse, L.; Mostarshed, S.; Hafeman, D. Minority-carrier diffusion length effects on light-addressable potentiometric sensor (LAPS) devices. *Sens. Actuators A* **1992**, *32*, 431–436. [CrossRef]
11. Bousse, L.; Mostarshed, S.; Hafeman, D.; Sartore, M.; Adami, M.; Nicolini, C. Investigation of carrier transport through silicon wafers by photocurrent measurements. *J. Appl. Phys.* **1994**, *75*, 4000–4008. [CrossRef]
12. Chen, C.-H.; Akuli, N.; Lu, Y.-J.; Yang, C.-M. Laser illumination adjustments for signal-to-noise ratio and spatial resolution enhancement in static 2D chemical images of NbOx/IGZO/ITO/glass light-addressable potentiometric sensors. *Chemosensors* **2021**, *9*, 313. [CrossRef]
13. Özsoylu, D.; Kizildag, S.; Schöning, M.J.; Wagner, T. Effect of plasma treatment on the sensor properties of a light-addressable potentiometric sensor (LAPS). *Phys. Status Solidi A* **2019**, *216*, 1900259. [CrossRef]
14. Guo, Y.; Miyamoto, K.; Wagner, T.; Schöning, M.J.; Yoshinobu, T. Device simulation of the light-addressable potentiometric sensor for the investigation of the spatial resolution. *Sens. Actuators B* **2014**, *204*, 659–665. [CrossRef]
15. Guo, Y.; Werner, C.F.; Handa, S.; Wang, M.; Ohshiro, T.; Mushiake, H.; Yoshinobu, T. Miniature multiplexed label-free pH probe in vivo. *Biosens. Bioelectron.* **2021**, *174*, 112870. [CrossRef] [PubMed]
16. Zhang, Q.; Wang, P.; Parak, W.J.; George, M.; Zhang, G. A novel design of multi-light LAPS based on digital compensation of frequency domain. *Sens. Actuators B* **2001**, *73*, 152–156. [CrossRef]
17. Miyamoto, K.; Kuwabara, Y.; Kanoh, S.; Yoshinobu, T.; Wagner, T.; Schöning, M.J. Chemical image scanner based on FDM-LAPS. *Sens. Actuators B* **2009**, *137*, 533–538. [CrossRef]
18. Werner, C.F.; Schusser, S.; Spelthahn, H.; Wagner, T.; Yoshinobu, T.; Schöning, M.J. Field-programmable gate array based controller for multi spot light-addressable potentiometric sensors with integrated signal correction mode. *Electrochim. Acta* **2011**, *56*, 9656–9660. [CrossRef]
19. Colalongo, L.; Verzellesi, G.; Passeri, D.; Lui, A.; Ciampolini, P.; Rudan, M.V. Modeling of light-addressable potentiometric sensors. *IEEE Trans. Electron. Devices* **1997**, *44*, 2083–2090. [CrossRef]

Communication

Simultaneous In Situ Imaging of pH and Surface Roughening during the Progress of Crevice Corrosion of Stainless Steel

Ko-ichiro Miyamoto [1,*], Rinya Hiramitsu [1], Carl Frederik Werner [2] and Tatsuo Yoshinobu [1,3]

[1] Department of Electronic Engineering, Tohoku University, Sendai 980-8579, Japan;
hiramitsu.h27.rinya@gmail.com (R.H.); tatsuo.yoshinobu.a1@tohoku.ac.jp (T.Y.)
[2] Department of Electronics, Kyoto Institute of Technology, Kyoto 606-8585, Japan; werner@kit.ac.jp
[3] Department of Biomedical Engineering, Tohoku University, Sendai 980-8579, Japan
* Correspondence: k-miya@ecei.tohoku.ac.jp; Tel.: +81-22-795-7075

Abstract: Stainless steel plays an important role in industry due to its anti-corrosion characteristic. It is known, however, that local corrosion can damage stainless steel under certain conditions. In this study, we developed a novel measurement system to observe crevice corrosion, which is a local corrosion that occurs inside a narrow gap. In addition to pH imaging inside the crevice, another imaging technique using an infrared light was combined to simultaneously visualize surface roughening of the test piece. According to experimental results, the lowering of local pH propagated inside the crevice, and after that, the surface roughening started and expanded due to propagation of corrosion. The real-time measurement of the pH distribution and the surface roughness can be a powerful tool to investigate the crevice corrosion.

Keywords: chemical sensor; light-addressable potentiometric sensor; stainless steel; crevice corrosion; pH imaging

Citation: Miyamoto, K.-i.; Hiramitsu, R.; Werner, C.F.; Yoshinobu, T. Simultaneous In Situ Imaging of pH and Surface Roughening during the Progress of Crevice Corrosion of Stainless Steel. *Sensors* **2022**, *22*, 2246. https://doi.org/10.3390/s22062246

Academic Editor: José Manuel Díaz-Cruz

Received: 30 January 2022
Accepted: 12 March 2022
Published: 14 March 2022

Publisher's Note: MDPI stays neutral with regard to jurisdictional claims in published maps and institutional affiliations.

1. Introduction

Stainless steel is an essential material in industries due to its superior anti-corrosion characteristic, which results from a passivation film of native oxide formed on the surface. The damage of the passivation film is immediately recovered by oxidization. Products made of stainless steel withstand corrosion, even in seawater. In an underwater environment, the passivation film is maintained by dissolved oxygen [1].

It is known, however, that local corrosion of stainless steel may occur under certain conditions. Crevice corrosion is an example of local corrosion, which occurs inside a narrow crevice with a gap on the order of microns. Inside a crevice, protons are produced by hydrolysis of eluted metal ions, and dissolved oxygen is consumed to recover the passivation film. Due to the narrow geometry, where diffusion from inside or outside of the crevice is restricted, the solution inside has a low pH and low oxygen. In addition, chloride ions are attracted to maintain charge neutrality, and it is known that these chloride ions attack the passivation film. The lowering of pH inside the crevice is accelerated by the interaction of the processes above. When the pH value becomes lower than the critical value, "depassivation pH", the passivation film can no longer be maintained, and the elution of the metal is further accelerated. The surface is then rapidly damaged by propagation of corrosion. It should be noted that the environment inside the crevice is not uniform, in general, reflecting the non-uniform geometry of the crevice, including in terms of its width and shape [2–6].

Although crevice corrosion has been extensively studied, experimental methods to probe inside such a narrow crevice have been limited due to the geometry. To overcome this limitation, Kaji et al. formed a gap between a test piece and a transparent glass plate, whose surface was modified with ion-sensitive dyes, and the spatial distributions of pH and chloride ions were optically monitored through the glass plate [7,8]. However, the measurable pH range of the dye is not wide enough, and the color change is almost invisible when the surface of the test piece is colored due to corrosion. There is a strong demand for an experimental method to observe the pH distribution inside crevices during the course of corrosion.

Recently, we proposed the label-free measurement of pH distribution inside a crevice [9–11] by applying a semiconductor-based chemical imaging sensor [12], which was based on the principle of the light-addressable potentiometric sensor (LAPS, [13,14]). Figure 1a shows a schematic view of a LAPS sensor plate. It has a flat sensor surface and a wide-range response to a pH change [9–11], which makes it an ideal sensor to monitor the pH change in the vicinity of the corroding surface of a test piece inside a crevice, as depicted in Figure 1b. Lowering of the local pH value was observed in the course of corrosion, which could be associated with the increase in the corrosion current. After the experiment, the corroded surface was optically inspected, and it was confirmed that the corroded area corresponded to the location where the lowering of pH was observed. However, the optical inspection was possible only ex situ, and it was not possible to identify the corroded area in the course of corrosion and to correlate it to the spatiotemporal change in pH distribution.

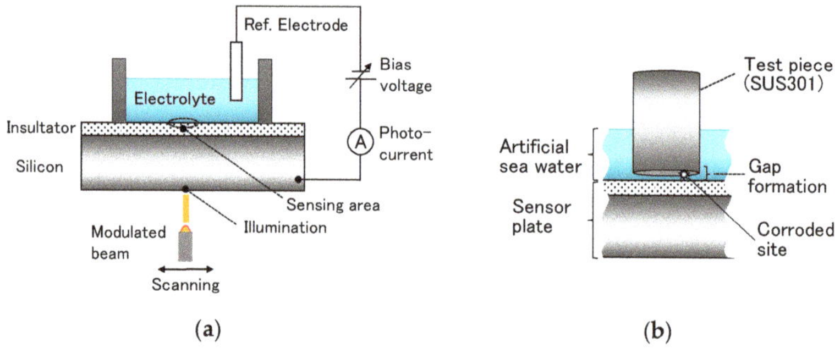

Figure 1. Schematic of the chemical imaging sensor system based on the LAPS principle: (**a**) the structure of the LAPS sensor plate and (**b**) the formation of a crevice with a narrow gap between the surface of the test piece and the sensor surface.

In this study, we developed a measurement system in which the pH distribution inside the crevice and the surface roughening of the test piece could be simultaneously visualized. Taking advantage of the fact that silicon is transparent to infrared light with a wavelength longer than ca. 1100 nm, additional optics was combined with the measurement system used in our previous study [10] to allow in situ inspection of the corroding surface of the test piece using an infrared light probe that penetrates the LAPS sensor plate.

2. Experiment

Figure 2 shows the measurement system developed in this study. The system consists of (1) a measurement cell with its bottom made of a LAPS sensor plate; (2) a SUS304 test piece; (3) a home-made potentiostat; (4) scanning optics for probing pH and surface roughness; and (5) a control PC and measurement software.

Sensor plate and measurement cell: The sensor plate was made of an *n*-type silicon substrate with double insulating layers of silicon dioxide (intermediate layer, 50 nm) and silicon nitride (pH-sensitive layer, 50 nm) on top. These insulating layers were formed by thermal oxidation and low-pressure chemical vapor deposition, respectively. The size and thickness of the sensor chip were 36×36 mm^2 and 200 μm. The electrochemical cell was made of PVC, which was pressed onto the sensor surface with a rubber seal in between. The sensor plate was fixed on a metal plate that contacted the gold electrode evaporated on the back surface of the sensor plate. These are intrinsically the same as those used in our previous studies [9,10].

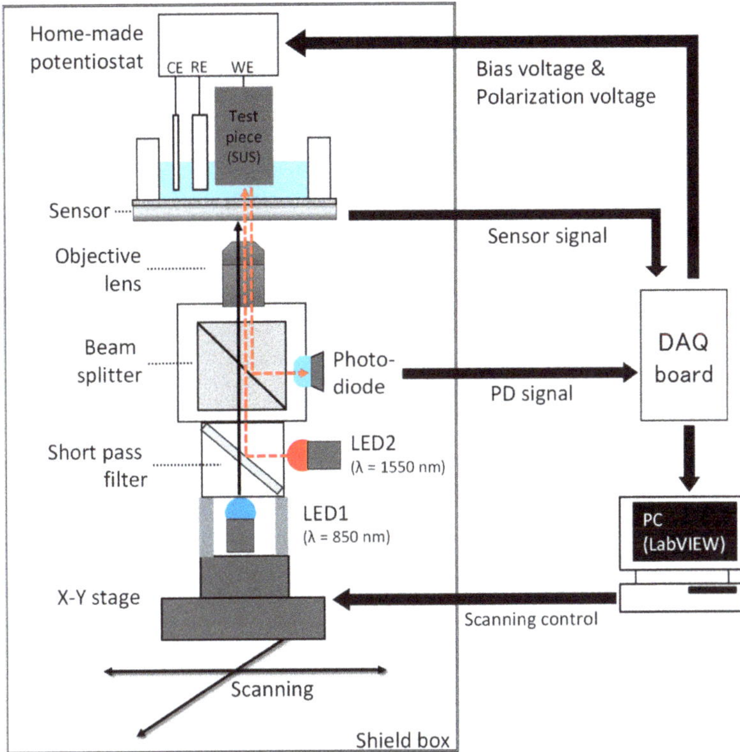

Figure 2. Measurement system.

Test piece: A rod-shaped piece of JIS SUS304 (18–8 stainless steel) with a length of 25–35 mm and a diameter of 12 mm was used as a test piece. The composition of SUS304 used in this study was as follows: C 0.068%, Si 0.48%, Mn 1.84%, P 0.029%, S 0.027%, Ni 8.11%, Cr 18.65%. The surface of the test piece was first passivated by immersion in 35% nitric acid (FUJIFILM Wako Pure Chemicals Corp., Osaka, Japan) for 1 h at room temperature. The bottom surface was then polished with abrasive paper (200 mm in diameter, P600, KOVAX, Yokohama, Japan) directly before the experiment. The measurement cell was filled with artificial seawater (pH 7.8), in which the test piece was immersed and placed on the sensor surface to form a narrow crevice.

Electrochemical system: The test piece is connected as a working electrode (WE) to the homemade potentiostat shown in Figure 3, together with an Ag/AgCl reference electrode (RE) and a platinum wire (CE). The LAPS sensor plate is virtually grounded via a transimpedance amplifier, and the potentials of the RE and the WE (test piece) with respect to the ground are set at V_1 and V_2, respectively. The circuit thereby polarizes the test piece at $V_{pol} = V_2 - V_1$ (vs. Ag/AgCl). The LAPS signal is obtained from the transimpedance amplifier, and the corrosion current I_{corr} is obtained by monitoring the output voltage, $V_2 - R_f I_{corr}$.

Scanning optics: As shown in Figure 2, two light beams are employed in the measurement system. A light beam from LED1 (L3989-01, Hamamatsu Photonics K.K.) has a wavelength of 850 nm, which excites the photocurrent signal in the LAPS sensor plate. The other light beam from LED2 (L12509-0155L, Hamamatsu Photonics K.K., Shizuoka, Japan) has a wavelength of 1550 nm, which penetrates the LAPS sensor plate and is reflected by the corroding surface of the test piece. The intensities of light beams from LED1 and LED2 are modulated at different frequencies of 2000 and 1500 Hz, respectively, so that they can be separated by lock-in detection to avoid cross-talk in the same manner described in [15].

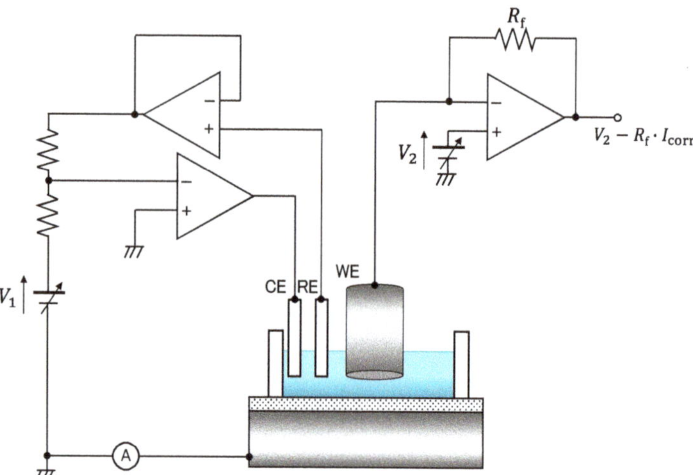

Figure 3. Circuit configuration of the homemade potentiostat.

The two light beams are mixed by a short pass filter with a cut-off wavelength of 1200 nm (Cat. #86-687, Edmund optics, Barrington, NJ, USA), which transmits and reflects the shorter and longer wavelengths, respectively. The two light beams are focused together by an objective lens (ULWD MIRPlan50, Olympus, Tokyo, Japan).

The light beam from LED2 specular reflected at the surface of the test piece penetrates the LAPS sensor plate again, and a part of it is guided to a photodiode (FCI-InGaAs-1000, OSI optoelectronics, Camarillo, CA, USA) by a beam splitter (#47-235, Edmund optics). The intensity of the specular reflected light beam, which decreases as a result of scattering by roughness or corrosion products, is measured by the photodiode as an indicator of the degree of local corrosion.

The optics is mounted on an X-Y stage to allow the light beams to two-dimensionally scan the LAPS sensor plate and the test piece. The photocurrent signal of the LAPS sensor plate and the current signal of the photodiode are converted into voltage signals by transimpedance amplifiers and recorded by PC via a data acquisition device (USB-6361, National Instruments, Austin, TX, USA). Maps of the pH distribution and corrosion are simultaneously obtained from these two signals.

3. Results and Discussion

As a preliminary experiment, visualization of the corroded surface using the infrared light beam from LED2 was tested using a test piece which underwent crevice corrosion in advance. Figure 4a shows an optical image of the surface, in which a dark area shows the corroded area. Figure 4b shows a map of the intensity of the reflected light beam, which clearly correlates with the optical image in Figure 4a. In addition, the non-uniformity inside the corroded area in these two images match each other, as indicated by arrows in the insets. This result proves the possibility of infrared imaging of the surface through the LAPS sensor plate.

In situ measurement of crevice corrosion: The polished surface of a test piece was pressed against the sensor surface by its weight, leaving a narrow gap of approximately 12.3 μm [10]. The test piece was then potentiostatically polarized at 150 mV vs. Ag/AgCl in artificial seawater for 3400 s. The time course of the anodic current is shown in Figure 5a. The anodic current started to increase at around 760 s and continued to increase until the end of the experiment. Figure 5b shows the visible light image of the test piece after the experiment. Three areas on the surface were corroded; two were at the edge of the sample, and the other one was near the center of the sample.

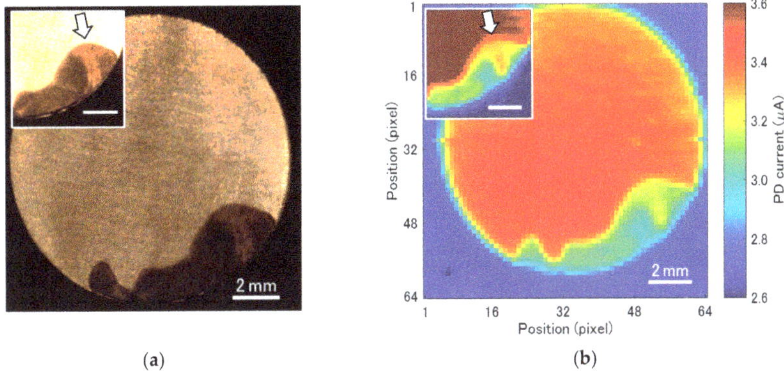

(a)   (b)

Figure 4. (**a**) An optical image of the corroded test piece and (**b**) the infrared reflection image observed through the LAPS sensor plate. Inset figures show the corroded area with enhanced contrast.

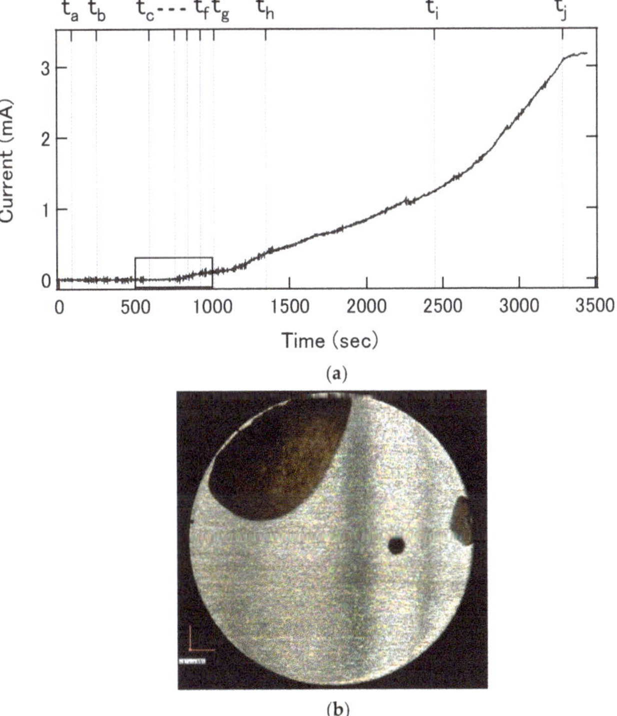

(b)

Figure 5. (**a**) The temporal change in the anodic current during the potentiostatic polarization of the test piece at 150 mV vs. Ag/AgCl in artificial seawater. The time stamps t_a to t_j correspond to the times at which the pH and roughness images in Figure 6a were recorded. A part of the curve surrounded by a rectangle, which corresponds to the initial stage of corrosion, is enlarged in Figure 7b. (**b**) Visual light image of the corroded surface after the experiment.

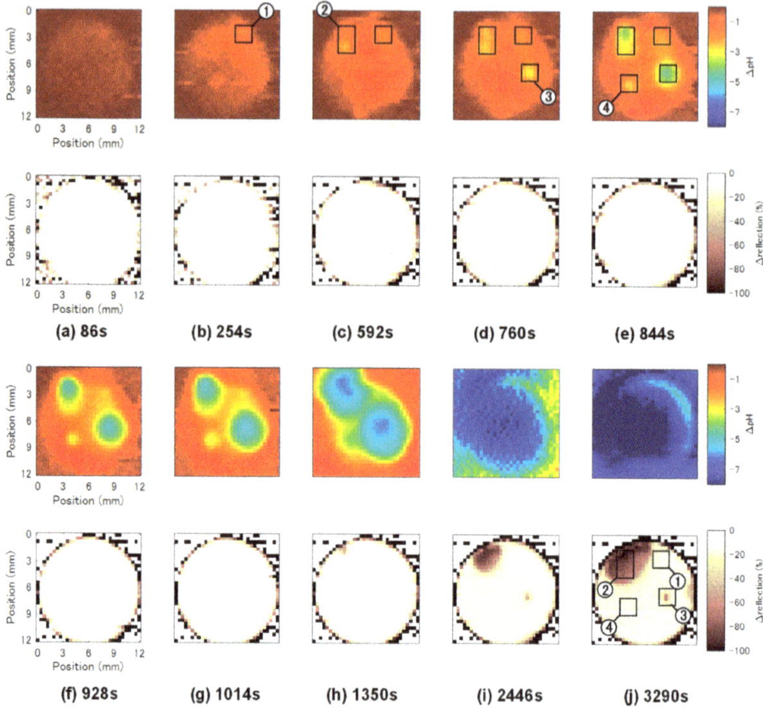

Figure 6. The change in pH distribution (**upper row**) and that of reflection (**lower row**). Areas 1 to 4 are the locations where the lowering of the local pH was observed.

During the potentiostatic polarization, two-dimensional distributions of pH and the intensity of reflected light, which indicates the roughness of the surface, were continuously acquired. The number of pixels, the scanning area, the sampling frequency, and the sampling number were 32 × 32, 12.8 mm × 12.8 mm, 100 kHz, and 2400, respectively. The pH and roughness images were recorded every 85 s. Images labelled with (a) to (j) in Figure 6 were collected at times indicated by the corresponding time stamps in Figure 5a. In this series of measurement, local pH changes were observed in four areas.

Figure 7a,b show the initial pH changes at Areas 1 to 4 and the increase in the corrosion current, respectively. The first obvious change in the local pH value was observed in Area 1 at 254 s (Figure 6b). At this stage, a major increase in the corrosion current was not yet detected in Figure 7b. Thereafter no further lowering of the local pH value was observed in Area 1, suggesting that this area was repassivated. Then, the local pH values started to lower successively in Area 2 at 592 s (Figure 6c) and Area 3 at 760 s (Figure 6d). In these two areas, the local pH values further continued to decrease, and the corroded areas expanded in the following pH images. The locations of these two areas correspond to the corroded areas observed in the optical image of the final surface shown in Figure 5b. Finally, the local pH value in Area 4 started to decrease at 844 s (Figure 6e), but the pH change in Area 4 was smaller than those in Areas 2 and 3. In Figure 7b, a major increase in the corrosion current was observed in synchronization with the acidification in Areas 2 and 3.

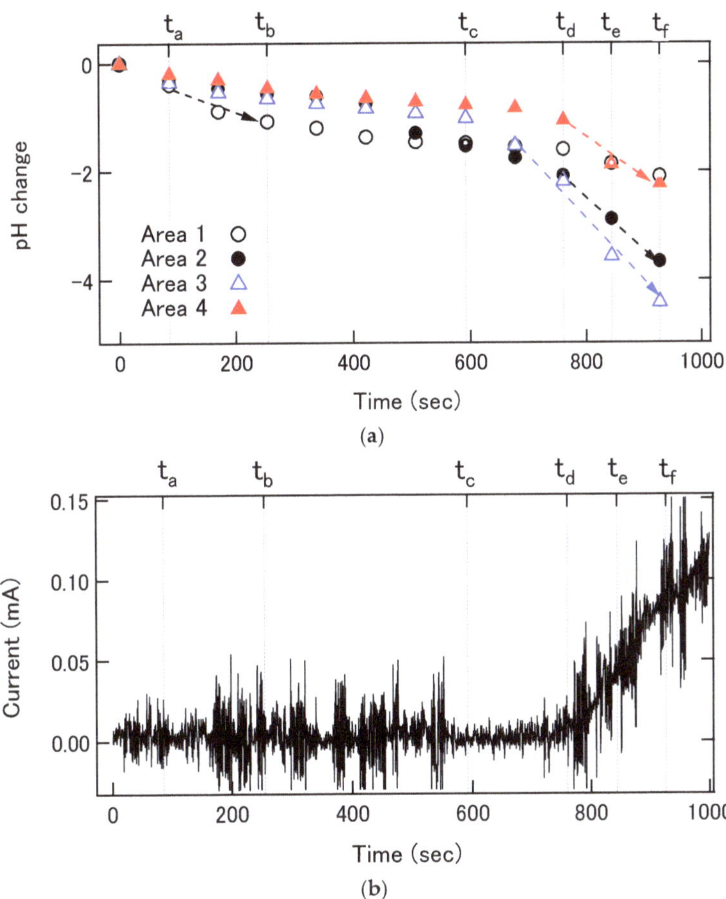

Figure 7. (**a**) pH changes in Areas 1 to 4 at the initial stage of corrosion. (**b**) Enlarged curve of the corrosion current shown in Figure 5a. (t_a = 86 s, t_b = 254 s, t_c = 592 s, t_d = 760 s, t_e = 844 s, t_f = 928 s.).

According to the mechanism of crevice corrosion, accelerated propagation of corrosion is triggered by elution of metal ions, which lowers the local pH value. Unless the elution is too much, the surface can be recovered by repassivation, consuming the remaining dissolved oxygen. The obtained results suggest the number of metal ions that were eluted at Areas 2 and 3 was greater than at Area 1. The position at which elution of metal ions occurs depends on the geometry of the crevice and the roughness of the sample surface. In addition, it is known that dissolution of nonmetallic inclusions at the surface can trigger pit corrosion, which will eventually initiate crevice corrosion. Manganese sulfide (MnS) is a typical inclusion in stainless steel [16–18]. Such inclusions included in the test piece may have determined the positions of pH reduction and corrosion.

Figure 8a shows the pH changes in these four areas for a longer period. The pH values in Areas 2 and 3 continued to decrease even after 1000 s, and eventually engulfed the minor pH changes in Areas 1 and 4, as observed in Figure 6.

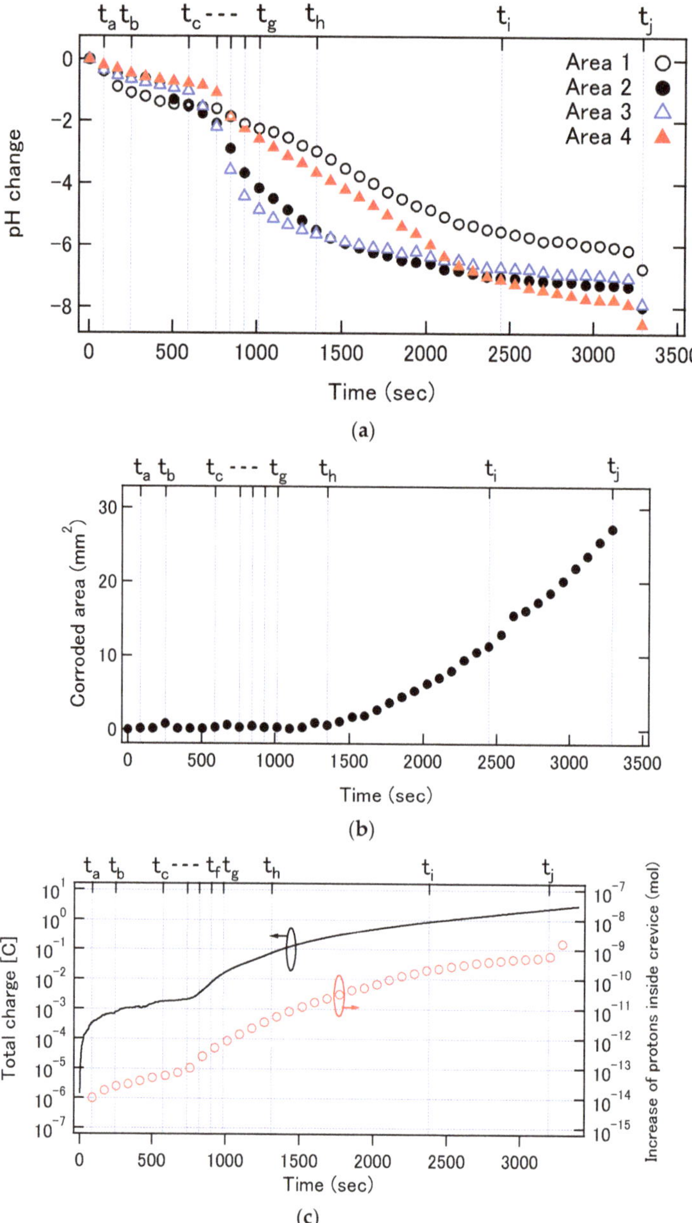

Figure 8. (**a**) Temporal changes in the averaged pH values in Areas 1 to 4. (**b**) Expansion of the corroded area detected by infrared reflection. (**c**) Time courses of the cumulative charge injected into the solution by the corrosion current and the increase in the number of protons inside the crevice calculated from the pH distribution.

According to the reflection images shown in Figure 6, the reflection in Area 2 decreased over 20% after 1350 s (Figure 6h), and then, the decrease propagated along the outer edge of the test piece. After 2446 s (Figure 6i), the reflection in Area 3 also decreased. After that, the reflection decreased at the right edge of the sample at 3290 s (Figure 6j). Figure 8b shows the temporal change in the total area of corrosion estimated from the reflection images by counting the pixels where the reflection decreased by more than 20%. Comparison of Figure 8a,b reveals that the onset of surface roughening at t_h is the time at which the local pH values at Areas 2 and 3 were approximately 2.2.

In order to illuminate the relationship between the corrosion current and the pH change, the following two quantities were calculated. The total increase in the number of protons inside the gap was calculated from the pH images by:

$$\sum_{\text{all pixels}} \left\{ \left(10^{-\text{pH}(t)} - 10^{-\text{pH}(t=0)} \right) \times \text{pixel area} \times \text{gap} \right\} \text{ [mol]}, \tag{1}$$

where the pixel area was $16 \times 10^4 \ \mu m^2$ and the gap of the crevice was assumed to be 12.3 μm [7]. The total charge of ions eluted from the test piece was calculated from the corrosion current by:

$$\int_0^t I_{\text{corr}}(\tau)d\tau \text{ [C]} \tag{2}$$

A proportional relationship was observed between the total charge injected by the corrosion current and the increase in the number of protons inside the crevice.

4. Conclusions

In this study, we developed a measurement system that is capable of visualizing both the pH distribution and the roughness inside a crevice. From the measurement results, the initial stage of crevice corrosion and its propagation were clearly observed. The lowering of the local pH value was synchronized to the increase in the corrosion current, and the surface roughening was observed after the local pH value was lowered. The combination of the chemical image sensor and the infrared light imaging through the sensor plate can be a powerful tool to study crevice corrosion.

Author Contributions: K.-i.M. and T.Y. conceived and led the research. R.H. and C.F.W. contributed to the implementation and experiment. All authors contributed to data interpretation and preparation of the manuscript text. All authors have read and agreed to the published version of the manuscript.

Funding: A part of this study was funded by JSPS KAKENHI (Grant Number 17H03074) and JFE 21st Century Foundation.

Institutional Review Board Statement: Not applicable.

Informed Consent Statement: Not applicable.

Acknowledgments: The authors gratefully acknowledge the advice and discussions of H. Kajimura, K. Matsuoka, R. Matsuhashi and K. Nose of Nippon Steel Corporation Group. A part of this research was supported by the Machine Shop Division of Fundamental Technology Center, Research Institute of Electrical Communication, Tohoku University.

Conflicts of Interest: The authors declare no conflict of interest. The funders had no role in the design of the study; in the collection, analyses, or interpretation of data; in the writing of the manuscript, or in the decision to publish the results.

References

1. Olsson, C.O.A.; Landolt, D. Passive films on stainless steelschemistry, structure and growth. *Electrochim. Acta* **2003**, *48*, 1093–1104. [CrossRef]
2. Lennox, T.J.; Peterson, M.H.; Groover, R.E. A study of crevice corrosion in type 304 stainless steel. *Mater. Prot.* **1970**, *9*, 23–26.
3. Szklarska-Smialowska, Z.; Mankowski, J. Crevice corrosion of stainless steels in sodium chloride solution. *Corros. Sci.* **1978**, *18*, 953–960. [CrossRef]
4. Alavi, A.; Cottis, R.A. The determination of pH, potential and chloride concentration in corroding crevices on 304 stainless steel and 7475 aluminium alloy. *Corros. Sci.* **1987**, *27*, 443–451. [CrossRef]
5. Watson, M.; Postlethwaite, J. Numerical simulation of crevice corrosion: The effect of the crevice gap profile. *Corros. Sci.* **1991**, *32*, 1253–1262. [CrossRef]

6. Nagaoka, A.; Matsuhashi, R.; Nose, K.; Matsuoka, K.; Kajimura, H. The effect of chloride ion and dissolved oxygen concentration on passivation pH of SUS304 stainless steel. *Zair. Kankyo* **2020**, *69*, 49–57. (In Japanese) [CrossRef]
7. Kaji, T.; Sekiai, T.; Muto, I.; Sugawara, Y.; Hara, N. Visualization of pH and pCl distributions: Initiation and propagation criteria for crevice corrosion of stainless steel. *J. Electrochem. Soc.* **2012**, *159*, C289–C297. [CrossRef]
8. Kaji, T.; Sekiai, T.; Muto, I.; Sugawara, Y.; Hara, N. Visualization of Solution Chemistry inside Crevice by pH and pCl Sensing Plates. *ECS Trans.* **2019**, *41*, 205–216. [CrossRef]
9. Miyamoto, K.-I.; Sakakita, S.; Wagner, T.; Schöning, M.J.; Yoshinobu, T. Application of chemical imaging sensor to in-situ pH imaging in the vicinity of a corroding metal surface. *Electrochim. Acta* **2015**, *183*, 137–142. [CrossRef]
10. Miyamoto, K.-I.; Sakakita, S.; Werner, C.F.; Yoshinobu, T. A Modified Chemical Imaging Sensor System for Real-Time pH Imaging of Accelerated Crevice Corrosion of Stainless Steel. *Phys. Status Solidi (A)* **2018**, *215*, 1700963. [CrossRef]
11. Miyamoto, K.-I.; Yoshinobu, T. Sensors and techniques for visualization and characterization of local corrosion. *Jpn. J. Appl. Phys.* **2019**, *58*, SB0801. [CrossRef]
12. Liang, T.; Qui, Y.; Gan, Y.; Sun, J.; Zhou, S.; Wan, H.; Wang, P. Recent developments of high-resolution chemical imaging systems based on light-addressable potentiometric sensors (LAPSs). *Sensors* **2019**, *19*, 4294. [CrossRef] [PubMed]
13. Hafeman, D.G.; Parce, J.W.; McConnell, H.M. Light-addressable potentiometric sensor for biochemical systems. *Science* **1988**, *240*, 1182–1185. [CrossRef] [PubMed]
14. Miyamoto, K.; Yoshinobu, T. Development and bio-imaging applications of a chemical imaging sensor. *Sens. Mater.* **2016**, *28*, 1091–1104. [CrossRef]
15. Miyamoto, K.; Kuwabara, Y.; Kanoh, S.I.; Yoshinobu, T.; Wagner, T.; Schöning, M.J. Chemical image scanner based on FDM-LAPS. *Sens. Actuators B Chem.* **2009**, *137*, 533–538. [CrossRef]
16. Eklund, S.G. On the initiation of crevice corrosion on stainless steel. *J. Electrochem. Soc.* **1976**, *123*, 170–173. [CrossRef]
17. Nagaoka, A.; Nose, K.; Nokami, K.; Kajimura, H. The role of micro pits in the initiation process of crevice corrosion of SUS304 stainless steel in an aqueous chloride solution. *Mater. Trans.* **2022**, *63*, 335–342. [CrossRef]
18. Chiba, A.; Muto, I.; Sugawara, Y.; Hara, N. Direct observation of pit initiation process on type 304 stainless steel. *Mater. Trans.* **2014**, *55*, 857–860. [CrossRef]

MDPI
St. Alban-Anlage 66
4052 Basel
Switzerland
Tel. +41 61 683 77 34
Fax +41 61 302 89 18
www.mdpi.com

Sensors Editorial Office
E-mail: sensors@mdpi.com
www.mdpi.com/journal/sensors